The Mouse in Animal Genetics and Breeding Research

The Mouse in Animal Genetics and Breeding Research

Eugene J. Eisen

Department of Animal Science
North Carolina State University, USA

ICP

Imperial College Press

Published by

Imperial College Press
57 Shelton Street
Covent Garden
London WC2H 9HE

Distributed by

World Scientific Publishing Co. Pte. Ltd.
5 Toh Tuck Link, Singapore 596224
USA office: 27 Warren Street, Suite 401-402, Hackensack, NJ 07601
UK office: 57 Shelton Street, Covent Garden, London WC2H 9HE

British Library Cataloguing-in-Publication Data
A catalogue record for this book is available from the British Library.

THE MOUSE IN ANIMAL GENETICS AND BREEDING RESEARCH

ISBN 1-86094-565-1

Printed in Singapore by B & JO Enterprise

To my marvelous family:

Jackie
Arri, Lisa H., Gabriel and Micah
Avram, Lisa B., Paul and Isabelle
Andrea, Daniel, Amos and Talia

TABLE OF CONTENTS

PREFACE

Genetics research with the mouse has expanded enormously in recent years, culminating with publication of the mouse genome sequence in 2002. Therefore, it is timely to present a comprehensive review of genetics research, which has used the mouse as a model organism. This volume is addressed to researchers and graduate students who are either actively involved in or contemplate using the mouse in genetics research. It is also valuable to those researchers and students who visualize how research results obtained with mice can assist in answering questions in their own area of expertise, be it human genetics, livestock genetics, veterinary genetics or evolutionary biology. Two notable examples are mouse research in the genetic control of obesity and in genetic resistance to disease. Other applications of the mouse model abound, including the genetics of behavior, genetics of maternal effects, searching for quantitative trait loci, use of mutagenesis to study mammalian development, and adaptation of embryo biotechnologies.

Since the scope of mouse genetics research is so broad, it was not realistic to incorporate every possible field of research. Of course, many topics have been covered in reviews elsewhere, and, certainly, opportunities exist for future volumes to include updated reviews on old and new topics.

My appreciation goes to all the contributing authors for preparing outstanding review articles in their respective areas of expertise. I gratefully acknowledge the assistance of several people who helped to make this volume possible. Marian Correll has formatted all of the articles, and I sincerely appreciate her expertise and patience. Ben Corl provided invaluable assistance with his word processing knowledge. My wife, Jackie, was of immeasurable help in all questions of grammar, making certain that the placement of commas, colons and semicolons were not random events.

Eugene (Gene) J. Eisen
January 2005

CHAPTER 1

THE BEGINNINGS: ODE TO A WEE MOUSE

E. J. Eisen

Animal Science Department
North Carolina State University, Raleigh, NC, USA
gene_eisen@ncsu.edu

1. Introduction

This chapter provides a brief overview and historical perspective of how the mouse came to be an important animal model in mammalian genetics research. More detailed reviews can be found in several sources.[1–5] Kenneth Paigen has recently published an excellent review of the first one hundred years of mouse genetics from 1902 to 2002.[6,7]

The house mouse became a well-recognized pest to humans with the introduction of plant and animal domestication. The transition of humans from hunter-gatherers to farmers began about 8,500 B.C. in the region of Southwest Asia called the Fertile Crescent.[8] The mouse found itself in a luxuriant environment with unlimited food in the form of stored grains and other food morsels in dwellings of the sedentary human population of this area.[5] Thus began the never-ending struggle of people to protect their stored food from what might be called the *bad mouse*. It is not surprising that the word *mouse* comes originally from the Sanskrit *mush* meaning to steal, which became *mus* in Latin and *mys* in Greek.[1] It has been suggested that the ancient Egyptians deified the cat because of its ability to control the mouse population. The mouse even receives some harsh treatment in the book of Leviticus:

> There also shall be an abomination to you among the creeping things that creep upon the earth: the weasel and the *mouse*, and the tortoise after his kind.... These are unclean to you among all that creep; whosoever doth touch them, when they be dead, shall be unclean until even.

At various times the Greeks and Romans worshipped mice, and physicians used mice in their medicinal formulas. The clergy of the Middle Ages in Europe considered mice as lustful creatures and instruments of the devil,[1] but the mouse had a more positive status in Asia.

The use of mice in pharmaceutical concoctions continued well into the 17th century and later in parts of Europe.[3] The Chinese and Japanese became enamored of mice.[1,3] Positive mouse symbolism in Asia became wide spread. In Japan, the mouse was given the special status as messenger of Daikoku, the God of Wealth.[9] In many parts of Asia, the mouse is recognized in several ways: a) one of every twelve years is known as the year of the mouse; b) the hours of the day between 11:00AM and 1:00PM are designated as the hours of the mouse; and c) multiplications by serial 2's are known as mouse numbers.[3]

Robert Burns, the great 18th century Scottish poet, immortalized the intimate relationship between the farmer and the mouse and their struggle for survival in his famous ode *To a Mouse.*

ON TURNING HER UP IN HER NEST,
WITH THE PLOUGH, NOVEMBER 1785

Wee, sleekit, cowrin, tim'rous beastie,
O, what a panic's in thy breastie!
Thou need na start awa sae hasty,
    Wi' bickering brattle!
I wad be laith to rin an' chase thee,
    Wi' murd'ring pattle!

I'm truly sorry Man's dominion
Has broken Nature's social union,
An' justifies that ill opinion,
    Which makes thee startle,
At me, thy poor, earth-born companion,
    An' fellow-mortal!...

...But, Mousie, thou art no thy lane,
In proving foresight may be vain:
The best-laid schemes o' mice an' men
    Gang aft a-gley,
An' lea'e us nought but grief an' pain,
    For promis'd joy!

Still though are blest, compar'd wi' me!
The present only toucheth thee:
But, Och! I backward cast my e'e

On prospects drear!
An' forward, tho' I canna see,
I guess an' fear!

2. Mouse Domestication

The Chinese and Japanese are believed to have been the first to domesticate mice.[3] They were the first to raise unusual mice, particularly with regard to coat color and waltzing mutants. The spotted mouse is mentioned in the Eh Yah lexicon in 1100 B.C., and the waltzing mouse is described as early as 80 B.C. in the annals of the Han Dynasty.[1,9]

The *mouse trade* brought fancy mice to Europe, so that by the 19th century the house mouse hobby became popular there and spread to the United States by the beginning of the 20th century.[5] One of these mouse fanciers, Ms. Abbie Lathrop, a retired schoolteacher, became the link between hobby and science for the house mouse. A detailed description of Abbie Lathrop's mouse breeding business in Granby, Massachusetts, USA from about 1900 until her death in 1918 at age 50, showed that she interacted with biologists of the day as well as mouse fanciers.[2] She first sold mice to hobbyists, but soon she received orders for mice from research laboratories, including the Bussey Institute at Harvard University. Many inbred mouse lines were derived from mice originally obtained from Lathrop's farm.[2] Thus, the wee mouse had come full circle from the *bad mouse*, a nemesis to humans, to the *good mouse*, an important biological model for biomedical research.

3. The Birth of Mouse Genetics

But for the intervention of a conservative bishop who forbade Gregor Mendel from continuing to study inheritance of coat color traits of mice,[10] mouse genetics may have had its beginnings in 1866 instead of 1902. The bishop in Mendel's district felt it was inappropriate for the monk to share his living quarters with critters that had sexual intercourse, and so Mendel was forced to turn his attention to making experimental crosses with the garden pea. Fortunately for Mendel, the bishop was apparently unaware that plants also had sex.

Upon the rediscovery of Mendel's laws in 1900 by Correns, De Vries and Tschermak working independently with plants, there was a question of whether the laws applied to animals as well.[6] In 1902 Lucien Cuénot in France demonstrated independent segregation of albino *vs* color and of yellow *vs* black coat color,[11] and in 1903 William Castle in the United States established segregation and independent assortment of albino *vs* colored, spotted *vs* solid colored, black *vs* brown, and yellow *vs* nonyellow.[12] Castle, together with a bright Harvard undergraduate, Clarence Cook Little, established nine genetic coat color loci.[13,14] As chronicled by Paigen,[6] Cuénot described the first lethal mutation, the A^y allele of the agouti locus,[15] which was verified in 1910 by Castle and Little.[14] The first genetic linkage group for mice was established by Haldane and co-workers in 1915.[16]

William Castle was the first co-director of the Bussey Institute at Harvard University, which opened in 1908. Castle was a catalyst for promoting research in mouse genetics. He attracted a covey of outstanding Ph.D. students to the Institute. Although only 13 of Castle's 246 publications were concerned primarily with mice, he had a pervasive influence on mouse genetics.[2] Most of the early American mouse and mammalian geneticists started out with Castle at the Bussey.[4] Numbered among Castle's students were Clarence Little (1914), developer of the first inbred mouse line; Sewell Wright (1915), co-founder of the field of population genetics; L. C. Dunn (1920), eminent developmental geneticist; and George Snell (1930), developer of coisogenic strains used to study histocompatability loci and recipient of the 1980 Nobel Prize in Physiology or Medicine.

Despite early indications that some human diseases might be mimicked by certain mouse mutants, progress in defining new genetic variants was slow.[3] In fact, Little and Bagg wrote in 1924 that "mice and rats are singularly free of morphologic variations."[17] We know that Little certainly did not heed his own advice. He went on to become the first director of the Jackson Laboratory in Bar Harbor, Maine in 1929, which eventually became the pre-eminent mouse research facility in the world.[18] Also, according to a personal communication to Morse[3] from C.E. Keeler, one of Castle's Ph.D. students, Castle had suggested to Keeler that "some species besides mice should occupy his time in the future as there were no new mutations to be discovered." Now here is a case where not paying attention to one's mentor was a smart decision. Of course, it is possible that Castle was simply trying to challenge Keeler.

The origin of mouse genetics research was, from its inception at the beginning of the 20th century, aimed at queries in human medicine.[4,18] The earliest of these studies involved the use of the first inbred line (DBA, denoting the coat color genes *d*, dilute; *b*, brown; and *a*, nonagouti) developed by Little while still an undergraduate at Harvard. In elegant tumor transplant experiments involving DBA and Japanese waltzing mice, Little and E.E. Tyzzer demonstrated that transplant acceptance depends upon polygenic inheritance; the successful recipient must carry the dominant gene at each histocompatibility locus carried by the donor tissue.[19] This research also demonstrated the value of the genetic uniformity of an inbred line and perhaps was the impetus for researchers to develop many more inbred lines for biomedical research.[2] The original inbred lines are thought to consist of a combination of four *Mus* species, *Mus musculus domesticus, M. m. musculus, M. m. castaneus and M. m. bactrianus*,[20] and are conventionally identified as *Mus musculus.*[5]

At the other end of the spectrum of early mouse genetics research was the interest by evolutionary biologists and animal breeders concerning the nature of genetic variation. The first experimental approach was to select for a specific quantitative trait and simply to see how far artificial selection could change a trait and how many generations were needed to reach a limit, if indeed there would be a limit. To insure maximum genetic variation, selection was initiated in a randombred strain or after crossing two or more inbred lines. The earliest experiments of this nature were by Goodale in 1931 for white hair on the face[20,21] and by Goodale in 1930 and MacArthur in 1939 for 60-day body weight.[22–26] The first modern selection experiments with the mouse, analyzed and interpreted by conventional quantitative genetics methods, were reported by Falconer at the University of Edinburgh.[27]

4. Conclusions

The relationship between the house mouse and humans became firmly established with the introduction of agriculture in the Fertile Crescent about 10,000 years ago. The mouse was well adapted to *steal* food from granaries following the harvest. As farming spread through Eurasia, the major defense against the insurgent mice was the use of cats, which may have led to deification of cats in Egypt. The mouse itself enjoyed periods of being protected and worshipped during the Greek and Roman eras, where they were also used as augurs. During these periods and at different times well into the 17th century, the mouse was employed in various pharmaceutical remedies. However, the mouse was scorned by the Catholic clergy as being libidinous and an instrument of the devil.

The domestication of the house mouse probably originated in China and Japan where the first mutant mice were maintained. Mouse fanciers from Asia brought their hobby to Europe in the 19th century, and fancy mice eventually reached the United States toward the latter part of the 19th century. At the beginning of the 20th century Ms. Abbie Lathrop started a mouse breeding business in Massachusetts, which supplied mice to mouse fanciers and later to laboratories for genetics research.

The rediscovery of Mendel's laws in 1900 launched a rush to verify these genetic principles in other organisms. Use of the mouse was a logical choice in studying mammalian genetics, the initial experiments being conducted by Cuénot in France and Castle in the United States. Early research efforts in mouse genetics were directed toward problems in human medicine, but were also applied to questions in evolutionary biology and breeding. Even more exciting mammalian genetics research was to be conducted with the wee mouse as it entered the world of genetic maps, transgenics, knockouts, positional cloning of mutants and quantitative trait loci affecting complex traits. A crowning milestone in mouse genetics was publication of the sequence of the mouse genome in *Nature* in December 2002.[28] We can only wonder what Castle, Little, Cuénot and the other pioneers in mouse genetics would have made of all this.

References

1. Keeler, C.E. 1931. *The Laboratory Mouse: Its Origin, Heredity, and Culture*. Harvard University Press, Cambridge.
2. Morse III, H.C. 1978. Introduction. In: *Origins of Inbred Mice*. ed. H.C. Morse III. pp. 3–21. Academic Press, Inc., New York.
3. Morse III, H.C. 1981. The laboratory mouse - historical perspective. In: *The Mouse in Biomedical Research, Vol. I. History, Genetics and Wild Mice*. ed. H.L. Foster, J.D. Small and J.G. Fox. pp. 1–16. Academic Press, Inc., New York.
4. Russell, E.S. 1985. A history of mouse genetics. *Ann. Rev. Genet.* **19**:1–28.
5. Silver, L.M. 1995. *Mouse Genetics*. Oxford University Press, Inc., New York.
6. Paigen, K. 2003. One hundred years of mouse genetics: An intellectual history. I. The classical period (1902-1980). *Genetics* **163**:1–7.
7. Paigen, K. 2003. One hundred years of mouse genetics: An intellectual history. II. The molecular revolution (1981-2002). *Genetics* **163**:1227–1235.
8. Diamond, J. 1999. *Guns, Germs, and Steel: The Fates of Human Society*. W.W. Norton & Co., New York.
9. Keeler, C.E. and S. Fuji. 1937. The antiquities of mouse varieties in the orient. *J. Heredity* **28**:92–96.
10. Henig, R.M. 2000. *The Monk in the Garden: The Lost and Found Genius of Gregor Mendel, the Father of Genetics*. Houghton Mifflin Co., Boston.
11. Cuénot, L. 1902. Notes et revues. *Arch. Zool. Exp. Gen.* XXVII.
12. Castle, W.E. 1903. Mendel's law of heredity. *Proc. Am. Acad. Arts Sci.* **38**:535–548.
13. Castle, W.E. and C.C. Little. 1909. The peculiar inheritance of pink eyes among colored mice. *Science* **30**:313–314.
14. Castle, W.E. and C.C. Little. 1910. On a modified Mendelian ratio among yellow mice. *Science* **32**:868–870.

15. Cuénot, L. 1905. Les races pures et leur combinaisons chez les souris. *Arch. Zool. Exp.* **3**:123–132.
16. Haldane, J.B.S., A.D. Sprunt and N.M. Haldane. 1915. Reduplication in mice. *J. Genet.* **5**:133–135.
17. Little, C.C. and H.J. Bagg. 1924. The occurrence of four inheritable morphological variations in mice and their possible relation to treatment with X-rays. *J. Exp. Zool.* **41**:45–91.
18. Russell, E.S. 1978. Origins and history of mouse inbred strains: Contributions of Clarence Cook Little. In: *Origins of Inbred Mice*. ed. H.C. Morse III. pp. 33–43. Academic Press, Inc., New York.
19. Little, C.C. and E.E. Tyzzer. 1916. Further studies on inheritance of susceptibility to a transplantable tumor of Japanese waltzing mice. *J. Med. Res.* **33**:393–398.
20. Bonhomme, F., J.-L. Guénet, B. Dod, K. Moriwaki and G. Bulfield. 1987. The polyphyletic origin of laboratory inbred mice and their rate of evolution. *J. Linnean. Soc.* **30**:51–58.
21. Goodale, H.D. 1942. Further progress with artificial selections. *Am. Nat.* **76**:515–519.
22. Kyle, W.H. and H.D. Goodale. 1963. Selection progress toward an absolute limit for amount of white hair in mice. In: *Genetics Today: Proc 11th Int. Congr. Genet. Vol 1*. ed. S.J. Geerts. pp. 154–155. Pergamon Press, Oxford.
23. Goodale, H.D. 1938. A study of the inheritance of body weight in the albino mouse by selection. *J. Hered.* **29**:101–112.
24. Wilson, S.P., H.D. Goodale, W.H. Kyle and E.F. Godfrey. 1971. Long term selection for body weight in mice. *J. Hered.* **62**: 228–234.
25. MacArthur, J.W. 1944. Genetics of body size and related characters. I. Selecting small and large races of the laboratory mouse. *Am. Nat.* **78**:142–157.
26. MacArthur, J.W. 1949. Selection for small and large body size in the house mouse. *Genetics* **34**:194–209.
27. Falconer, D.S. 1955. Patterns of response in selection experiments with mice. *Cold Spring Harbor Symp. Quant. Biol.* **20**:178–196.
28. Waterston, R.H. *et al.* 2002. Initial sequencing and comparative analysis of the mouse genome. *Nature* **420**: 520–562.

CHAPTER 2

TESTING QUANTITATIVE GENETIC SELECTION THEORY

E. J. Eisen

Animal Science Department
North Carolina State University, Raleigh, NC, USA
gene_eisen@ncsu.edu

1. Introduction

Quantitative genetics deals with the inheritance of complex traits that are controlled by many loci, each with relatively small effects, and by environmental influences such as diet and physical activity. Most of these traits are continuously distributed, although some like ovulation rate and litter size can only be whole numbers; nevertheless, these metric characters are assumed to have an underlying continuous distribution. Until the introduction of molecular marker technology, there was little success in identifying the location of individual genes controlling complex traits. Therefore, statistical methodology had to be developed based on variances and covariances among relatives to determine the amount and types of genetic variability affecting these traits.[1]

The theory of quantitative genetics is based on the supposition of Mendelian inheritance and allows the prediction of certain outcomes in a population. Theoretical developments found immediate application in plant and animal breeding because most economic traits are of a quantitative nature. However, quantitative genetics now makes important contributions to many other fields, including evolutionary biology, behavioral genetics and genetics of complex human diseases.

Experimental issues in quantitative genetics address three major goals: a) deducing the properties of genes associated with quantitative variation; b) testing validity of the theory by experimental breeding; and c) determining consequences of specific breeding procedures that cannot be determined by the theory.[1] The mouse was adopted as a pilot organism to test quantitative genetic theory, particularly beginning after 1945.[2] Fueled by the observation that mice share many metric traits analogous to those in livestock, tests of quantitative genetic theory involved traits such as body weight, growth rate, feed efficiency, feed intake, body composition and litter size, emphasis often being on the genetic architecture of the traits themselves rather than testing theory *per se*.[3–8]

The present review concentrates on several key topics in testing quantitative genetic selection theory with the mouse: a) relating genetic parameters estimated from covariances among relatives with realized direct and correlated responses from single-trait selection experiments; b) making use of variation among replicates in direct and correlated responses to selection; c) comparing observed and predicted responses to index selection; d) determining the basis for selection limits after long-term selection; and e) testing the validity of the infinitesimal model. Other aspects of experimental tests of quantitative genetic theory are given in this volume on maternal effects (Ch. 3), inbreeding and crossbreeding (Ch. 4) and genotype by environment interaction (Ch. 5).

2. Predicted and Realized Heritabilities

Prediction of selection response for a quantitative trait x is determined by the equation

$$\Delta G_{x\cdot x} = ih_x^2\sigma_{p_x} \tag{1}$$

where i = selection intensity, h_x^2 = heritability and σ_{p_x} = phenotypic standard deviation. Heritability is most often estimated from paternal half-sib or offspring-sire (offspring-dam in the case of traits limited to females only) covariances in the base population. The assumptions are that estimates are from a random mating population in Hardy-Weinberg and linkage equilibrium, no genotype by environment interaction, no epistasis and no sex-linkage. A further assumption for offspring-parent covariances is no nuclear maternal effects.[9] The prediction equation provides vital information to breeders of economic species as to expected rate of genetic gain. Therefore, breeders must have a high degree of confidence in base population heritability estimates. Additionally, the magnitude of the base heritability plays a critical role in determining the most efficient type of selection that should be applied to the trait; e.g., individual selection for highly heritable traits *versus* index selection for traits of low heritability.

Sheridan[10] has reviewed the literature on predicted (h_p^2) and realized (h_r^2) heritabilities in selection experiments with laboratory and farm animals. The most frequently used estimate of the realized heritability is the regression of response in trait x, usually deviated from a contemporary control line, on cumulative selection differential.[1] In those populations with a past history of long-term selection, the realized response was found to be poorly predicted. A summary across all species surveyed, where selection was initiated in a population not at a plateau, also showed poor agreement between predicted and realized heritability; approximately half had a disagreement of at least 30 percent and about 25 percent had realized heritabilities significantly different from predicted heritabilities.[2,10]

It is not unexpected that long-term selection experiments would show poor agreement between h_p^2 and h_r^2 because long-term selection can cause negative linkage disequilibrium, nonlinearity in the regression of offspring on parent, and changes in allele frequencies.[11] In addition, drift and mutations may become more important than in short-term selection.[11] Several factors can contribute to the poor prediction of short-term responses: a) genotype by environment interaction; b) linkage disequilibrium; c) genetic drift; d) inbreeding depression; e) maternal effects; f) environmental change; g) correlated environmental effects; and h) epistasis.[10,11]

There are, however, several design points indicating that the disagreements are not as serious as suggested in.[10] Hill and Caballero[2] noted that a) no account was taken of the Bulmer effect[12] in individual or other selection procedures using among-family information; b) for low heritability, a large percentage difference is small both in absolute terms and when compared to its standard error; c) agreement actually improved the further selection proceeded; d) the high percentage of significant statistical tests may be biased upward because most of the standard errors of realized responses were based on ordinary least squares, which underestimates the true standard errors; and e) analyses of plateaued populations are confounded by mutational effects and segregation of deleterious genes affecting fitness.

An updated summary of a comparison of predicted (h_p^2) and realized (h_r^2) heritabilities in mouse selection experiments suggests that, at least in mice, the disagreement between h_p^2 and h_r^2 is not great (Table 1). Standard errors of realized heritabilities were based on either variance among replicates or approximations derived in.[13] Where standard errors were available for both h_p^2 and h_r^2, only one of 24 comparisons (4.2%) of the differences between h_p^2 and h_r^2 was greater than twice the standard error of the difference, approximately what would be expected by random sampling alone. This result contrasts to 18.8% significant differences reported in.[10] For 12-day litter weight, higher estimates of h_p^2 compared to h_r^2 [14,15] may be related to less reliable variance component estimates because there was good agreement in h_r^2 between studies and among replicates within studies. Omitting these two reports, the correlation

Table 1. Comparison of predicted (h_p^2) and realized (h_r^2) heritabilities

Trait[a]	$h_p^2 \pm SE$	$h_r^2 \pm SE$	Replicates[e]	Generations	Reference
LW12	0.20[b]	0.12 ± 0.03 H[d]	4	10	(14)
LW12	0.25[b]	0.08 ± 0.04 H	2	9	(15)
BW6	0.47 ± 0.06	0.40 ± 0.03 H	6	10	(16)
		0.33 ± 0.05 L			
		0.37 ± 0.03 D			
BW6	0.36 ± 0.10	0.25 ± 0.01 H	2	7	(17)
BW6	0.42 ± 0.02	0.55 ± 0.07 H	U	12	(18)
PWG	0.28[b]	0.35 ± 0.04 H	2	14	(19)
PWG	0.19 ± 0.08	0.25 ± 0.03 H	4		(20)
TL6	0.44 ± 0.15	0.44 ± 0.03 H	2	7	(17)
BW8[c]	0.27 ± 0.07	0.29 ± 0.02 H	2	18	(21)
		0.18 ± 0.02 L	2	18	
		0.23 ± 0.02 D	2	18	
BW8	0.33 ± 0.05	0.24 ± 0.02 H	2	18	(21)
		0.22 ± 0.01[f] L	2	18	
		0.23 ± 0.02 D	2	18	
EFP	0.50 ± 0.09	0.63 ± 0.09 H	2	10	(22)
		0.66 ± 0.13 L	2	10	
		0.66 ± 0.05 D	2	10	
HC	0.42 ± 0.09	0.47 ± 0.08 H	2	10	(22)
		0.34 ± 0.08 L	2	10	
		0.40 ± 0.04 D	2	10	
LS	0.17 ± 0.04	0.19 ± 0.03 H	U	12	(18)
NBA	0.04 ± 0.10	0.10 ± 0.06 H	U	7	(23)
LWW	0.24 ± 0.10	0.11 ± 0.07 H	U	7	(23)
NBAD	0.21 ± 0.10	0.22 ± 0.04 H	U	7	(23)
LWWD	0.15 ± 0.12	0.22 ± 0.08 H	U	7	(23)

[a] LW12, 12-day litter weight; BW6, 6-week body weight; PWG, 3- to 6-week weight gain; TL6, 6-week tail length; BW8, 8-week body weight; EFP, epididymal fat pad weight/body weight; HC, hind carcass weight/body weight; LS, total number born; NBA, number born alive; LWW, litter weight weaned; NBAD, number born alive/dam body weight; LWWD, litter weight weaned/dam body weight.

[b] Standard error was not available.

[c] Estimates were from crossbred and randombred populations, respectively; realized heritabilities for individual and within-family selection were homogeneous and pooled within each base population.

[d] Indicates direction of selection: H, high; L, low; D, divergence.

[e] Number of replicates; U, denotes only one selected line.

[f] Difference between predicted and realized heritability exceeded twice the standard error of the difference.

between h_p^2 and h_r^2 was 0.81 ± 0.12 ($P < 0.01$), and there were about as many estimates of h_p^2 that were greater than h_r^2 as those that were less. It was concluded that for the mouse experiments reviewed, there was no evidence of disagreement between predicted and realized heritabilities.

Of the five bi-directional replicated selection experiments summarized in Table 1, only the study of 8-week body weight in crossbreds[21] showed asymmetric realized heritabilities, with a greater upward than downward estimate. The authors[21] attributed the asymmetry to a high frequency of genes for low body weight. However, with only two replicates, genetic drift cannot be ruled out as a cause of the asymmetry of h_r^2.[16] Earlier reports on selection for body weight were reversed with smaller realized heritabilities in the upward selected lines,[24–26] but failure to replicate these studies as well leaves open the possibility that asymmetry may have been due to genetic drift.

3. Predicted and Realized Genetic Correlations

Correlated responses to selection provide key information on pleiotropy between selected and correlated traits. The predicted correlated response in trait y when truncation selection is applied to x is

$$\Delta G_{y \bullet x} = i h_x h_y r_G \sigma_{P_y} \qquad (2)$$

where r_G = genetic correlation between x and y, and the other parameters are as defined previously. The magnitude and sign of genetic correlations determine the size and direction of correlated responses, many of which have been reported for mice.[3–6,8] However, relatively few studies have compared realized genetic correlations (r_{G_r}) with those genetic correlations estimated from covariance among relatives in the base population (r_{G_p}).[2]

Realized genetic correlations estimated from selection for a single trait are often reported. But these correlations are not, strictly speaking, realized estimates because they depend on the validity of the base population heritability of the correlated trait. Ideally, a realized genetic correlation should be based on selection for a different trait in each of two lines. Formulas for estimates of r_{G_r} using this method are given in.[1,17] Falconer[16] emphasized the importance of replication to minimize the likelihood that r_{G_r} is due to genetic drift as opposed to pleiotropy. An additional complication is asymmetric correlated responses, defined as selection in trait x yielding a response different from that expected when selection is for trait y.[27]

Few paired single-trait selection experiments have been reported with mice.[17, 18,22] A summary of realized (r_{G_r}) and base population (r_{G_p}) genetic correlations from these studies is given in Table 2. Standard errors of r_{G_r} were calculated from the variance among replicates or, in the absence of replication, by the method of Hill.[28] In general, there was good agreement between r_{G_p} and r_{G_r}, with only one of 13 differences exceeding twice the standard error of the difference. This difference was caused by an exceedingly large estimate of r_{G_r} in one replicate selected for high epididymal fat pad weight as a proportion of body weight.[22] However, the pooled estimates of r_{G_r} agreed closely with those of r_{G_p} in this study.[22] Over all studies, the correlation between r_{G_p} and r_{G_r} was 0.82 ± 0.17 ($P < 0.01$).

Table 2. Comparison of predicted (r_{Gp}) and realized (r_{Gr}) genetic correlations in paired single-trait studies

Trait x[a]	Trait y[a]	r_{Gp} ± SE	r_{G_r} ± SE	Criterion	Reference[d]
BW6	TL6	0.29 ± 0.09	0.31 ± 0.09 H[b]	Selection for BW6	(17)
			0.38 ± 0.06 H	Selection for TL6	
			0.33 ± 0.02 H	Pooled	
BW6	LS	0.63 ± 0.14	0.53 ± 0.10 H	Selection for BW6	(18)
			0.50 ± 0.10 H	Selection for LS	
			0.52 ± 0.10 H	Pooled	
EFP	HC	−0.61 ± 0.09	−0.72 ± 0.18 H	Selection for EFP	(22)
			−0.46 ± 0.05 L	Selection for HC	
			−0.57 ± 0.10	Pooled	
			−1.13 ± 0.24[c] L	Selection for EFP	
			−0.51 ± 0.28 H	Selection for HC	
			−0.69 ± 0.14	Pooled	
			−0.67 ± 0.12 D		

[a] BW6, 6-week body weight; TL6, 6-week tail length; LS, total number born; EFP, epididymal fat pad weight/body weight; HC, hind carcass weight/body weight.
[b] Indicates direction of selection: H, high; L, low; D, divergence.
[c] Difference between predicted and realized genetic correlation exceeded twice the standard error of the difference.
[d] See Table 1 for number of replicates and number of generations.

Asymmetry of correlated responses may be caused by a breakdown of the infinitesimal model due to linkage disequilibrium.[2,27] Single-trait selection for the positively correlated traits BW6 and TL6[17] and BW6 and LS,[18] respectively, showed no asymmetry (Table 2). For the negatively correlated traits, EFP and HC, there was a suggestion that high and low selection for EFP yielded a higher r_{G_r} than low and high selection for HC, but the standard errors were too large, due to the high variation among replicates, to draw any firm conclusion.[22] More replicates in this study may have clarified the basis for the tendency toward asymmetry.

4. Variation Among Replicates

In a series of classic studies, Hill[13,28,29] showed that standard errors of realized heritabilities and genetic correlation were biased downward when estimated by ordinary least squares regression, a consequence of positive correlated errors associated with genetic drift. The consequences of this finding are far reaching because it says that h_r^2, r_{G_r} and asymmetry of these parameter estimates would be declared significant more often than they should be. The same situation holds for the tests of significance of the differences between h_r^2 and h_p^2 and between r_{G_r} and r_{G_p}, respectively, as mentioned earlier.

One approach proposed to avoid this pitfall is to replicate the selection experiment and use the empirical standard error based on the variance among replicates, which includes drift effects.[16] Ten generations of replicated selection for large and small 6-week body weight in mice are summarized in Table 3.[16] The mean empirical standard errors of h_r^2 were 55% (large lines), 228% (small lines) and 135% (divergence) greater than the pooled standard error based on ordinary least squares regression. Also apparent is the absence of significant asymmetry of h_r^2 in the pooled data, whereas pairing of selected replicates would give a false impression that asymmetry was present. Similar findings were reported in another study involving divergent selection for three other traits.[30] Falconer[16] also showed that correlated responses in weaning weight varied considerably among replicates, emphasizing the need for replication in estimating realized correlated responses due to pleiotropy without the confounding effect of genetic drift.

In selection studies where replication was not used, it is still possible to obtain approximately unbiased standard errors of h_r^2 as outlined by Hill.[13,28,29] An example is shown in Table 4 for single-trait selection for 6-week body weight and litter size.[18] The approximate unbiased standard errors of h_r^2 for these respective traits were 55% and 89% larger than the biased standard errors.

Table 3. Realized heritabilities (h_r^2) for 6-week body weight in six replicates: data from[16]

	Large		Small		Divergence	
Replicate	h_r^2	SE	h_r^2	SE	h_r^2	SE
A	0.390	0.066	0.501	0.058	0.434	0.042
B	0.438	0.047	0.301	0.058	0.375	0.024
C	0.251	0.025	0.159	0.024	0.212	0.020
D	0.457	0.041	0.288	0.041	0.392	0.037
E	0.385	0.051	0.365	0.039	0.379	0.033
F	0.444	0.043	0.376	0.046	0.420	0.032
Pooled[a]	0.398	0.020	0.328	0.014	0.369	0.014
Mean[b]	0.395	0.031	0.331	0.046	0.369	0.033

[a] Regression of mean response of lines on mean selection differential with SE of least squares regression.
[b] Arithmetic mean h_r^2 with empirical SE based on variance among replicates.
Table adapted from[16] with permission from Cambridge University Press.

Table 4. Drift and error variances, realized heritability (h_r^2) and biased and approximately unbiased SE (h_r^2): data from[18]

	Trait[a]	
Item	BW6	LS
Drift Variance	0.0825	0.0425
Error Variance	0.0384	0.1263
Biased SE (h_r^2)[b]	0.042	0.019
Approximately unbiased SE (h_r^2)[c]	0.065	0.036
h_r^2	0.55	0.19

[a] Single-trait selection: BW6, 6-week body weight; LS, number born.
[b] Estimated by ordinary least squares regression.
[c] Estimated by a modification of Hill.[13]

5. Response to Index Selection

The selection index was developed to select simultaneously for two or more quantitative traits,[31,32] its immediate application being apparent for plant and animal genetic improvement. Efficiency of the selection index was shown to be greater than or equal to independent culling whose efficiency, in turn, is greater than or equal to tandem selection, the relative efficiencies being dependent on the genetic and phenotype variance-covariance matrices and economic weights of the traits.[33] Selection index experiments have been used to: a) test the theory of selection index and its efficiency compared with other multitrait selection methods; b) estimate realized heritabilities and genetic correlations; and c) test the validity of restricted and desired gains index theory.[34,35] Restricted and desired gains indexes are particularly useful for conducting "antagonistic" selection where the desired selection goals are contrary to the sign of the genetic correlation ,[17] a situation often encountered in livestock breeding.

Only one study with mice has compared index selection with independent culling and tandem selection; two replicates were selected for 3- to 6-week postweaning weight gain and litter size at birth using the multitrait methods.[35] Response of the aggregate genotype was greatest for the selection index followed by independent culling and tandem selection in that order, but these differences were not significant. One difficulty with this study was that predicted responses based on base population parameters, economic weights and selection intensities were not reported so it is not possible to compare predicted with observed responses.

Procedures have been developed for estimating realized genetic correlations and heritabilities when a pair of lines has been selected for two traits using different selection indexes.[37,38] Precision of these estimated genetic parameters has been derived.[39,40] This methodology has been applied to several antagonistic index selection studies.

Antagonistic index selection for 6-week body weight and tail length resulted in much higher realized genetic correlations (0.70, 0.82 and 1.19) than were obtained from the base population (0.29 ± 0.09) or from single-trait selection (0.33 ± 0.02) whereas the realized heritabilities were in fair agreement with base estimates.[17] In the high weight-low tail length line, observed response was 50% of the expected response while in the low weight-high tail length line, observed response was 75% of the expected response.

Desired gains index selection was applied to selection for high 6-week body weight-low litter size and *vice versa*; the genetic correlation used to calculate the index was 0.45.[18] Divergence in litter size was about one-half of the expected divergence, while that for 6-week body weight was only slightly less than expected.

Eisen[41] used a restricted selection index to select divergently for epididymal fat pad weight, holding body weight constant. Realized responses in the high fat-restricted body weight line agreed with expectation, but responses in the low fat-restricted body weight line were contrary to expectation with no significant response in epididymal fat pad weight and an increase in body weight. The realized genetic correlation of 0.93 ± 0.01 was higher than the base population estimate of 0.78 ± 0.05.

Several other antagonistic selection experiments with mice have yielded observed responses less than expected.[38,42–44] There are, however, several exceptions. Successful selection was attained for high and low lean body mass while holding fat content unchanged.[30] Selection for high or low early weight gain while holding late weight gain constant and for high or low late weight gain while holding early weight gain constant resulted in significant divergence among lines, and the changes in the growth trajectories were close to expectation.[45]

Collectively, these studies indicate that realized responses to index selection often do not agree well with expectation when antagonisms exist between the signs of the genetic correlations and the direction of selection for component traits. Possible explanations for these discrepancies include: a) rapid changes in gene frequency with selection, causing changes in genetic parameters;[46] b) genetic drift; c) linkage disequilibrium introduced as a consequence of selection; d) high correlation among traits in the index causing multicollinearity; and e) imprecise estimates of genetic parameters. The selection index experiments suggest the need to closely monitor the correlated responses of component traits, particularly when using restricted or desired gains indexes or when there are antagonisms between the signs of the genetic correlations and the economic weights. A review of restricted and desired gains indexes in chickens, turkeys and *Tribolium castaneum*, as well as some of the mouse studies reported here, prompted the conclusion that these indexes are generally less effective than predicted in reducing undesirable correlated responses.[11]

6. Individual Selection Compared to Other Methods

Surprisingly, only one mouse experiment has compared the response to individual selection with other selection methods. Short-term response to within-family selection is expected to yield a smaller response than individual selection, but it has the advantage of approximately doubling the effective population size and eliminating maternal effects,[1] apart from a fraction due to additive direct-maternal covariance.[19] Short-term, within-family divergent selection for 8-week body weight was nearly 60 percent as efficient as individual selection, which was expected based on the full-sib intraclass correlation.[21] Across 18 generations of selection, the disadvantage of within-family over individual selection was less, chiefly because of reduced genetic drift and the lower rate of inbreeding in within-family selection,[21] a result predicted from computer simulation studies.[47]

7. Long-term Responses and Selection Limits

The mouse has played an important role in characterizing the nature of long-term selection response.[48–52] Apart from moderate perturbations in generation means, typical long-term responses are characterized by an initial linear response followed by a deceleration of response until finally a plateau is reached.[1,11] The genetic bases for selection limits of a quantitative trait are as follows:[1,48,50] a) exhaustion of additive genetic variance due to fixation of loci; b) overdominance at many loci; c) opposing natural selection; d) undesirable recessive genes segregating at low frequency; e) negative genetic correlations among component characters; f) genotype by environment interactions; and g) tight linkages. The conflict between natural and artificial selection can arise in several ways:[11] a) alleles favored by artificial selection can have deleterious effects on fitness; b) loci favored by artificial selection can be in linkage disequilibrium with loci having deleterious effects on fitness; and c) selection increases the amount of inbreeding, which can increase the frequency of deleterious recessives. Roberts[49] noted that the limits reached in a specific selection experiment are a function of three factors: a) the specific environment in which selection was conducted; b) genetic properties of the trait; and c) how selection was practiced, including selection intensity, effective population size and method of selection.

Most experiments with mice on the nature of selection limits and how to overcome the plateau have concentrated on one of two traits, litter size or body weight. Falconer[53] originally proposed a method to overcome a selection limit thought to be caused by deleterious recessive alleles by forming partially inbred sublines from the selected line and then selecting among the sublines and crossing the best ones. An apparent plateau for increased litter size was overcome by this method in two independent studies.[53,54] It was, therefore, concluded that the cause of the selection limit in both cases was the segregation of undesirable recessive alleles at low frequency.

Attempts to overcome the selection limits in two other lines selected for large litter size have been reported.[55,56] One line had been selected with litters not standardized[57] and reached a plateau at about generation 30, even though fertility was 93%.[53] Divergent selection for litter size using standardized litters was initiated at generation 48 and continued for 12 generations.[55] The mean litter size in the standardized line was 0.7 pups higher than the original unstandardized line.

This increase suggests that the plateau may have been partially caused by the negative maternal environmental effect arising because females born in large litters tend to produce smaller litters and *vice versa*. Thus, a partial explanation of the selection limit may have been a genotype by environment interaction, the environmental treatments being unstandardized and standardized litters. However, continued upward selection yielded no further significant response although the granddaughter-grandam estimate of heritability was 0.31 ± 0.11. Reverse selection was successful, indicating that additive genetic variance was still present at the selection limit; the response may have been hindered by recessive alleles present at low frequency. This line was crossed with another line that had plateaued at about generation 20, and selection for large litter size initiated after two generations of random mating.[56] The rationale for this approach was that the synthetic line should respond if there were different alleles fixed in each parental line. In addition, upward selection was continued in both original lines. Averaged over generations, the synthetic line had 2.39 and 1.11 more pups than the average of the parental lines in generations 22 to 31 and 32 to 41, respectively. However, the realized heritability in the synthetic line was not significantly different from zero, suggesting that the two lines had not reached fixation for different alleles. Again, the lack of response was attributed to the presence of recessive alleles at low frequency.

The Falconer design[53] was employed in two studies to try to overcome an apparent selection limit for increased growth.[58, 59] Although recessive alleles were still segregating at the limit for 6-week body weight in one of the replicate lines described in,[16] total gain obtained by their elimination was too small to significantly overcome the plateau.[58] A significant response to reverse selection indicated the presence of additive genetic variance and suggested the possibility of reproductive difficulties causing the plateau. In a second study, the Falconer design failed to overcome the selection limit for 3- to 6-week weight gain.[59] Reversed and relaxed selection led to significant responses and improvement in net reproductive rate. Therefore, the conclusion in both studies was that the selection limit was likely due to a negative association between the selected trait and fitness.[58, 59]

Roberts[60] reported on two lines selected for 3- to 6-week weight gain that had reached an apparent limit. The high growth line failed to respond to reversed and relaxed selection. This result indicated that effectively, the additive genetic variance in this line had been exhausted. In contrast, the small line at the limit regressed slightly when selection was relaxed, and reverse selection yielded a response. The conclusion was that natural selection affecting viability explained the lack of continued response in the small line.[60]

An analysis of Goodale's large body weight line[61] showed a plateau despite a persistent positive regression of offspring on sire through 84 generations.[62] There was a negative correlation between body weight and litter size. Possible explanations for the plateau were exhaustion of additive genetic variance or a negative association between body weight and fitness.

A logical approach to overcoming selection limits in two or more lines is to cross the lines and initiate selection in the synthetic line after a few generations of random selection, as noted earlier. This procedure should prove successful if the selected lines were fixed for different alleles. Success of this approach has been reported for crossing large body weight lines.[63,64] However, crossing of low body weight lines[64] and crossing the aforementioned high litter size lines[56] failed to elicit a response. One possible explanation other than fixation for the same alleles is extremely tight linkages.[64]

In summary, indications are that the reason for the selection plateaus for high litter size, a trait with a relatively low heritability, was segregation of undesirable recessives. In contrast, selection limits for large body weight, a trait with a moderate heritability, was due to exhaustion of genetic variance and/or a negative impact on reproductive fitness.

A key finding of the theory of selection limits for a quantitative trait is that the total response is approximately twice the effective population size (N_e) times the response in the first generation.[65] The generations required to reach the limit and total response tend to increase approximately linearly with N_e i.[50] Realized heritabilities for 3- to 6-week weight gain increased with N_e at each of two levels of i, and genetic variation was more quickly depleted in later generations with smaller N_e.[66] Sixteen generations of selection for high postweaning gain resulted in increased rates of response at larger values of N_e [67] (Table 5). In a long-term selection study for large 6-week body weight using a larger $N_e = 60$ than previous reports, the total selection response was higher and continued for more generations than observed in earlier reports with smaller N_e.[68] The reduction in response at lower N_e is a consequence of loss of desirable alleles because of genetic drift and/or inbreeding depression. Theoretically, mutational effects proportional to N_e may contribute to the response observed in long-term selection experiments.[69] Fifty generations of divergent selection for 6-week body weight from an inbred base population of mice led to a difference of about three phenotypic standard deviations between the high and low lines.[70] Therefore, mutational events have likely contributed to long-term selection experiments with mice.[52]

Table 5. Relationship between population size and response to 16 generations of selection for postweaning gain in mice[67]

Parental Pairs [a]	F_{16} [b]	N_e [c]	Response (g)
10	0.326	21	1.9
20	0.262	27	2.6
50	0.165	45	4.8
100	0.099	77	4.7
150	0.086	89	5.1
200	0.078	99	5.5

[a]Three replicates at each population size.
[b]Cumulative inbreeding coefficient at generation 16.
[c]Approximated by the formula $1 - F_t = (1 - \Delta F)^t$ where $\Delta F = 1/2N_e$ and $t = 16$.
Table adapted from[50]: ©1980. Blackwell Publishing Limited.

8. Testing Validity of the Infinitesimal Model

The infinitesimal model assumes that traits are determined by an infinite number of unlinked additive loci each with small effect, and gene frequencies are assumed to be unchanged by directional selection.[12] While use of the animal model in restricted maximum likelihood (REML) assumes the infinitesimal model,[11] its validity has been questioned.[71]

Using REML methodology, it was shown that long-term divergent selection for food intake[72] and lean mass,[73] respectively, reduced additive genetic variance and heritability in later generations. The magnitude of the reductions was too large to be accounted for by selection causing linkage disequilibrium of unlinked loci. Therefore, the results suggest that the assumptions of the infinitesimal model were not valid for these data sets.[72,73] A possible explanation for the failure of the infinitesimal model to hold for lean mass is evidence that sex-linked effects influenced selection response[73] whereas autosomal loci are assumed for the infinitesimal model. The REML methodology also was applied to a long-term selection experiment for high and low gonadal fat pads as a percentage of body weight.[74] Despite a four-fold difference between the high fat and low fat lines, there was no indication that an infinitesimal model could not adequately describe the data.[74]

In a contrasting study, 20 generations of divergent selection for 6-week body weight led to an increase in phenotypic variance in the low and high lines, which was attributable to increases in environmental and additive genetic variances.[75]

Possible explanations for the failure of the infinitesimal model in this case include dominance, epistasis and linkage disequilibrium.[75] Analyses of quantitative trait loci (QTL) indicate that the distribution of QTL effects approximates an exponential distribution for body weight in mice[76] and for many other quantitative traits in other species,[77] as predicted by Robertson.[78] It is apparent that further studies are needed to determine the effects of failure of the assumptions of the infinitesimal model on estimates of genetic parameters obtained from selection experiments using the animal model.

9. Conclusions

Heritability of a quantitative trait calculated from the covariance among relatives is an estimate of additive genetic variance as a proportion of phenotypic variance. The base population heritability estimate is used to predict direct selection response and to determine the most efficient selection method to improve the trait. A survey of selection experiments in mice for growth, body composition and reproductive traits showed reasonably good agreement between predicted and realized heritabilities.

Genetic correlation obtained from the covariance among relatives in an equilibrium population reflects the degree of pleiotropy between two quantitative traits and is used to predict the magnitude and direction of correlated responses. For the few paired single-trait selection experiments reported with mice, there is reasonably good agreement between predicted and realized genetic correlations.

Use of the mouse in selection experiments has confirmed the importance of replication in accounting for genetic drift effects. Replicated selection experiments have verified the theory that standard errors of response and realized heritability obtained from ordinary least squares are biased downward. As a consequence, replication is shown to be essential in testing for asymmetry and in minimizing the likelihood that correlated responses are caused by genetic drift. However, even without replication, methods are available to account for drift effects when calculating standard errors of these estimates.

In contrast to single-trait selection experiments, selection index studies with mice have not often agreed closely with prediction, particularly where antagonistic index selection has been applied. This observation has been noted in studies with other species and merits further theoretical investigation.

Long-term selection studies with mice are represented mainly by litter size, a lowly heritable trait, and body weight, a moderately heritable trait. The plateaus for litter size were primarily associated with segregation of undesirable recessives at low frequency. In contrast, the basis for the selection limit for body weight was exhaustion of additive genetic variance and poor reproductive fitness. In general, the results with mice verify the theory of selection limits, in that the total response was increased with greater effective population size.

The few analyses of selection experiments for quantitative traits with mice based on the animal model and REML suggest that the assumptions of the infinitesimal model were not valid. More research is needed to determine the causes and consequences for failure of the infinitesimal model to hold and to develop analytical paradigms to deal with this issue.

References

1. Falconer, D.S. and T.F.C. Mackay. 1996. *Introduction to Quantitative Genetics*. Fourth Edition. Pearson Education Limited, Essex.
2. Hill, W.G. and A. Caballero. 1992. Artificial selection experiments. *Annu. Rev. Ecol. Syst.* **23**:287–310.
3. Roberts, R.C. 1965. Some contributions of the laboratory mouse to animal breeding research. Part I. *Anim. Br. Abst.* **33**:339–353.
4. Roberts, R.C. 1965. Some contributions of the laboratory mouse to animal breeding research. Part II. *Anim. Br. Abst.* **33**:515–526.
5. Eisen, E.J. 1974. The laboratory mouse as a mammalian model for the genetics of growth. *Proc. First World Congr. Genet. Appl. Livest. Prod.* **1**:467–492.
6. Roberts, R.C. 1979. Side effects of selection for growth in laboratory animals. *Livest. Prod. Sci.* **6**:93–104.
7. McCarthy, J.C. 1982. The laboratory mouse as a model for animal breeding: A review of selection for increased body weight and litter size. *Proc. Second World Congr. Genet. Appl. Livest. Prod.* **5**:66–83.
8. Eisen, E.J. 1989. Selection experiments for body composition in mice and rats: A review. *Livest. Prod. Sci.* **23**:17–32.
9. Willham, R.L. 1963. The covariance between relatives for characters composed of components contributed by related individuals. *Biometrics* **19**:18–27.
10. Sheridan, A.K. 1988. Agreement between estimated and realized genetic parameters. *Anim. Br. Abst.* **56**:877–889 .
11. Walsh, B. and M. Lynch. 2000. *Selection and Evolution of Quantitative Traits.* http://nitro.biosci.arizona.edu/zbook/volume_2/vol2.html
12. Bulmer, M.G. 1971. The effect of selection on genetic variability. *Am. Nat.* **105**:201–211.
13. Hill, W.G. 1972. Estimation of realized heritabilities from selection experiments, II. Selection in one direction. *Biometrics* **28**:767–780.
14. Eisen, E.J., J.E. Legates and O.W. Robison. 1970. Selection for 12-day litter weight in mice. *Genetics* **64**:511–532.
15. Robinson, Jr., W.A., J.M. White and W.E. Vinson. 1974. Selection for increased 12-day litter weight in mice. *Theor. Appl. Genet.* **44**:337–344.

16. Falconer, D.S. 1973. Replicated selection for body weight in mice. *Genet. Res.* **22**:291–321.
17. Rutledge, J.J., E.J. Eisen and J.E. Legates. 1973. An experimental evaluation of genetic correlation. *Genetics* **75**:709–726.
18. Eisen, E.J. 1978. Single-trait and antagonistic index selection for litters size and body weight in mice. *Genetics* **88**:781–811.
19. Hanrahan, J.P., E.J. Eisen and J.E. Legates. 1973. Effects of population size and selection intensity on short-term response to selection for postweaning gain in mice. *Genetics* **73**:513–530.
20. LaSalle, T.J., J.M. White and W.E. Vinson. 1974. Direct and correlated responses to selection for increased postweaning gain in mice. *Theor. Appl. Genet.* **44**:272–277.
21. von Butler, I., H. Willeke and F. Pirchner. 1984. Two-way within-family and mass selection for 8-week body weight in different mouse populations. *Genet. Res.* **43**:191–200.
22. Eisen, E.J. 1987. Selection for components related to body composition in mice: Direct responses. *Theor. Appl. Genet.* **74**:793–801.
23. Luxford, B.G. and R.G. Beilharz. 1990. Selection response for litter size at birth and litter weight at weaning in the first parity in mice. *Theor. Appl. Genet.* **80**:625–630.
24. Falconer, D.S. 1953. Selection for large and small size in mice. *Genetics* **51**:470–501.
25. Legates, J.E. 1969. Direct and correlated responses to selection in mice. In: *Genetics Lectures,* Volume 1. ed. R. Bogart. pp. 149–165. Oregon State University Press. Corvallis.
26. Sutherland, T.M., P.E. Biondini and C.H. Haverland. 1968. Selection under assortative mating in mice. *Genet. Res.* **11**:171–178.
27. B.B. Bohren, W.G. Hill and A. Robertson. 1966. Some observations on asymmetrical correlated responses to selection. *Genet. Res.* **7**:44–57.
28. Hill, W.G. 1971. Design and efficiency of selection experiments for estimating genetic parameters. *Biometrics* **27**:293–311.
29. Hill, W.G. 1972. Estimation of realized heritabilities from selection experiments. I. Divergent selection. *Biometrics* **28**:747–765.
30. Sharp, G.S., W.G. Hill and A. Robertson. 1984. Effects of selection on growth, body composition and food intake in mice. I. Responses in selected traits. *Genet. Res.* **43**:75–92.
31. Smith, H.F. 1936. A discriminant function for plant selection. *Ann. Eugenics* **7**:240–250.
32. Hazel, L.N. 1943. The genetic basis for constructing selection indexes. *Genetics* **28**:476–490.
33. Young, S.S.Y. 1961. A further examination of the relative efficiency of three methods of selection for genetic gains under less restricted conditions. *Genet. Res.* **2**:106–121.
34. Cunningham, E.P., R.A. Moen and T.Gjedrem. 1970. Restriction of selection indexes. *Biometrics* **26**:67–74.
35. Yamada, Y., K. Yokouchi and A. Nishida. 1975. Selection index when genetic gains of individual traits are of primary concern. *Japan. J. Genetics* **50**:33–41.
36. Doolittle, D.P., S.P. Wilson and L.L. Hulbert. 1972. A comparison of multiple trait selection methods in the mouse. *J. Heredity.* **63**:366–372.
37. Harvey, W.R. 1972. Direct and indirect response in two-trait selection experiments in mice. *Proc. 21st National Breeders Round Table.* Kansas City, MO.
38. Berger, P.J. and W.R. Harvey. 1975. Realized genetic parameters from index selection in mice. *J. Anim. Sci.* **40**:38–47.
39. Gunsett, F.C., K.N. Andriano and J.J. Rutledge. 1982. Estimating the precision of estimates of genetic parameters realized from multiple-trait selection experiments. *Biometrics* **38**:981–989.

40. Cameron, N.D. and R. Thompson. 1986. Design of multivariate selection experiments to estimate genetic parameters. *Theor. Appl. Genet.* **72**:466–476.
41. Eisen, E.J. 1992. Restricted index selection in mice designed to change body fat without changing body weight: Direct responses. *Theor. Appl. Genet.* **83**:973–980.
42. McCarthy, J.C. and D.P. Doolittle. 1977. Effects of selection for independent changes in two highly correlated body weight traits in mice. *Genet. Res.* **29**:133–145.
43. Eisen, E.J. 1977. Restricted selection index: An approach to selecting for feed efficiency. *J. Anim. Sci.* **44**:958–972.
44. Eisen, E.J. 1993. Multitrait restricted and desired gains selection indices designed to change growth and body composition in mice. *J. Anim. Breed. Genet.* **110**:13–29.
45. Atchley, W.R., S. Xu and D.E. Cowley. 1997. Altering developmental trajectories in mice by restricting index selection. *Genetics* **146**:629–640.
46. Mortimer, S.I. and J.W. James. 1987. Changes in genetic parameters under restricted index selection. *Genet. Res.* **49**:129–134.
47. Dempfle, L. 1975. A note on increasing the limit of selection through selection within families. *Genet. Res.* **24**:127–135.
48. Al-Murrani, W.K. 1974. The limits to artificial selection. *Anim. Br. Abst.* **42**:587–592.
49. Roberts, R.C. 1974. Selection limits in the mouse and their relevance to animal breeding. *First World Congr. Genet. Appl. Livest. Prod.* **1**:493–509.
50. Eisen, E.J. 1980. Conclusions from long-term selection experiments with mice. *Z. Tierzüchtg. Züchtgsbiol.* **97**:305–319.
51. Bünger, L., U. Renne and R.C. Buis. 2001. Body weight limits in mice-long-term selection and single genes. In: *Encyclopedia of Genetics.* ed. E.C.R. Reeve. pp. 337–360. Fitzroy Dearborn Publ., London.
52. Hill, W.G. and L. Bünger. 2004. Inference on the genetics of quantitative traits from long-term selection in laboratory and domestic animals. In: *Plant Breeding Reviews.* Vol. 24, Part 2. ed. J. Janick. pp. 169–210. John Wiley & Sons, Inc., New York.
53. Falconer, D.S. 1971. Improvement of litter size in a strain of mice at a selection limit. *Genet. Res.* **17**:215–235.
54. Eklund, J. and G.E. Bradford. 1977. Genetic analysis of a strain of mice plateaued for litter size. *Genetics* **85**:529–542.
55. Buis, R.C. 1988. Investigation of a selection limit for litter size in mice. *Livest. Prod. Sci.* **20**:161–172.
56. Vangen, O. 1993. Results from 40 generations of divergent selection for litter size in mice. *Livest. Prod. Sci.* **37**:197–211.
57. Bakker, H., J.H. Wallinga and R.D. Politiek. 1978. Reproduction and body weight of mice after long-term selection for large litter size. *J. Anim. Sci.* **46**:1572–1580.
58. Al-Murrani, W.K. and R.C. Roberts. 1974. Genetic variation in a line of mice selected to its limit for body weight. *Anim. Prod.* **19**:273–289.
59. Barria, N. and G.E. Bradford. 1981. Long-term selection for rapid gain in mice. I. Genetic analysis at the limit of response. *J. Anim. Sci.* **52**:729–738.
60. Roberts, R.C. 1966. The limits to artificial selection for body weight in the mouse. II. The genetic nature of the limits. *Genet. Res.* **8**:361–375.
61. Goodale, H.D. 1938. A study of the inheritance of body weight in the albino mouse by selection. *J. Heredity* **29**:101–112.

62. Wilson, S.P., H.D. Goodale, W.H. Kyle and E.F. Godfrey. 1971. Long term selection for body weight in mice. *J. Heredity*. **62**:228–234.
63. Falconer, D.S. and J.W.B. King. 1953. A study of selection limits in the mouse. *J. Genetics*. **51**:561–581.
64. Roberts, R.C. 1967. The limits to artificial selection for body weight in the mouse. III. Selection from crosses between previously selected lines. *Genet. Res.* **9**:73–85.
65. Robertson, A. 1960. A theory of limits in artificial selection. *Proc. Royal Soc.* **B 153**:234–249.
66. Eisen, E.J. 1975. Population size and selection intensity effects on long term selection response in mice. *Genetics* **79**:305–323.
67. Kownacki, M. 1979. Effect of reciprocal crossing of selected lines of mice. *Theor. Appl. Genet.* **54**:169–179.
68. Renne, U., M. Langhammer, E. Wytrwat, G. Dietl and L. Bünger. 2003. Genetic-statistical analysis of growth in selected and unselected mouse lines. *J. Expt. Anim.* **42**:216–232.
69. Hill, W. G. 1982. Predictions of response to artificial selection from new mutations. *Genet. Res.* **40**:255–278.
70. Keightley, P.D. 1998. Genetic basis of response to 56 generations of selection on body weight in inbred mice. *Genetics*. **148**:1931–1939.
71. Fairfull, R.W., I.M. McMillan and W.M. Muir. 1998. Poultry breeding: Progress and prospects for genetic improvement of egg and meat production. *Proc. Sixth World Congr. Genet. Appl. Livest. Prod.* **24**:271–278.
72. Meyer, K., and W.G. Hill. 1991. Mixed model analysis of a selection experiment for food intake in mice. *Genet. Res.* **57**:71–81.
73. Beniwal, B.K., I.M. Hastings, R.Thompson and W.G. Hill. 1992. Estimation of changes in genetic parameters in selected lines of mice using REML with an animal model. I. Lean mass. *Heredity* **69**:352–360.
74. Martinez, V., L. Bünger, and W.G. Hill. 2000. Analysis of response to 20 generations of selection for body composition in mice: fit to infinitesimal model assumptions. *Genet. Sel. Evol.* **32**:3–21.
75. Heath, S.C., G. Bulfield, R. Thompson and P.D. Keightley. 1995. Rates of change of genetic parameters of body weight in selected mouse lines. *Genet. Res.* **66**:19–25.
76. Rocha, J.L., E. J. Eisen., L.D. Van Vleck and D. Pomp. 2004. A large-sample QTL study in mice: I. Growth. *Mamm. Genome* **15**:83–99.
77. Hayes, B. and M.E. Goddard. 2001. The distribution of the effects of genes affecting quantitative traits in livestock. *Genet. Sel. Evol.* **33**:209–229.
78. Robertson, A. 1967. The nature of quantitative genetic variation. In: *Heritage from Mendel* ed. A. Brink. pp. 265–280. Wisconsin University Press, Madison.

CHAPTER 3

MATERNAL EFFECTS, GENOMIC IMPRINTING AND EVOLUTION

Jhondra Funk-Keenan[1, 3] and William R. Atchley[1, 2, 4]

[1]*Department of Genetics*
[2]*Center for Computational Biology*
North Carolina State University
Raleigh, NC, USA
[3]*jfunkk@ncsu.edu*
[4]*bill@atchleylab.org*

1. Introduction

In mammals, both mother and father contribute a haploid component of their genes to their progeny. However, the mother makes additional contributions through so-called "maternal effects," as well as through the egg cytoplasm, which includes mitochondrial genes and other cytoplasmic effects. Maternal effects are defined as contributions over and above the direct effects of a mother's own genes that she contributes to her progeny. These effects arise from both heritable and non-heritable maternal attributes and can be thought of as "epigenetic effects."[1-3] The developmental consequences of these maternal effects can be substantial during ontogeny and may persist until adulthood. Maternal effects may significantly impact the progeny's growth, development and reproduction. The effect can be transgenerational because the mother is conditioning the expression of her progeny's genes and may influence the action of natural selection acting on the phenotypes in the next generation (Figure 1).

Maternal effects originate from an individual whose phenotype is not being observed and are received as environmental effects by the offspring. A plethora of maternal attributes may influence progeny development including maternal age, maternal body size, uterine size, placental exchange of nutrients, quality and quantity of milk, growth factors, hormones, nesting behavior and thermal regulation.

Maternal effects should not be confused with the *Drosophila* maternal genes, such as Bicoid and Nanos. These maternal genes are transcribed by the

Drosophila mother during oogenesis and are deposited as messenger RNAs in the offspring to start the process of development and body pattern formation. While they do condition the expression of the offspring's genes, maternal genes originate from the mother's direct effects and are not maternal effects as defined above.

Mammalian maternal effects vary in their underlying causes and ramifications. Among maternal effects with significant human impact are the effects of alcohol, drug use and smoking. These non-heritable maternal effects will impact the progeny's development during ontogeny via intrauterine growth retardation, birth defects, and altered behavior. For example, exposure to cigarette smoke either pre- or postnatally decreases weight and growth in rodents.[4] Fetal exposure to alcohol and cocaine can lead to decreased postnatal growth and altered neurological function in progeny.[5]

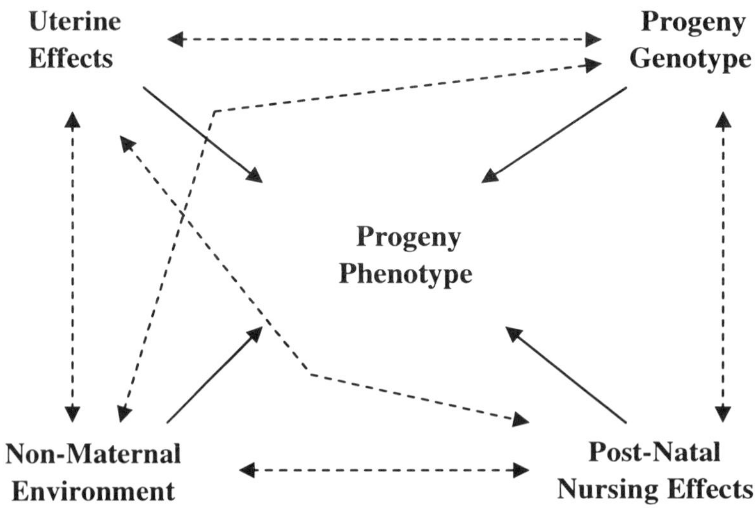

Figure 1. Developmental quantitative genetic model of mammalian development. Solid lines denote causal components, while dashed lines denote interactions between components. Modified from references.[1–3,14]

Maternal effects of greatest interest to evolutionary biologists and animal breeders are those that arise from heritable attributes in the mother, i.e., attributes arising by expression of the mother's own genes. Unlike typical environmental effects, heritable maternal effects can respond to selection and undergo systematic evolutionary changes themselves.[6] In this review, the importance of these heritable maternal effects from a developmental and evolutionary standpoint will be discussed.

Model selection is fundamental for studying evolution of complex traits like growth and development. Unfortunately, many commonly applied models use oversimplifications for modeling complex traits. Direct effects models, for example, include only classical Mendelian inheritance; they ignore real-life factors influencing the variation of traits and consequently may be inadequate to describe the evolution of traits over the course of ontogeny.[7] Use of more complex models, such as epigenetic effects models, permit inclusion of temporal and spatial interactions and provide a better description of the evolution of complex traits. The epigenetic effects model[7,8] allows us to look at both intrinsic variation (variation due to additive genetic variation) and heritable extrinsic variation (sources of variation outside the progeny's genome) such as maternal effects. Further, epigenetic models permit estimation of the covariation between intrinsic and extrinsic or maternal effects and thus represent more accurately how natural selection operates on phenotypic variation.[7]

Research on mammalian maternal effects has focused on the effect on growth and development, especially body weight.[9] Table 1 shows a partial list of mammalian traits known to be influenced by maternal effects. While this review discusses only mammalian maternal effects, maternal influences are not restricted to mammalian organisms but rather are wide-spread. They occur for seed dormancy in plants,[10] reproductive traits and diapause in fish and insects,[11] growth in cockroaches,[12] and growth in birds,[13] to name a few.

Table 1. A partial list of offspring traits influenced by mammalian maternal effects

Trait	Species	Reference
Body weight and growth rate	Horse (*Equus*), human, mouse (*Mus*, *Phyllotis darwini*), red squirrel (*Tamiasciurus hudsonicus*)	14, 15, 16, 17, 18, 19, 20, 21
Onset of puberty/vaginal opening	Mouse (*Mus*), rat (*Rattus*), guinea pig (*Cavia porcellus*)	22, 23, 24
Fertility and fecundity	Human, rat (*Rattus*), mouse (*Mus*)	25, 26, 27, 28
Bone strength	Horse (*Equus*), human, rat (*Rattus*)	29, 30, 31
Organ weight	Mouse (*Mus*), rat (*Rattus*), sheep (*Ovis*)	19, 32, 33, 34
Disease state/onset	Human, mouse (*Mus*), rat (*Rattus*)	35, 36, 37, 38
Endocrine function and activity	Human, mouse, (*Mus*), rat (*Rattus*)	22, 39, 40, 41, 42, 43

2. Temporal Aspects of Maternal Effects

Use of a developmental quantitative genetic model for studying the architecture of complex traits requires the recognition that there is a dynamic ontogenetic component. The variance and covariance components for a trait, as well as interactions between traits, change significantly during ontogeny. Only by understanding and documenting these ontogenetic changes can one expect to accurately model the genetic control of complex traits. There have been a number of studies that document the dynamics of variance and covariance components during ontogeny.[44-46] In each of these, the maternal component of variation has shown a clear temporal dynamic. One of the obvious temporal aspects of mammalian maternal effects is development in different types of maternal environments, e.g., prenatal uterine and postnatal nursing (Figure 2).

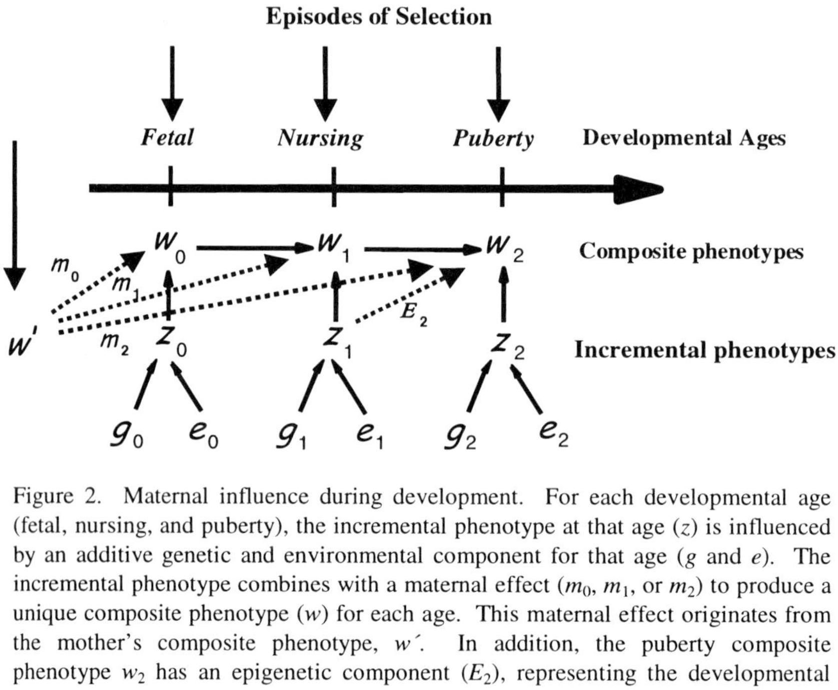

Figure 2. Maternal influence during development. For each developmental age (fetal, nursing, and puberty), the incremental phenotype at that age (z) is influenced by an additive genetic and environmental component for that age (g and e). The incremental phenotype combines with a maternal effect (m_0, m_1, or m_2) to produce a unique composite phenotype (w) for each age. This maternal effect originates from the mother's composite phenotype, w'. In addition, the puberty composite phenotype w_2 has an epigenetic component (E_2), representing the developmental cascade of traits, where early phenotypes will condition development of later phenotypes. Arrows indicate episodes of natural selection influencing both progeny and mother's phenotype. Modified from.[7]

Thus, in actuality, development of typical mammalian organisms, like the mouse, involves the expression of two genomes (mother and progeny) in three environments (uterine, nursing, and postnatal non-nursing). Maternal effects may directly impact growth in one phase such as prenatal growth or may affect growth in multiple phases. Maternal effects such as maternal age and maternal nutrition are illustrations of multi-phasic maternal effects, affecting both pre- and postnatal growth.

After parturition, the phenotypic variance explained by maternal effects for a quantitative trait typically increases until it is highest near weaning. After weaning, the phenotypic variance due to maternal effects may either plateau or decrease.[47-49] Phenotypic variation explained by maternal effects can be greater than variance from an individual's own genotype, especially right after birth.[50,51] Maternal effects will continue to influence the offspring's phenotypes long after the offspring has left the mother.[47,48] Typically, maternal effects impact adult behaviors such as maternal care performed by female offspring[26,52] and adult size of progeny.[3,14,44,46,47]

As mentioned above, a plethora of maternal attributes may influence progeny development. Consider several of these pre- and postnatal maternal effects and how they influence the progeny's fitness and phenotype.

3. Effects of Maternal Age

Both early motherhood (immediately after puberty onset) and late motherhood (near menopause) can negatively impact offspring growth and increase disease incidence. Adult offspring of mice bred in adolescence (soon after puberty onset) or at middle age (near menopause) weighed less than adults born to females bred at early adulthood.[53] These lighter mice also had delayed onset of puberty, relative to offspring from early adulthood females.[53] Eisen[54] found that females bred soon after puberty onset showed decreased litter size, littering rate, and pup birth weight with increased pup mortality compared to females bred at older ages. However, this is not consistent as other research has found females bred soon after puberty showed the same lactational performance as females bred at older ages, suggesting an interaction with genetic background.[55] Increased maternal age is also associated with decreased birth weight in humans and rats.[56–58] The reduced prenatal growth may be due to a decrease in uterine and placental quality affecting nutrients received by the fetus[59] while decreased postnatal growth may be due to decrease in quality and quantity of lactation.

Advanced maternal age is also associated with increased risks for diseases such as diabetes,[60] kidney disease,[61] leukemia,[62] and hypertension in adults.[35] In humans, offspring of young mothers have increased incidence in psychological and physiological disorders such as schizophrenia and cancer.[63–65] Mouse pups born to older mothers may also show diminished mental function relative to mice born to younger mothers.[66] In humans, increased maternal age is correlated with increases in fetal systolic blood pressure.[67]

4. Effects of Maternal Nutrition

Prenatal maternal nutrition influences disease prevalence and onset, growth and fat deposition in the progeny, puberty onset, and gene expression in mammals.[23,35,68,69] We expect caloric restriction during gestation or lactation would lead to decreased fetal growth and maternal weight loss due to the increased energy requirements on the mother. However, caloric restriction during these two phases may not act via the same mechanism, as restricted diet during postnatal development shows greater effects on adult body weight in mice than restricted diet during prenatal development.[70] The progeny of rodent mothers fed calorically-restricted diets during gestation have smaller weights early in ontogeny but show compensatory growth via increases in fat composition after birth.[71,72] Accordingly, mice born to mothers on restricted diets during gestation show increased levels of gene expression in fat synthesis genes.[73] In humans, maternal nutrition during early pregnancy also influences fetal growth.[74] However, it is frequently not possible to estimate the effect of maternal nutrition on human fetal growth, due to geographic and socioeconomic variation in maternal diet.

Some debate exists as to whether maternal nutrition differentially affects the two sexes. Sons of female mice starved during gestation will show decreased adult weight while females have the same weight as control mice.[27,75] However, the daughters of these food-deprived mothers have fewer pups in their own litters, suggesting a multi-generational effect on fertility.[27] Prenatal malnutrition also differentially affects male and female adult brain function and adult blood pressure in rats.[76,77]

5. Prenatal Maternal Effects

The uterine environment influences the embryos in any of several ways: uterine size, nutrient exchange, uterine position, and prenatal exposure to toxins. Prenatal maternal effects may be expected to affect only gestational development, but research shows this is not correct.[14] In inbred mouse lines, prenatal maternal effects had a greater impact on progeny body weight at some ages than the progeny's own genotype did.[14,15] In mice, the overall uterine environment interacts with the progeny's genotype to influence growth and response to selection.[14–16,49,78] Uterine environment also had a significant and long-lasting impact on skeletal development in inbred mice and mice undergoing artificial selection.[3,14,16,79,80]

Variation in size or weight of the mother (independent of nutritional variation) can also impact progeny development. In both inbred mice and mice undergoing artificial selection, weight of the gestational mother significantly influenced both prenatal and postnatal growth in mice; heavier mothers produced heavier and faster growing pups, regardless of the pup's genotype.[14,16,49] Nonhuman primates and several livestock species also show a similar proportional relationship between maternal weight or size and growth in offspring.[81,82] Weight of the uterine mother also affects skeletal growth in mouse pups.[3,83]

5.1. *Effects of Litter Attributes*

Litter attributes such as intrauterine position, uterine size, gestational litter size, and placental efficiency are known to significantly influence progeny growth and development. Uterine position impacts several aspects in the offspring: growth, hormone level, hormonal activity, morphology and behavior.[84] Rat and mouse fetuses developing near male fetuses have increased birth weight compared to fetuses developing near female fetuses.[85,86] However, this pattern in mice is not consistent[87,88] and may not continue after parturition.[89] Fetuses frequently show intrauterine growth retardation when they develop in the middle of the uterus, being 0.85 times the weight of normal fetuses at birth.[90,91] Pigs also show a uterine position effect on weight of the fetus, with lighter fetuses developing closer to the cervix.[92]

Uterine position may lead to altered reproduction and behavior for both sexes. Male and female rodents developing beside male fetuses show increases in aggression, territoriality and novelty-seeking incidences; this is presumably due to increased exposure to androgens during gestation.[84,86,93] Female rodents developing between two males also show a decrease in "female-like" behaviors such as lordosis (a measure of female receptivity to males) and have later puberty onset and shorter estrus compared to other females.[86,94] Uterine position may also affect fetal response to stimulants such as cocaine.[95,96]

Gestational litter size also affects fetal body weight and morphological traits in mice and other species. Unlike uterine size or uterine position, prenatal litter size can be manipulated via embryo transfer, allowing more experimental access on its effect on progeny phenotypes. Prenatal litter size has a well-documented effect on pre- and postnatal growth of offspring, as well as skeletal growth and organogenesis.[3,14,97–100] Depending on the strain of mouse and the time during ontogeny, litter size explains between 10–50% of the phenotypic variance for body size.[14,100] Traditionally, mice born to larger litters are smaller and have higher mortality.[3,98,101,102]

This effect of litter size will also impact subsequent generations, as a female's fertility is correlated with her growth. Selection for increased growth has typically increased gestational litter size as a correlated response to selection; selection for decreased growth will decrease litter size.[103–106] However, the increase in fertility is offset by higher pre- and postnatal mortality of fetuses.[102,105] Increased fetal death probably occurs around the time of parturition, since the number of live fetuses during gestation is greater than or equal to that of control mice.[105,107] This positive genetic correlation between litter size and growth is especially interesting because there is a negative environmental correlation between litter size and growth in female mice. Inbred female mice raised in larger litters tend to be smaller and produce smaller litters.[14] Thus, the effect of litter size on growth and growth's subsequent effect on litter size is cyclic and transgenerational.

5.2. *Effects of Prenatal Toxin Exposure*

Prenatal exposure to toxins typically leads to altered brain development and organogenesis in mammals and birds, but can also affect the immune and endocrine system in chickens.[108] The fetus can be exposed to several different toxins while in utero: alcohol, radiation, environmental toxins such as lead and mercury compounds, and bacterial toxins. Prenatal exposure to bacterial toxins, alcohol and some environmental toxins can alter neuron levels in adult mice and may lead to neurological disease in humans and rats.[109–111] Exposure to bacterial toxins during gestation can also result in altered heart growth.[112]

Fetuses of grazing mammals can be exposed to toxins from plants and forage; these toxins may affect the fetus in a narrow time frame during development or over the whole pregnancy. These toxins typically lead to increases in spontaneous abortion, alterations in fetal growth and morphology, and decreased fertility.[113] Mice fed an endophyte-infected fescue diet during gestation had decreased littering rate compared to mice fed either control or fescue non-infected diet.[114] The pups exposed to endophytes during prenatal development weighed less and had decreased incidence of vaginal opening at normal puberty onset compared to control or non-infected diet mice.[114] There is evidence for a genotype by environment interaction with regard to the mouse's physiological response to toxins, as discussed in Ch. 5.

6. Postnatal Maternal Effects

The postnatal nursing environment influences the neonate in a variety of ways, most notably through lactation and maternal behavior, including nest quality, thermoregulation and general care of offspring. Postnatal maternal effects influence offspring body weight and morphology in rodents,[46,47,77,115] as well as other mammals. Postnatal maternal effects have a larger impact on growth in females than males, probably due to the effect of postnatal nutrition on puberty onset in females.[46,116,117] These postnatal maternal effects also impact adult behaviors such as alcohol consumption, stress response, and maternal care.[51,118]

6.1. *Effect of Lactation*

Milk composition and quantity can vary significantly across mothers and will produce variation in weight gain and growth among pups. Lactation performance of the mother will depend on her growth and parity, as well as the nursing litter size.[49] When nursing larger litters, females produce more milk, but offspring with more littermates still get less milk than offspring with few littermates.[119] In addition, the concentration of solids in the milk is decreased in large litters relative to small litters.[119] Therefore, individuals in large litters get less and poorer quality milk. Milk composition also varies with the number of lactations. Mouse pups can have decreased weight gain as the number of lactations increased, but this is strain-specific.[120] In addition, temperature during nursing can also affect lactation quality. Females nursing at cooler temperatures produce more milk with higher energy content, more solids and more fat than females nursing at higher temperatures.[121]

Hormones and proteins involved in milk production and letdown will also influence growth and reproduction in the offspring; for example, the hormone prolactin and its receptor are involved in carbohydrate metabolism. Consequently, prolactin level in milk is correlated with offspring growth, and mice lacking the prolactin receptor have decreased weight gain in offspring.[117,122] The *hub* mutation in mice may also affect the prolactin pathway.[123] Female *hub* mice have decreased milk yield and lower epidermal growth factor (EGF) levels in milk, with associated increased pup mortality and decreased pup growth compared to normal females.[124,125] Levels of other hormones in milk such as growth hormone and progesterone are also correlated with offspring growth and vaginal opening in mice.[122] Administration of an oxytocin antagonist in pregnant females will lead to decreased growth in offspring starting at 3 days of age, only after lactation is established.[126] Females lacking the *Peg3* gene are deficient in milk ejection and have a reduced number of oxytocin neurons in the brain; pups nursed by these females have decreased postnatal growth compared to pups nursed by control mice.[127]

6.2. *Effect of Maternal Care*

Postnatal maternal behavior or care affects pup mortality and growth, especially of the brain development and the associated development of adult behaviors, e.g., stress response and expression of associated genes.[18,128] Maternal care is a transgenerational maternal effect since maternal care influences maternal behavior of female offspring.[26,51] Females lacking the *FosB* and *Dbh* genes showed maternal care defects such as pup cleaning and retrieval, as well as lactation problems.[129] The *hub* mutant mothers have slower pup retrieval and decreased pup grooming compared to control mothers.[125] *Mest1*-deficent females are poor in general maternal care, pup retrieval and nest building, as are *Peg3*-deficent females.[127,130] Female mice lacking one or both copies of the prolactin receptor gene are slow to retrieve pups and defensively crouch over nests, taking longer than control females.[131]

7. Maternal Effects and Role in Selection

The majority of theoretical research on selection response and the evolution of traits have used direct effect genetic models. As mentioned above, direct effect models assume genetic variation and covariation arise only from genes within an individual's genome and ignore other causes of phenotypic variance such as maternal effects and sex effects. Under this model, an individual's phenotypic variance for a single trait will be equal to its additive genetic variance, environmental variance and variance from an interaction between the two.

If we expand this model to one with "n" traits under selection, the relationship becomes

$$\mathbf{P} = \mathbf{G} + \mathbf{E} \tag{1}$$

where **P**, **G**, and **E** are the *n x n* phenotypic, additive genetic, and environmental variance/covariance matrices, respectively. The response to selection with this model is

$$\mathbf{\Delta z} = \mathbf{G}\,\mathbf{P}^{-1}\,\mathbf{s} \tag{2}$$

where **Δz** is the vector of changes in trait means and **s** is the vector of selection differentials or traits under selection.[132]

As noted above, direct effects models are oversimplifications of the underlying causes of genetic variation in mammalian species. Maternal effects influence expression of their offspring's genes, alter phenotypes, possibly changing the correlation between genotype and phenotype, and thus influence the efficacy of selection response.[8] Therefore, response to selection models must include a maternal component to accurately estimate the underlying causes of variation and to accurately predict response to selection.

Under a model with maternal effects, an individual's phenotypic variance for a single trait will now be a function of its additive genetic and environmental effects and a maternal effect coefficient "m" times the maternal phenotype for that trait.[133] This "m" coefficient measures the strength with which a mother's phenotype can alter the progeny's phenotype.[133] These authors point out that this maternal effect coefficient can be negative or positive.

In multi-dimensional biological systems, maternal effects rarely affect only one trait. In the more realistic multivariate case, the phenotypic variance for a single trait in "n" maternally influenced traits is a function of the additive genetic and environmental effects for that trait plus a multivariate maternal effect coefficient, measuring the strength of the maternal effect in trait 2 on trait 1. The response for one generation of selection is modeled by

$$\mathbf{\Delta z}(t) = (\mathbf{C_{az}} + \mathbf{MP})\boldsymbol{\beta}(t) + \mathbf{M\Delta z}(t-1) - \mathbf{MP}\boldsymbol{\beta}(t-1) \tag{3}$$

where $\mathbf{C_{az}}$ is a symmetric matrix representing covariances between each additive genetic value and phenotypic value, $\boldsymbol{\beta}$ is a matrix of selection gradients (= $\mathbf{P}^{-1}\mathbf{s}$) and **M** is a matrix of maternal effect coefficients.

Thus, response to selection in generation "t" is a function of the force of selection in generation "t", based on the midparent and offspring resemblance,[133]

$$(\mathbf{C_{az}} + \mathbf{MP})\boldsymbol{\beta}(t) \tag{4}$$

However, response in generation "t" also depends on the force of selection and evolutionary change in generation "t – 1," which is the key difference between response in maternally influenced traits and response in Mendelian traits.

Direct effects models assume no maternal effects; thus **M** is a matrix where all elements are equal to zero, and $\mathbf{C_{az}} = \mathbf{G}$, where $\mathbf{C_{az}}$ shows how changes in gene frequency lead to changes in phenotype. If a trait is maternally influenced, $\mathbf{C_{az}} \neq \mathbf{G}$ and an individual phenotype may appear "better" or "worse" than its genotype suggests. Thus, maternal effects will alter the genotype/phenotype correlation, impacting the efficiency of selection.[1,8,133]

Maternally influenced traits often show a time-lag in response to selection, since selection response in one generation depends on the force of selection in two generations.[130] Such time lags can lead to continued selection response after selection has ceased (evolutionary momentum) and a variable selection response across multiple generations even when selection strength is constant.[134] This leads to evolutionary change in traits with no additive genetic variance, since no genetic variance is required; rather, only maternal variance is needed to respond to selection.[6]

Even when using a maternal effects model, estimating response of maternally influenced traits is difficult. The evolutionary change for previous generations ($\mathbf{M}\Delta\mathbf{z}(t - 1)$) is difficult to measure since evolutionary change for a given generation can fluctuate around its expectation.[133] In addition, one cannot assume maternal coefficients remain constant as evolution continues, making estimation more difficult.[16, 135]

When maternal effects influence a trait, they create a direct-maternal covariance between direct (effect of genotype) and indirect effects (effect of maternal effect) due to the pleiotropic nature of some genes. Selection for a trait under maternal influence can act on the direct genetic variance, the indirect genetic variance, or the direct-maternal covariance. If this covariance is negative, response to selection will be diminished, since changes in direct effects will be offset by a large (negative) change in maternal effects. This covariance can cause some traits to evolve away from the true fitness optimum ("maladaptive evolution"). As Kirkpatrick and Lande[133] point out, there is a tendency to assume a positive correlation of the same trait between mother and offspring, but this is not necessarily true. Negative covariances between direct and maternal effects are common for growth and fitness traits,[136] such as the negative correlation between maternal litter size and subsequently progeny litter size.[137]

This covariance may also change over the course of ontogeny, further complicating selection response modeling. Cheverud[136] found an initially negative covariance between maternal effects and body weight in mice that became positive after weaning. Hanrahan and Eisen[138] also found similar results. Other traits showed negative covariances between maternal effects and offspring traits later in ontogeny, especially near onset of puberty.[136] In pigs, there is a positive covariance between direct and maternal effects for body weight after birth which becomes negative as an animal ages.[139, 140] Other research has found a positive covariance between direct and maternal effects for body weight in mice throughout ontogeny. However, the magnitude of these positive covariances increases later in ontogeny, suggesting direct effects at later ages are more correlated with maternal effects than earlier direct effects.[141]

8. Maternal Effects and Familial Relationships

Maternal effects will also affect two familial relationships, complicating parameter estimation.[142] The first is the resemblance among offspring raised in a common environment; however, the relationship between offspring and mother remains the same.[142] This is related to variation in maternal performance; maternal performance is a composite phenotype, comprised of all the maternal attributes in the mother influencing her offspring.[50] An example of this is variation in an aspect of maternal performance such as pup retrieval leads to an environmental component in sibling covariation for a trait such as body weight.[142] The variation among offspring results from variation in maternal performance. Such variation in maternal performance can be due to genetic variation in behavioral and lactation genes.[143] The second impact is the mother's phenotype for a trait and its conditioning of the offspring's phenotype for the same trait. This will modify the resemblance between full sibs and maternal half sibs, as well as the relationship between offspring and mother.[50,142]

The tendency of sibs to resemble each other due to the common maternal environment is one of the more troublesome environmental aspects to model and assess via experimental design.[142] To separate common maternal effect from other sources of variation, it is necessary to crossfoster full sibs to different nurse mothers Breeding designs to estimate causal components of variance without crossfostering will lead to inaccurate estimation of genetic parameters due to the correlation of genetic covariances between relatives.

Eisen[144] suggested that traditional mating designs used to estimate causal components of variance can lead to doubling the contribution of indirect genetic effects (both maternal effects and the direct effects–maternal effects covariance). This leads to inflated estimates of additive genetic variance and heritability, and in some cases, can double the heritability estimate.[144,145] Estimating realized heritability via regression of offspring phenotype on midparent phenotype for a trait with maternal effects can also produce a biased estimate, sometimes greater than 1 or less than 0.[133,136] The direction and magnitude of the heritability bias will depend on the magnitude and sign of the direct-maternal covariance.[1]

9. Maternal Effects Across Multiple Generations

Maternal effects can extend across multiple generations, i.e., grandmaternal effects. As expected, they are not as pronounced as typical maternal effects, but so-called "grandmaternal effects" may still condition gene expression and impact the phenotype of the grandchild. Grandmaternal effects influence birth and weaning weight and may interact with nutrition in certain cattle breeds.[146–149] In humans, grandmaternal effects may influence fetal birth weight, although these results are inconsistent.[150] Some stress-responsive behaviors in adult rats also show grandmaternal effects.[151]

10. Maternal Effects and Interactions with Other Epigenetic Phenomena

Maternal effects are one of several epigenetic phenomena involved in development. These maternal effects can interact with other epigenetic phenomena of DNA modification, such as methylation and acetylation, to impact the progeny's growth and development. One such DNA modification is genomic imprinting, or the differential expression of one parent's allele via methylation for a subset of genes. At an imprinted gene, one parental allele is active (transcribed) and the other allele is silent (not transcribed). Which parental allele is silent or active is determined on a gene-by-gene basis. This differential expression of parental alleles is achieved via differential DNA methylation at the two alleles by the enzyme Dmnt1. The methylation pattern for a gene is established in the germline and is maintained until after implantation.[152]

Previously, it was thought epigenetic patterns were erased during gametogenesis, thereby eliminating transmission from one generation to the next. However, Rakyan *et al.*[153] documented that epigenetic patterns at some genes can be inherited due to incomplete erasure of the epigenetic pattern. This fact, coupled with the fact that DNA replication has a 95–97% fidelity of transferring methylation patterns,[154,155] means there is great potential for introducing epigenetic variation, even within the same individual and certainly to subsequent generations.

Unlike maternal effects, DNA methylation involves only one genome, the progeny's. However, evidence suggests some maternal effects, e.g., maternal nutrition may interact with genomic imprinting at certain loci to condition their expression. While less than 0.5% of rodent genes are known to be imprinted,[156] a subset of imprinted genes are involved in regulation of fetal growth, placental development and maternal behaviors. This makes imprinted genes a possible source of both quantitative variation for growth and maternal effects.

The link between methylation variation and expression variation at imprinted genes, which will then impact phenotypic variation in a population, has been well documented at several imprinted genes. Methylation levels of both the maternal and paternal *IGF2* gene vary temporally and spatially in prenatal ontogeny.[157] The methylation level of the expressed paternal allele is correlated with *IGF2* expression levels.[157] Deletion of the upstream sequences of the maternal *IGF2* allele can lead to altered methylation patterns and polymorphic expression of the silent maternal allele in mouse embryos.[158] The methylation patterns at imprinted genes can change due to environmental conditions, and these methylation changes are correlated with changes in expression levels.[159-163] In addition, a small percentage of embryos and adults in natural populations express both alleles of some imprinted genes, confirming the potential for expression variation at imprinted genes.[164-167]

10.1. *Maternal Effects and Its Effect on Genomic Imprinting*

The best supported link between methylation variation at imprinted genes and maternal effects is through maternal nutrition. Nutrients involved in methyl metabolism such as folate, methionine, and choline are required to maintain DNA methylation,[168] and variations in dietary methyl amount can lead to differential methylation at the *agouti* locus in mice.[169] The imprinted gene *H19* shows differential methylation patterns and expression levels when mouse embryos are raised in different environments.[170] The maternally expressed *H19* allele had increased methylation and an associated decrease in expression when mouse embryos were cultured in fetal calf serum, compared to fetuses cultured in medium without fetal calf serum.[171] These embryos also had decreased expression of the paternal *IGF2* allele, resulting in lower birth weights than control pups.[171]

There is also a postnatal nutritional influence on methylation at imprinted genes. Poor diet during early postnatal development can alter the expression of the *IGF2* gene and other imprinted genes; this expression change and its associated phenotypic effect can persist long after nutrition has improved.[172] Adult rats nursed in small litters showed differential expression at two imprinted genes, compared to rats nursed in litters with more pups, showing the effect of litter size on methylation.[173] Inhibition of DNA methylation in young mice led to changes in *IGF2* and *H19* expression in some tissues.[174] This modified *IGF2* expression was not the same for all tissues; some tissues expressed both alleles, while others expressed only one allele (either maternal or paternal).[174]

10.2. *Genomic Imprinting and Its Effect on Maternal Effects*

Conversely, genomic imprinting will influence maternal effects such as maternal behavior and placental quality. Loss of the paternally expressed imprinted gene *Mest1* produces female mice deficient in pup retrieval and nest building, leading to increased postnatal pup mortality.[130] *Mest1*-deficent mice also have uterine growth retardation and had decreased pre- and postnatal growth compared to non-deficient mice.[130] Loss of another paternally expressed gene *Peg3* also leads to growth retardation in pups and abnormal lactation in mothers.[127] Interestingly, the methylation pattern of the imprinted gene *Mest1* does not seem sensitive to nutrient variation as other imprinted genes, such as *IGF2* and *H19*.[171]

Imprinting genes also influence growth of the placenta, impacting nutrient exchange to the fetus. Knockout mice for the paternally expressed genes *IGF2*, *Peg3*, and *Mest1* will have reduced placental size in gestation; knockouts of maternally expressed genes such as *H19* and *IGF2R* will increase placental size.[175] However, this placental growth is dependent on both progeny and maternal IGF2 expression. Both *IGF2*-deficent mothers[176] and normal mothers with embryos lacking a paternal *IGF2* allele have smaller placentas compared to control mothers.[177] The decreased placental size does decrease diffusion of nutrients to the fetus but does not affect the growth of the fetus until late in gestation, possibly due to a compensation mechanism in amino acid transport.[178] Imprinted genes also influence the transfer of nutrients over the placenta: the *Slc22*, *Impt1*, and *Ata3* genes are transporter genes.

11. Evolution of Maternal Effects and Genomic Imprinting

In order for genomic imprinting to evolve, it must convey a selective advantage. In general, paternally expressed genes increase fetal growth, thus requiring more maternal resources. This helps ensure the offspring will survive and reproduce, at the cost of the mother's (and other males') future offspring, to the father's advantage. Conversely, maternally-expressed genes tend to decrease offspring growth, thereby shortening the maternal care period and increasing her chance of bearing subsequent litters.[179] The balance between the best interests of the two parents can be seen in the *IGF2/IGF2R* gene pair. Increased expression of paternally imprinted *IGF2* leads to increased fetal growth, increasing the likelihood of a large offspring. The maternally expressed *IGF2R* gene works to "balance" the expression of the *IGF2* gene and minimize the postnatal care and nursing needed, spreading the mother's resources equally over all her offspring. Classic maternal effects such as postnatal lactation and care can also be explained by a similar hypothesis, the parent-offspring conflict hypothesis. Increased maternal care and lactation are advantageous to the offspring due to increased fitness but deleterious to the mother due to decreases in fertility from nursing. Like the father, it is to the fetus's advantage to obtain more maternal resources at the expense of other (and future) siblings.

Interestingly, genes involved in kin recognition such as genes in the major histocompatability complex are also imprinted.[180] In rodent species that form communal nests, these loci can serve as an advantage to a polygamous male. Without kin recognition, mothers will be willing to nurse any offspring, including ones that are not her own, to the father's advantage.

Both maternally-expressed genomic imprinting and maternal effects serve the same function: to equally distribute maternal resources to all her offspring. Placentally imprinted genes function to serve the same purpose: to decrease the supply of nutrients to the increasingly nutrient-demanding fetuses.[175]

12. Conclusion

Maternal effects arise from both prenatal and postnatal maternal-by-fetal interactions and can significantly influence offspring gene expression patterns, growth, reproduction, behavior and disease incidence. These effects may alter genotype-phenotype correlations in the offspring, thus affecting the efficacy of selection response and influencing evolutionary change.

While some of the well-characterized maternal effects have been discussed, more research is needed on phenomena such as the causal relationship between maternal effects and gene methylation and gene expression. The mouse is well-suited for such research because it is one of the few mammals with well-characterized maternal effects that are known to show temporal methylation variation at imprinted genes. In addition, with the use of mouse microarray technology, the causative relationship between maternal effects and gene expression can be better documented.

Acknowledgements

The authors would also like to express our gratitude to our mothers for their genetic and epigenetic contributions to our phenotypes (for better or for worse). We are indebted to our colleague Gene Eisen for his helpful comments on the manuscript. This work was supported by NIH Program Project Grant number GM045344 to W.R.A. and NIH Training Grant in Quantitative Genetics number GM08443 to J.F.K.

References

1. Atchley, W.R. and S. Newman. 1989. A quantitative-genetics perspective on mammalian development. *Am. Nat.* **134:**486–512.
2. Atchley, W.R. and B.K. Hall. 1991. A model for development and evolution of complex morphological structures. *Bio. Rev. Camb. Philos. Soc.* **66:**101–157.
3. Atchley, W.R., T. Logsdon, D.E. Cowley, and E.J. Eisen. 1991. Uterine effects, epigenetics, and postnatal skeletal development in the mouse. *Evolution* **45:**891–909.
4. Gaworski, C.L., E.L. Carmines, A.S. Faqi, and N. Rajendran. 2004. In utero and lactation exposure of rats to 1R4F reference cigarette mainstream smoke: effect on prenatal and postnatal development. *Toxicol. Sci.* **79:**157–169.
5. Chiriboga, C.A. 2003. Fetal alcohol and drug effects. *Neurologist* **9:**267–279.
6. Wolf, J.B., E.D. Brodie, J.M. Cheverud, A.J. Moore, and M.J. Wade. 1998. Evolutionary consequences of indirect genetic effects. *Trends. Ecol. Evol.* **13:**64–69.
7. Cowley, D.E. and W.R. Atchley. 1992. Quantitative genetic models for development, epigenetic selection, and phenotypic evolution. *Evolution* **46:**495–518.
8. Atchley, W.R., S. Xu, and C. Vogl. 1994. Developmental quantitative genetic models of evolutionary change. *Dev. Genet.* **15:**92–103.
9. Reinhold, K. 2002. Maternal effects and the evolution of behavioral and morphological characters: a literature review indicates the importance of extended maternal care. *J. Hered.* **93:**400–405.
10. Roach, D.A. and R.D. Wulff. 1987. Maternal effects in plants. *Annu. Rev. Ecol. Syst.* **18:**209–235.
11. Mousseau, T.A. 1991. Geographic-variation in maternal-age effects on diapause in a cricket. *Evolution* **45:**1053–1059.
12. Holbrook, G.L. and C. Schal. 2004. Maternal investment affects offspring phenotypic plasticity in a viviparous cockroach. *Proc. Natl. Acad. Sci. USA* **101:**5595–5597.
13. Hayward, L.S. and J.C. Wingfield. 2004. Maternal corticosterone is transferred to avian yolk and may alter offspring growth and adult phenotype. *Gen. Comp. Endocrinol.* **135:**365–371.
14. Cowley, D.E., D. Pomp, W.R. Atchley, E.J. Eisen, and D. Hawkins-Brown. 1989. The impact of maternal uterine genotype on postnatal growth and adult body size in mice. *Genetics* **122:** 193–203.

15. Pomp, D., D.E. Cowley, E.J. Eisen, W.R. Atchley, and D. Hawkins-Brown. 1989. Donor and recipient genotype and heterosis effects on survival and prenatal growth of transferred mouse embryos. *J. Reprod. Fertil.* **86:**493–500.
16. Ernst, C.A., B.K. Rhees, C.H. Miao, and W.R. Atchley. 2000. Effect of long-term selection for early postnatal growth rate on survival and prenatal development of transferred mouse embryos. *J. Reprod. Fertil.* **118:**205–210.
17. MacNeil, M.D. 2003. Genetic evaluation of an index of birth weight and yearling weight to improve efficiency of beef production. *J. Anim. Sci.* **81:**2425–2433.
18. Nespolo, R.F., L.D. Bacigalupe, and F. Bozinovic. 2003. Heritability of energetics in a wild mammal, the leaf-eared mouse (Phyllotis darwini). *Evolution* **57:**1679–1688.
19. Vonnahme, K.A., B.W. Hess, T.R. Hansen, R.J. McCormick, D.C. Rule, G.E. Moss, W.J. Murdoch, M.J. Nijland, D.C. Skinner, P.W. Nathanielsz, and S.P. Ford. 2003. Maternal undernutrition from early- to mid-gestation leads to growth retardation, cardiac ventricular hypertrophy, and increased liver weight in the fetal sheep. *Biol. Reprod.* **69:**133–140.
20. McAdam, A.G. and S. Boutin. 2004. Maternal effects and the response to selection in red squirrels. *Proc. R. Soc. Lond., B, Biol. Sci.* **271:**75–79.
21. Thame, M., C. Osmond, F. Bennett, R. Wilks, and T. Forrester. 2004. Fetal growth is directly related to maternal anthropometry and placental volume. *Eur. J. Clin. Nutr.* **58:**894–900.
22. Leonhardt, M., J. Lesage, D. Croix, I. Dutriez-Casteloot, J.C. Beauvillain, and J.P. Dupouy. 2003. Effects of perinatal maternal food restriction on pituitary-gonadal axis and plasma leptin level in rat pup at birth and weaning and on timing of puberty. *Biol. Reprod.* **68:**390–400.
23. da Silva Faria, T., C. da Fonte Ramos, and F.J. Sampaio. 2004. Puberty onset in the female offspring of rats submitted to protein or energy restricted diet during lactation. *J. Nutr. Biochem.* **15:**123–127.
24. Hines, M. and R.W. Goy. 1985. Estrogens before birth and development of sex-related reproductive traits in the female guinea pig. *Horm. Behav.* **19:**331–347.
25. Joffe, M. and I. Barnes. 2000. Do parental factors affect male and female fertility? *Epidemiology* **11:**700–705.
26. Lovic, V., A. Gonzalez, and A.S. Fleming. 2001. Maternally separated rats show deficits in maternal care in adulthood. *Dev. Psychobiol.* **39:**19–33.
27. Meikle, D. and M. Westberg. 2001. Maternal nutrition and reproduction of daughters in wild house mice (Mus musculus). *Reproduction* **122:**437–442.
28. Andrade, A.J., S. Araujo, G.M. Santana, M. Ohi, and P.R. Dalsenter. 2002. Reproductive effects of deltamethrin on male offspring of rats exposed during pregnancy and lactation. *Regul. Toxicol. Pharmacol.* **36:**310–317.
29. Glade, M.J. 1993. Effects of gestation, lactation, and maternal calcium intake on mechanical strength of equine bone. *J. Am. Coll. Nutr.* **12:**372–377.
30. Ronis, M.J., J. Aronson, G.G. Gao, W. Hogue, R.A. Skinner, T.M. Badger, and C.K. Lumpkin, Jr. 2001. Skeletal effects of developmental lead exposure in rats. *Toxicol. Sci.* **62:**321–329.
31. Javaid, M.K. and C. Cooper. 2002. Prenatal and childhood influences on osteoporosis. *Best. Pract. Res. Clin. Endocrinol. Metab.* **16:**349–367.
32. Pausova, Z., T. Paus, L. Sedova, and J. Berube. 2003. Prenatal exposure to nicotine modifies kidney weight and blood pressure in genetically susceptible rats: a case of gene-environment interaction. *Kidney Int.* **64:**829–835.

33. Lisle, S.J., R.M. Lewis, C.J. Petry, S.E. Ozanne, C.N. Hales, and A.J. Forhead. 2003. Effect of maternal iron restriction during pregnancy on renal morphology in the adult rat offspring. *Br. J. Nutr.* **90:**33–39.
34. Piasek, M., M. Blanusa, K. Kostial, and J.W. Laskey. 2004. Low iron diet and parenteral cadmium exposure in pregnant rats: the effects on trace elements and fetal viability. *Biometals* **17:**1–14.
35. Waterland, R.A. and C. Garza. 1999. Potential mechanisms of metabolic imprinting that lead to chronic disease. *Am. J. Clin. Nutr.* **69:**179–197.
36. Bassett, S.S., D. Avramopoulos, and D. Fallin. 2002. Evidence for parent of origin effect in late-onset Alzheimer disease. *Am. J. Med. Genet.* **114:**679–686.
37. Fernandez-Twinn, D.S., S.E. Ozanne, S. Ekizoglou, C. Doherty, L. James, B. Gusterson, and C.N. Hales. 2003. The maternal endocrine environment in the low-protein model of intra-uterine growth restriction. *Br. J. Nutr.* **90:**815–822.
38. Jones, R.D., A.H. Morice, and C.J. Emery. 2004. Effects of perinatal exposure to hypoxia upon the pulmonary circulation of the adult rat. *Physiol. Res.* **53:**11–17.
39. El-Khattabi, I., F. Gregoire, C. Remacle, and B. Reusens. 2003. Isocaloric maternal low-protein diet alters IGF-I, IGFBPs, and hepatocyte proliferation in the fetal rat. *Am. J. Physiol. Endocrinol. Metab.* **285:**E991–E1000.
40. Oyama, L.M. and C.M. Oller Do Nascimento. 2003. Effect of ethanol intake during lactation on male and female pups' liver and brain metabolism during the suckling-weaning transition period. *Nutr. Neurosci.* **6:**183–188.
41. Ren, J., Z.K. Roughead, L.E. Wold, F.L. Norby, S. Rakoczy, R.L. Mabey, and H.M. Brown-Borg. 2003. Increases in insulin-like growth factor-1 level and peroxidative damage after gestational ethanol exposure in rats. *Pharmacol. Res.* **47:**341–347.
42. Harding, J.E. 2003. Nutrition and growth before birth. *Asia. Pac. J. Clin. Nutr.* **12 Suppl:** S28.
43. Nef, S., T. Shipman, and L.F. Parada. 2000. A molecular basis for estrogen-induced cryptorchidism. *Dev. Biol.* **224:**354–361.
44. Cheverud, J.M., L.J. Leamy, W.R. Atchley, and J.J. Rutledge. 1983. Quantitative genetics and the evolution of ontogeny .1. Ontogenetic changes in quantitative genetic variance-components in randombred mice. *Genet. Res.* **42:**65–75.
45. Atchley, W.R. 1984. Ontogeny, timing of development, and genetic variance-covariance structure. *Am. Nat.* **123:**519–540.
46. Riska, B., W.R. Atchley, and J.J. Rutledge. 1984. A genetic-analysis of targeted growth in mice. *Genetics* **107:**79–101.
47. Atchley, W.R. and J. Zhu. 1997. Developmental quantitative genetics, conditional epigenetic variability and growth in mice. *Genetics* **147:**765–776.
48. Rutledge, J.J., O.W. Robison, E.J. Eisen, and J.E. Legates. 1972. Dynamics of genetic and maternal effects in mice. *J. Anim. Sci.* **35:**911–918.
49. Moore, R. W., E. J. Eisen, and L. C. Ulberg, 1970. Prenatal and postnatal maternal influences on growth in mice selected for body weight. *Genetics* **64:**59–68.
50. Cheverud, J. M., and A.J. Moore. 1994. Quantitative genetics and the role of the environment provided by relatives in behavioral evolution. In: *Quantitative Genetic Studies of Behavioral Evolution.* ed. C. R. B. Boake. pp. 67–100. University of Chicago Press, Chicago.
51. Wolf, J.B. 2000. Gene interactions from maternal effects. *Evolution* **54:**1882–1898.

52. Francis, D.D., F.C. Champagne, and M.J. Meaney. 2000. Variations in maternal behaviour are associated with differences in oxytocin receptor levels in the rat. *J. Neuroendocrinol.* **12:** 1145–1148.
53. Wang, M.H. and F.S. vom Saal. 2000. Maternal age and traits in offspring. *Nature* **407:**469–470.
54. Eisen, E.J. 1980. Influence of early pregnancy on reproductive rates in lines of mice selected for litter size. *Theor. Appl. Genet.* **57**:209–220.
55. Eisen, E. J. W.R. Williams, and J. F. Hayes. 1977. Effects of early pregnancy on postnatal maternal performance of mice. *Anim. Prod.* **25**:1–10.
56. Vorhees, C.V. 1988. Maternal age as a factor in determining the reproductive and behavioral outcome of rats prenatally exposed to ethanol. *Neurotoxicol. Teratol.* **10:**23–34.
57. Bottini, E., G.F. Meloni, J. MacMurray, M. Ammendola, T. Meloni, and G. Gloria-Bottini. 2001. Maternal age and traits of offspring in humans. *Placenta* **22:**787–789.
58. Tarin, J.J., V. Gomez-Piquer, C. Manzanedo, J. Minarro, C. Hermenegildo, and A. Cano. 2003. Long-term effects of delayed motherhood in mice on postnatal development and behavioural traits of offspring. *Hum. Reprod.* **18:**1580–1587.
59. Hansen, J.P. 1986. Older maternal age and pregnancy outcome: a review of the literature. *Obstet. Gynecol. Surv.* **41:**726–742.
60. Loos, R.J., D.I. Phillips, R. Fagard, G. Beunen, C. Derom, C. Mathieu, J. Verhaeghe, and R. Vlietinck. 2002. The influence of maternal BMI and age in twin pregnancies on insulin resistance in the offspring. *Diabetes Care* **25:**2191–2196.
61. Dyck, R., H. Klomp, L. Tan, and M.R. Stang. 2003. An association of maternal age and birth weight with end-stage renal disease in Saskatchewan. Sub-analysis of registered Indians and those with diabetes. *Am. J. Nephrol.* **23:**395–402.
62. Hemminki, K., P. Kyyronen, and P. Vaittinen. 1999. Parental age as a risk factor of childhood leukemia and brain cancer in offspring. *Epidemiology* **10:**271–275.
63. Fergusson, D.M. and L.J. Woodward. 1999. Maternal age and educational and psychosocial outcomes in early adulthood. *J. Child. Psychol. Psychiatry* **40:**479–489.
64. Kirchengast, S. and B. Hartmann. 2003. Impact of maternal age and maternal somatic characteristics on newborn size. *Am. J. Human Biol.* **15:**220–228.
65. El-Saadi, O., C.B. Pedersen, T.F. McNeil, S. Saha, J. Welham, E. O'Callaghan, E. Cantor-Graae, D. Chant, P.B. Mortensen, and J. McGrath. 2004. Paternal and maternal age as risk factors for psychosis: findings from Denmark, Sweden and Australia. *Schizophr. Res.* **67:** 227–236.
66. Tarin, J.J., V. Gomez-Piquer, C. Manzanedo, J. Minarro, C. Hermenegildo, and A. Cano. 2003. Long-term effects of delayed motherhood in mice on postnatal development and behavioural traits of offspring. *Hum. Reprod.* **18:**1580–1587.
67. Gillman, M.W., J.W. Rich-Edwards, S.L. Rifas-Shiman, E.S. Lieberman, K.P. Kleinman, and S.E. Lipshultz. 2004. Maternal age and other predictors of newborn blood pressure. *J. Pediatr.* **144:**240–245.
68. Gambling, L., S. Dunford, and H.J. McArdle. 2004. Iron deficiency in the pregnant rat has differential effects on maternal and fetal copper levels. *J Nutr. Biochem.* **15:**366–372.
69. Senoh, D., T. Hata, and M. Kitao. 1995. Effect of maternal hyperglycemia on fetal regional circulation in appropriate for gestational age and small for gestational age fetuses. *Am. J. Perinatol.* **12:**223–226.

70. Rogers, E.H., E.S. Hunter, M.B. Rosen, J.M. Rogers, C. Lau, P.C. Hartig, B.M. Francis, and N. Chernoff. 2003. Lack of evidence for intergenerational reproductive effects due to prenatal and postnatal undernutrition in the female CD-1 mouse. *Reprod. Toxicol.* **17:**519–525.
71. Jones, A.P., E.L. Simson, and M.I. Friedman. 1984. Gestational undernutrition and the development of obesity in rats. *J Nutr.* **114:**1484–1492.
72. Bayol, S., D. Jones, G. Goldspink, and N.C. Stickland. 2004. The influence of undernutrition during gestation on skeletal muscle cellularity and on the expression of genes that control muscle growth. *Br. J. Nutr.* **91:**331–339.
73. Maloney, C.A., A.K. Gosby, J.L. Phuyal, G.S. Denyer, J.M. Bryson, and I.D. Caterson. 2003. Site-specific changes in the expression of fat-partitioning genes in weanling rats exposed to a low-protein diet in utero. *Obes. Res.* **11:**461–468.
74. Neufeld, L.M., J.D. Haas, R. Grajeda, and R. Martorell. 2004. Changes in maternal weight from the first to second trimester of pregnancy are associated with fetal growth and infant length at birth. *Am. J. Clin. Nutr.* **79:**646–652.
75. Meikle, D.B. and M.W. Thornton. 1995. Premating and gestational effects of maternal nutrition on secondary sex ratio in house mice. *J. Reprod. Fertil.* **105:**193–196.
76. Khan, I.Y., P.D. Taylor, V. Dekou, P.T. Seed, L. Lakasing, D. Graham, A.F. Dominiczak, M.A. Hanson, and L. Poston. 2003. Gender-linked hypertension in offspring of lard-fed pregnant rats. *Hypertension* **41:**168–175.
77. Palmer, A.A., D.J. Printz, P.D. Butler, S.C. Dulawa, and M.P. Printz. 2004. Prenatal protein deprivation in rats induces changes in prepulse inhibition and NMDA receptor binding. *Brain Res.* **996:**193–201.
78. Rhees, B.K., C.A. Ernst, C.H. Miao, and W.R. Atchley. 1999. Uterine and postnatal maternal effects in mice selected for differential rate of early development. *Genetics* **153:**905–917.
79. Nonaka, K., Y. Sasaki, K. Yanagita, T. Matsumoto, Y. Watanabe, and M. Nakata. 1993. Intrauterine effect of dam on prenatal development of craniofacial complex of mouse embryo. *J. Craniofac. Genet. Dev. Biol.* **13:**206–212.
80. Sasaki, Y., K. Nonaka, and M. Nakata. 1994. The effect of four strains of recipients on the intrauterine growth of the mandible in mouse fetuses. *J. Craniofac. Genet. Dev. Biol.* **14:**118–123.
81. Tardif, S.D. and K.L. Bales. 2004. Relations among birth condition, maternal condition, and postnatal growth in captive common marmoset monkeys (Callithrix jacchus). *Am. J. Primatol.* **62:**83–94.
82. Allen, W.R., S. Wilsher, C. Tiplady, and R.M. Butterfield. 2004. The influence of maternal size on pre- and postnatal growth in the horse: III Postnatal growth. *Reproduction* **127:**67–77.
83. Nonaka, K., Y. Sasaki, K. Yanagita, Y. Watanabe, T. Matsumoto, and M. Nakata. 1994. The effect of dam's strain on the intrauterine craniofacial growth of mouse fetuses. *J. Assist. Reprod. Genet.* **11:**359–366.
84. Ryan, B.C. and J.G. Vandenbergh. 2002. Intrauterine position effects. *Neurosci. Biobehav. Rev.* **26:**665–678.
85. Kinsley, C., J. Miele, C.K. Wagner, L. Ghiraldi, J. Broida, and B. Svare. 1986. Prior intrauterine position influences body weight in male and female mice. *Horm. Behav.* **20:**201–211.
86. Hernandez-Tristan, R., C. Arevalo, and S. Canals. 1999. Effect of prenatal uterine position on male and female rats sexual behavior. *Physiol. Behav.* **67:**401–408.

87. Lee, M. 1987. Early fetal growth and development in mice chronically exposed to ethanol during gestation. *Growth* **51:**146–153.
88. Palanza, P., S. Morley-Fletcher, and G. Laviola. 2001. Novelty seeking in periadolescent mice: sex differences and influence of intrauterine position. *Physiol. Behav.* **72:**255–262.
89. Nagao, T., K. Wada, M. Kuwagata, M. Nakagomi, C. Watanabe, S. Yoshimura, Y. Saito, K. Usumi, and J. Kanno. 2004. Intrauterine position and postnatal growth in Sprague-Dawley rats and ICR mice. *Reprod. Toxicol.* **18:**109–120.
90. Argente, M.J., M.A. Santacreu, A. Climent, and A. Blasco. 2003. Relationships between uterine and fetal traits in rabbits selected on uterine capacity. *J. Anim. Sci.* **81:**1265–1273.
91. Cellini, C., J. Xu, A. Arriaga, and T.L. Buchmiller-Crair. 2004. Effect of epidermal growth factor infusion on fetal rabbit intrauterine growth retardation and small intestinal development. *J. Pediatr. Surg.* **39:**891–897.
92. Wise, T., A.J. Roberts, and R.K. Christenson. 1997. Relationships of light and heavy fetuses to uterine position, placental weight, gestational age, and fetal cholesterol concentrations. *J. Anim. Sci.* **75:**2197–2207.
93. Gandelman, R., F.S. vom Saal, and J.M. Reinisch. 1977. Contiguity to male foetuses affects morphology and behaviour of female mice. *Nature* **266:**722–724.
94. vom Saal, F.S. 1989. The production of and sensitivity to cues that delay puberty and prolong subsequent oestrous cycles in female mice are influenced by prior intrauterine position. *J. Reprod. Fertil.* **86:**457–471.
95. Lipton, J.W., H.C. Robie, Z. Ling, D.E. Weese-Mayer, and P.M. Carvey. 1998. The magnitude of brain dopamine depletion from prenatal cocaine exposure is a function of uterine position. *Neurotoxicol. Teratol.* **20:**373–382.
96. Lipton, J.W., H.S. Robie, Z. Ling, D.E. Weese-Mayer, and P.M. Carvey. 1998. Uterine position determines the extent of dopamine reduction after chronic prenatal cocaine exposure. *Ann. N. Y. Acad. Sci.* **844:**314–323.
97. Cowley, D.E. 1991. Genetic prenatal maternal effects on organ size in mice and their potential contribution to evolution. *J. Evolution Biol.* **4:**363–381.
98. Reading, A.J. 1966. Effects of parity and litter size on birth weight of inbred mice. *J. Mammal.* **47:**111–114.
99. Leamy, L. 1992. Morphometric studies in inbred and hybrid house mice. VIII. Effects of litter size on brain size and body size. *Growth Dev. Aging* **56:**35–43.
100. Leamy, L. and Z. Zhang. 1993. Effects of prenatal litter size in inbred mice on morphometric characters with different developmental patterns. *Growth Dev. Aging* **57:**13–23.
101. Falconer, D.S. 1960. The genetics of litter size in mice. *J. Cell. Comp. Physiol.* **56(Suppl 1):** 153–167.
102. Rauw, W.M., P.W. Knap, L. Gomez-Raya, L. Varona, and J.L. Noguera. 2003. Reallocation of body resources in lactating mice highly selected for litter size. *J. Anim. Sci.* **81:**939–944.
103. Falconer, D.S. 1955. Patterns of response in selection experiments with mice. *Cold Spring Harb. Symp. Quant. Biol.* **20:**178–196.
104. Rahnefeld, G.W., W.J. Boylan, and R.E. Comstock. 1962. Genetic correlation between growth rate and litter size in mice. *Can. J. Genet. Cytol.* **4:**289–295.
105. Cargill, S.L., J.F. Medrano, T.R. Famula, and G.B. Anderson. 2000. Effects of the high growth (hg) mutation on reproduction in high growth (HG) female mice. *Growth Dev. Aging* **64:**21–31.

106. Bunger, L., A. Laidlaw, G. Bulfield, E.J. Eisen, J.F. Medrano, G.E. Bradford, F. Pirchner, U. Renne, W. Schlote, and W.G. Hill. 2001. Inbred lines of mice derived from long-term growth selected lines: unique resources for mapping growth genes. *Mamm. Genome* **12:**678–686.
107. Naar, E.M., A. Bartke, S.S. Majumdar, F.C. Buonomo, J.S. Yun, and T.E. Wagner. 1991. Fertility of transgenic female mice expressing bovine growth hormone or human growth hormone variant genes. *Biol. Reprod.* **45:**178–187.
108. Schrott, L.M., M.E. Getty, P.W. Wacnik, and S.B. Sparber. 1998. Open-field and LPS-induced sickness behavior in young chickens: effects of embryonic cocaine and/or ritanserin. *Pharmacol. Biochem. Behav.* **61:**9–17.
109. Castoldi, A.F., T. Coccini, and L. Manzo. 2003. Neurotoxic and molecular effects of methylmercury in humans. *Rev. Environ. Health* **18:**19–31.
110. Cambonie, G., H. Hirbec, M. Michaud, J.M. Kamenka, and G. Barbanel. 2004. Prenatal infection obliterates glutamate-related protection against free hydroxyl radicals in neonatal rat brain. *J. Neurosci. Res.* **75:**125–132.
111. Ling, Z.D., Q. Chang, J.W. Lipton, C.W. Tong, T.M. Landers, and P.M. Carvey. 2004. Combined toxicity of prenatal bacterial endotoxin exposure and postnatal 6-hydroxydopamine in the adult rat midbrain. *Neuroscience* **124:**619–628.
112. Rounioja, S., J. Rasanen, V. Glumoff, M. Ojaniemi, K. Makikallio, and M. Hallman. 2003. Intra-amniotic lipopolysaccharide leads to fetal cardiac dysfunction. A mouse model for fetal inflammatory response. *Cardiovasc. Res.* **60:**156–164.
113. McEvoy, T.G., J.J. Robinson, C.J. Ashworth, J.A. Rooke, and K.D. Sinclair. 2001. Feed and forage toxicants affecting embryo survival and fetal development. *Theriogenology* **55:**113–129.
114. Godfrey, V.B., S.P. Washburn, E.J. Eisen, and B.H. Johnson. 1994. Effects of consuming endophyte-infected tall fescue on growth, reproduction and lactation in mice selected for high fecundity. *Theriogenology* **41:**1393–1409.
115. Gomez-Serrano, M., L. Tonelli, S. Listwak, E. Sternberg, and A.L. Riley. 2001. Effects of cross fostering on open-field behavior, acoustic startle, lipopolysaccharide-induced corticosterone release, and body weight in Lewis and Fischer rats. *Behav. Genet.* **31:**427–436.
116. Kurnianto, E., A. Shinjo, and D. Suga. 1998. Prenatal and postnatal maternal effects on body weight in cross-fostering experiment on two subspecies of mice. *Exp. Anim.* **47:**97–103.
117. Freemark, M., D. Fleenor, P. Driscoll, N. Binart, and P. Kelly. 2001. Body weight and fat deposition in prolactin receptor-deficient mice. *Endocrinology* **142:**532–537.
118. Meaney, M.J. 2001. Maternal care, gene expression, and the transmission of individual differences in stress reactivity across generations. *Annu. Rev. Neurosci.* **24:**1161–1192.
119. Konig, B., J. Riester, and H. Markl. 1988. Maternal-Care in House Mice (Mus-Musculus) .2. The Energy-Cost of Lactation as a Function of Litter Size. *J. Zool.* **216:**195–210.
120. Nagasawa, H. and U. Koshimizu. 1989. Difference in reproductivity and offspring growth between litter numbers in four strains of mice. *Lab. Anim.* **23:**357–360.
121. Krol, E. and J.R. Speakman. 2003. Limits to sustained energy intake. VI. Energetics of lactation in laboratory mice at thermoneutrality. *J. Exp. Biol.* **206:**4255–4266.
122. Nagasawa, H., T. Naito, H. Namiki, T. Inaba, and J. Mori. 1988. Relationships between milk levels of hormones and growth or puberty of offspring in mice. *Exp. Clin. Endocrinol.* **91:** 119–122.

123. Alston-Mills, B., A. C. Parker, E. J. Eisen, R. Wilson, and S. Fletcher. 1999. Factors influencing maternal behavior in the *hubb/hubb* mutant mouse. *Physiol. Behav.* **68**:3–8.
124. Saxton, A. M., E. J. Eisen, B. H. Johnson, and J. G. Burkhart. 1985. New mutation causing jaundice in mice. *J. Hered.* **76**:441–446.
125. Alston-Mills, B., E. J. Eisen, S. Anderson, and D. Brauns. 1995. Structure and biochemistry of normal and mutant (hub/hub) mouse mammary tissue during gestation and lactation. *Comp. Biochem. Phys.* **112A**:527–536.
126. Lipschitz, D.L., W.R. Crowley, and S.L. Bealer. 2003. Central blockade of oxytocin receptors during late gestation disrupts systemic release of oxytocin during suckling in rats. *J. Neuroendocrinol.* **15:**743–748.
127. Li, L., E.B. Keverne, S.A. Aparicio, F. Ishino, S.C. Barton, and M.A. Surani. 1999. Regulation of maternal behavior and offspring growth by paternally expressed Peg3. *Science* **284:**330–333.
128. Bredy, T.W., R.J. Grant, D.L. Champagne, and M.J. Meaney. 2003. Maternal care influences neuronal survival in the hippocampus of the rat. *Eur. J. Neurosci.* **18:**2903–2909.
129. Thomas, S.A. and R.D. Palmiter. 1998. Examining adrenergic roles in development, physiology, and behavior through targeted disruption of the mouse dopamine beta-hydroxylase gene. *Adv. Pharmacol.* **42:**57–60.
130. Lefebvre, L., S. Viville, S.C. Barton, F. Ishino, E.B. Keverne, and M.A. Surani. 1998. Abnormal maternal behaviour and growth retardation associated with loss of the imprinted gene Mest. *Nat. Genet.* **20:**163–169.
131. Lucas, B.K., C.J. Ormandy, N. Binart, R.S. Bridges, and P.A. Kelly. 1998. Null mutation of the prolactin receptor gene produces a defect in maternal behavior. *Endocrinology* **139:**4102–4107.
132. Lande, R. 1979. Quantitative genetic-analysis of multivariate evolution, applied to brain - body size allometry. *Evolution* **33:**402–416.
133. Kirkpatrick, M. and R. Lande. 1989. The evolution of maternal characters. *Evolution* **43:** 485–503.
134. Lande, R. and M. Kirkpatrick. 1990. Selection response in traits with maternal inheritance. *Gen. Res.* **55:**189–197.
135. Arnold, S.J. 1988. Quantitative genetics and selection in natural populations: Microevolution of vertebral numbers in the garter snake *Thamnophis elegans*. In: *Proceedings of the Second International Conference on Quantitative Genetics.* ed. B.S. Weir, E.J. Eisen, M.M. Goodman, and G. Namkoong. pp. 619–636. Sinauer, Sunderland, MA.
136. Cheverud, J.M. 1984. Evolution by kin selection - a quantitative genetic model illustrated by maternal performance in mice. *Evolution* **38:**766–777.
137. Falconer, D.S. 1965. Maternal effects and selection response. In: *Genetics Today, Proceedings of the XI International Congress on Genetics.* ed. S. J. Geerts. pp. 763–774. Pergamon, Oxford.
138. Hanrahan, J.P. and E.J. Eisen. 1973. Sexual dimorphism and direct and maternal genetic effects on body-weight in mice. *Theor. Appl. Genet.* **43:**39–45.
139. Ahlschwede, W.T. and O.W. Robison. 1971. Maternal effects on weights and backfat of swine. *J. Anim. Sci.* **33:**1206–1210.
140. Kuhlers, D.L., A.B. Chapman, and N.L. First. 1977. Estimates of maternal and grandmaternal influences on weights and gains of pigs. *J. Anim. Sci.* **44:**181–188.

141. Riska, B., J.J. Rutledge, and W.R. Atchley. 1985. Covariance between direct and maternal genetic effects in mice, with a model of persistent environmental influences. *Genet. Res.* **45:** 287–297.
142. Falconer, D.S. and T.F.C. Mackay. 1996. *Introduction to Quantitative Genetics.* Fourth Edition. Pearson Education Limited, Essex.
143. Peripato, A.C., R.A. De Brito, T.T. Vaughn, L.S. Pletscher, S.R. Matioli, and J.M. Cheverud. 2002. Quantitative trait loci for maternal performance for offspring survival in mice. *Genetics* **162:**1341–1353.
144. Eisen, E.J. 1967. Mating designs for estimating direct and maternal genetic variances and direct-maternal genetic covariances. *Can. J. Genet. Cytol.* **9:**13–22.
145. Clement, V., B. Bibe, E. Verrier, J.M. Elsen, E. Manfredi, J. Bouix, and E. Hanocq. 2001. Simulation analysis to test the influence of model adequacy and data structure on the estimation of genetic parameters for traits with direct and maternal effects. *Genet. Sel. Evol.* **33:**369–395.
146. Dodenhoff, J., L.D. Van Vleck, S.D. Kachman, and R.M. Koch. 1998. Parameter estimates for direct, maternal, and grandmaternal genetic effects for birth weight and weaning weight in Hereford cattle. *J. Anim. Sci.* **76:**2521–2527.
147. Davis, K.C., D.D. Kress, D.E. Doornbos, and D.C. Anderson. 1998. Heterosis and breed additive effects for Hereford, Tarentaise, and the reciprocal crosses for calf traits. *J. Anim. Sci.* **76:**701–705.
148. Brown, M.A., A.H. Brown, W.G. Jackson, and J.R. Miesner. 2000. Genotype x environment interactions in Angus, Brahman, and reciprocal-cross cows and their calves grazing common bermudagrass, endophyte-infected tall fescue pastures, or both forages. *J. Anim. Sci.* **78:**546–551.
149. Cole, N.A., M.A. Brown, and W.A. Phillips. 2001. Genetic x environment interactions on blood constituents of Angus, Brahman, and reciprocal-cross cows and calves grazing common bermudagrass or endophyte-infected tall fescue. *J. Anim. Sci.* **79:**1151–1161.
150. McCarron, P., G. Davey Smith, and A.T. Hattersley. 2004. Type 2 diabetes in grandparents and birth weight in offspring and grandchildren in the ALSPAC study. *J. Epidemiol. Community Health* **58:**517–522.
151. Ahmadiyeh, N., G.A. Churchill, K. Shimomura, L.C. Solberg, J.S. Takahashi, and E.E. Redei. 2003. X-linked and lineage-dependent inheritance of coping responses to stress. *Mamm. Genome* **14:**748–757.
152. Kafri, T., M. Ariel, M. Brandeis, R. Shemer, L. Urven, J. McCarrey, H. Cedar, and A. Razin. 1992. Developmental pattern of gene-specific DNA methylation in the mouse embryo and germ line. *Genes Dev.* **6:**705–714.
153. Rakyan, V.K., J. Preis, H.D. Morgan, and E. Whitelaw. 2001. The marks, mechanisms and memory of epigenetic states in mammals. *Biochem. J.* **356:**1–10.
154. Pollack, Y., R. Stein, A. Razin, and H. Cedar. 1980. Methylation of foreign DNA sequences in eukaryotic cells. *Proc. Natl. Acad. Sci. USA* **77:**6463–6467.
155. Silva, A.J., K. Ward, and R. White. 1993. Mosaic methylation in clonal tissue. *Dev. Biol.* **156:** 391–398.
156. Murphy, S.K. and R.L. Jirtle. 2000. Imprinted genes as potential genetic and epigenetic toxicologic targets. *Environ. Health. Perspect.* **108 Suppl 1:**5–11.

157. Weber, M., L. Milligan, A. Delalbre, E. Antoine, C. Brunel, G. Cathala, and T. Forne. 2001. Extensive tissue-specific variation of allelic methylation in the Igf2 gene during mouse fetal development: relation to expression and imprinting. *Mech. Dev.* **101:**133–141.

158. Reed, M.R., C.F. Huang, A.D. Riggs, and J.R. Mann. 2001. A complex duplication created by gene targeting at the imprinted -* locus results in two classes of methylation and correlated Igf2 expression phenotypes. *Genomics* **74:**186–196.

159. Abdollahi, A., D. Roberts, A.K. Godwin, D.C. Schultz, G. Sonoda, J.R. Testa, and T.C. Hamilton. 1997. Identification of a zinc-finger gene at 6q25: a chromosomal region implicated in development of many solid tumors. *Oncogene* **14:**1973–1979.

160. Adollahi, A., D. Pisarcik, D. Roberts, J. Weinstein, P. Cairns, and T.C. Hamilton. 2003. LOT1 (PLAGL1/ZAC1), the candidate tumor suppressor gene at chromosome 6q24-25, is epigenetically regulated in cancer. *J. Biol. Chem.* **278:**6041–6049.

161. Baqir, S. and L.C. Smith. 2003. Growth restricted in vitro culture conditions alter the imprinted gene expression patterns of mouse embryonic stem cells. *Cloning Stem Cells* **5:** 199–212.

162. Nezer, C., C. Collette, L. Moreau, B. Brouwers, J.J. Kim, E. Giuffra, N. Buys, L. Andersson, and M. Georges. 2003. Haplotype sharing refines the location of an imprinted quantitative trait locus with major effect on muscle mass to a 250-kb chromosome segment containing the porcine IGF2 gene. *Genetics* **165:**277–285.

163. Van Laere, A.S., M. Nguyen, M. Braunschweig, C. Nezer, C. Collette, L. Moreau, A.L. Archibald, C.S. Haley, N. Buys, M. Tally, G. Andersson, M. Georges, and L. Andersson. 2003. A regulatory mutation in IGF2 causes a major QTL effect on muscle growth in the pig. *Nature* **425:**832–836.

164. Xu, Y., C.G. Goodyer, C. Deal, and C. Polychronakos. 1993. Functional polymorphism in the parental imprinting of the human IGF2R gene. *Biochem. Biophys. Res. Commun.* **197:**747–754.

165. Bunzel, R., I. Blumcke, S. Cichon, S. Normann, J. Schramm, P. Propping, and M.M. Nothen. 1998. Polymorphic imprinting of the serotonin-2A (5-HT2A) receptor gene in human adult brain. *Brain Res. Mol. Brain Res.* **59:**90–92.

166. Croteau, S., C. Polychronakos, and A.K. Naumova. 2001. Imprinting defects in mouse embryos: stochastic errors or polymorphic phenotype? *Genesis* **31:**11–16.

167. Sakatani, T., M. Wei, M. Katoh, C. Okita, D. Wada, K. Mitsuya, M. Meguro, M. Ikeguchi, H. Ito, B. Tycko, and M. Oshimura. 2001. Epigenetic heterogeneity at imprinted loci in normal populations. *Biochem. Biophys. Res. Commun.* **283:**1124–1130.

168. Cooney, C.A. 1993. Are somatic cells inherently deficient in methylation metabolism? A proposed mechanism for DNA methylation loss, senescence and aging. *Growth Dev. Aging* **57:**261–273.

169. Cooney, C.A., A.A. Dave, and G.L. Wolff. 2002. Maternal methyl supplements in mice affect epigenetic variation and DNA methylation of offspring. *J. Nutr.* **132:**2393S–2400S.

170. Doherty, A.S., M.R. Mann, K.D. Tremblay, M.S. Bartolomei, and R.M. Schultz. 2000. Differential effects of culture on imprinted H19 expression in the preimplantation mouse embryo. *Biol. Reprod.* **62:**1526–1535.

171. Khosla, S., W. Dean, D. Brown, W. Reik, and R. Feil. 2001. Culture of preimplantation mouse embryos affects fetal development and the expression of imprinted genes. *Biol. Reprod.* **64:** 918–926.

172. R.A. Waterland and R.L. Jirtle. 2003. Developmental relaxation of insulin-like growth factor 2 imprinting in kidney is determined by weanling diet. *Pediatr. Res.* **53 suppl**, p. 5A.
173. Waterland, R.A. and C. Garza. 2002. Early postnatal nutrition determines adult pancreatic glucose-responsive insulin secretion and islet gene expression in rats. *J. Nutr.* **132:**357–364.
174. Hu, J.F., P.H. Nguyen, N.V. Pham, T.H. Vu, and A.R. Hoffman. 1997. Modulation of Igf2 genomic imprinting in mice induced by 5-azacytidine, an inhibitor of DNA methylation. *Mol. Endocrinol.* **11:**1891–1898.
175. Reik, W., M. Constancia, A. Fowden, N. Anderson, W. Dean, A. Ferguson-Smith, B. Tycko, and C. Sibley. 2003. Regulation of supply and demand for maternal nutrients in mammals by imprinted genes. *J. Physiol.* **547:**35–44.
176. DeChiara, T.M., E.J. Robertson, and A. Efstratiadis. 1991. Parental imprinting of the mouse insulin-like growth factor II gene. *Cell* **64:**849–859.
177. Gardner, R.L., S. Squire, S. Zaina, S. Hills, and C.F. Graham. 1999. Insulin-like growth factor-2 regulation of conceptus composition: effects of the trophectoderm and inner cell mass genotypes in the mouse. *Biol. Reprod.* **60:**190–195.
178. Constancia, M., M. Hemberger, J. Hughes, W. Dean, A. Ferguson-Smith, R. Fundele, F. Stewart, G. Kelsey, A. Fowden, C. Sibley, and W. Reik. 2002. Placental-specific IGF-II is a major modulator of placental and fetal growth. *Nature* **417:**945–948.
179. Moore, T. and D. Haig. 1991. Genomic imprinting in mammalian development: a parental tug-of-war. *Trends Genet.* **7:**45–49.
180. Isles, A.R., M.J. Baum, D. Ma, A. Szeto, E.B. Keverne, and N.D. Allen. 2002. A possible role for imprinted genes in inbreeding avoidance and dispersal from the natal area in mice. *Proc. R. Soc. Lond. B: Biol. Sci.* **269:**665–670.

CHAPTER 4

INBREEDING AND CROSSBREEDING

Gudrun A. Brockmann

Institute for Animal Sciences, Unit Breeding Biology and Molecular Genetics, Humboldt-University of Berlin, Invalidenstraße 42, 10115 Berlin, Germany
gudrun.brockmann@agrar.hu-berlin.de

1. What is Inbreeding?

1.1. *Definition of Inbreeding*

Inbreeding is the mating of related individuals. Inbreeding leads to the creation of animals that are homozygous for random alleles at each locus. These gene loci are said to be fixed in the population. Consequent mating of full sibs over many generations eventually generates groups of offspring that are homozygous for the whole genome, except for the difference between males and females carrying the X or Y chromosome. Such animal groups are called inbred strains. These strains are said to be isogenic because all individuals are genetically identical.

The genetic variance within a strain decreases with increasing inbreeding and approaches zero in nearly completely homozygous strains. However, there is always residual heterozygosity because of the statistical probabilities of inbreeding and because of new mutations in the strain. As the fixation of alleles during inbreeding is a random process, each inbred strain that has been produced from a common founder population has been fixed for a different set of alleles among those occurring in the founder population. Thus, every inbred strain is genetically unique. Inbreeding reduces the genetic variation within a strain, but variation between inbred strains remains high, even if the strains descended from the same founder population. The differences exhibited by the resulting inbred types depend directly upon the number of alleles present in the population and, in particular, upon the initially randomly chosen individuals. Figure 1 depicts the changes of genetic and phenotypic variation during the process of inbreeding.

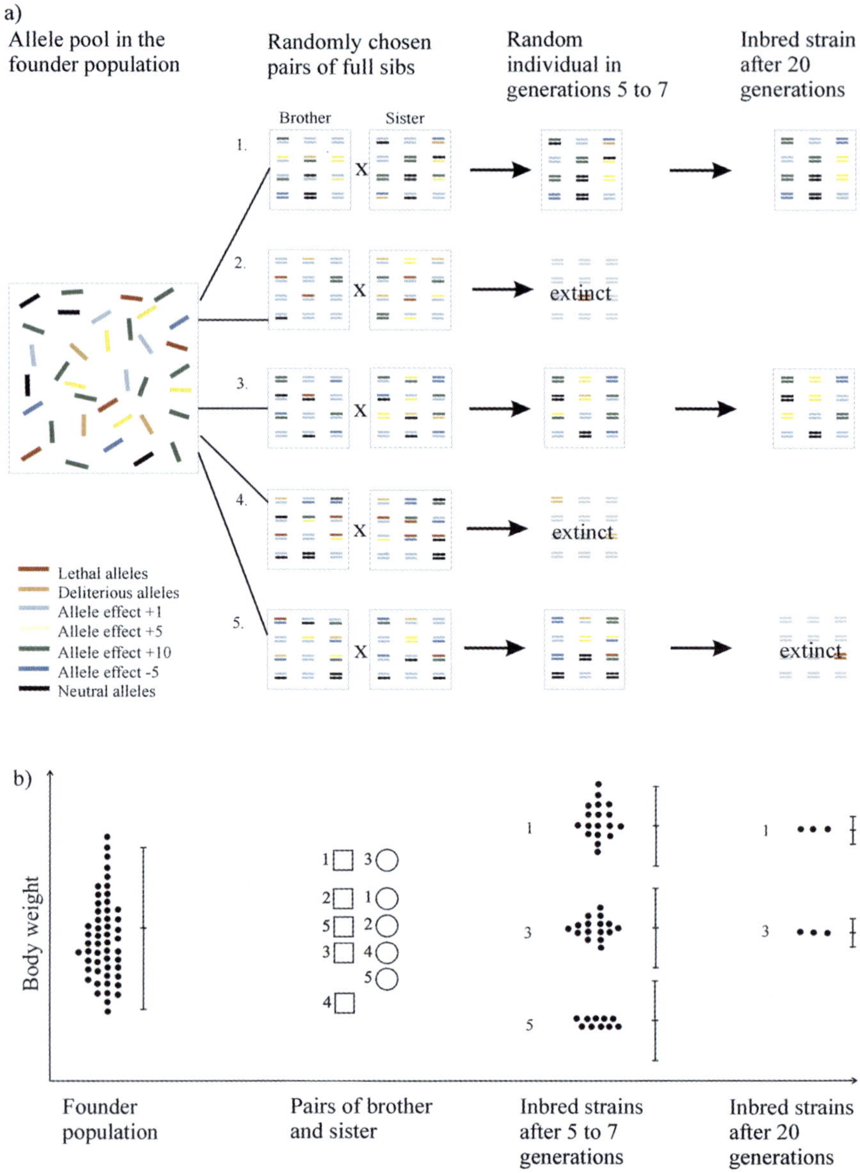

Figure 1. Genetic and phenotypic changes during inbreeding. (a) Inbreeding fixes genome variants of a heterogeneous base population. During the process of recurrent full-sib mating, several strains become extinct as a result of fixation of "hidden" recessive lethal and deleterious alleles. Twenty subsequent full sib matings generate nearly homogenous inbred strains. As the fixation of alleles is a random process, every inbred strain is genetically unique. (b) Inbreeding leads to loss of genetic variation within a strain, but genetic variation among different strains remains high, which might result in distinct phenotypes. The combination of inbreeding with selection of full sib pairs of favored phenotypes is used to generate inbred strains for specific traits.[84]

1.2. *History of Inbreeding*

Inbreeding is a phenomenon that is widely distributed in natural populations. It is of paramount importance in small populations, but it also occurs to a greater extent than biologists have thought, in many species with variable effects on individual fitness (independently of population size).

Before inbreeding was systematically used in laboratory animals, it was applied during domestication of many animal species. In the process of animal breeding, inbreeding has been applied in the improvement of livestock by the introgression of favorable traits into established breeds, but in most cases inbreeding has been strictly avoided to prevent reduction of fitness and fecundity.[1] Early herd books of valuable breeds showed unambiguously that mating of close relatives could fix characteristics in the resulting progeny. First applications of systematic inbreeding to animals were made by Robert Bakewell (1725–1795), who was a pioneer in showing the value of mating close relatives as a means of fixing a uniform type. His principle work was the foundation of the Leicester breed of sheep and the bringing of it to a high degree of desirable qualities. Bakewell's practice of mating brother with sister and parent with offspring horrified his neighbors, but they soon appreciated the practical results he achieved, because the yearly revenue from the services of his rams rose from about $4 to $2000 in thirty years.[2]

The first controlled experiments on inbreeding with laboratory animals were carried out on mice, rats, and guinea pigs at the end of the 19th century after Mendel had pioneered the theory of inheritance. Weismann inbred a stock of white mice for 29 generations. He observed that the average number of young obtained in three successive ten-generation periods decreased from 6.1 to 5.6 to 4.2. The average number of litters raised in the first part of the experiment was 22, in the latter part only 3. There was a greater opportunity to select healthy breeding stock in the earlier period, which might account for a part of the remarked reduction in fertility.[3] These first mouse inbred lines were lost for unknown reasons.

At the beginning of the 20th century, scientists were taking early steps towards understanding genetics. They began inbreeding mice from exotic pet mice, which had a wide range of coat colors, behaviors and other conspicuous characteristics, including resistance and susceptibility to cancer. In particular, Castle's group began a systematic analysis of inheritance and genetic variation.[4] The oldest inbred mouse strain is the DBA strain, which carries the three mutant coat color alleles *d*ilute, *b*rown and non-*a*gouti. This inbred strain was founded by Clarence

Little in 1909. Only a few years later, Castle and colleagues inbred the lines C57BL/6, C57BL/10, C3H, CBA, and BALB/c. Since then, many inbred strains became a powerful tool for basic and biomedical research. In this way, a variety of specialized strains predisposed to developing cancer has been developed. In 1921, Leonell Strong established many inbred strains that developed cancer frequently and spontaneously. These strains serve as a virtually unlimited source of many types of tumors and have made it possible to study their growth and general characteristics. The genetic homogeneity of inbred strains permitted reproducible experiments to be carried out on standardized mice around the world.

Inbred mouse strains became powerful models for research in human diseases and behavior when they were used to investigate the heritability of many different traits with relevance to human health and behavior. The first genetic study establishing the linkage between two coat-color mutations was published by Haldane in 1915.[5] The recognition that variations in DNA could be assayed directly and used as genetic markers in linkage studies was a milestone on the road towards gene discovery by linkage analysis.[6] Since then, comprehensive gene and marker maps have been developed in mice.[7] The dissection of complex traits into single genetic entities was consequently applied in the search for genetic loci controlling complex traits in linkage analyses. As a highlight in mouse genetics at the beginning of the 21st century, the genome sequences of the four most widely used mouse inbred strains C57BL/6J, DBA/2J, 129S1, and A/J were published. The mouse genome sequence is available at the Ensembl Web Site.[8]

1.3. *Advantages of Inbreeding*

Many breeders use inbreeding as a means to fix specific gene variants and to create strains of mice with particular characteristics. Inbreeding standardizes and simplifies the genotype. Therefore, inbred animals have substantial advantages for genetic and biometric research. Inbred strains of mice are a source for the identical reproduction of experimental mammals. Thus, experiments can be repeated with the same genetic material worldwide. However, attention should be paid to slight changes of environmental conditions among laboratories, which might cause phenotypic effects. Particularly sensitive to environment is the behavior of a mouse.[9]

Corresponding with the genetic homogeneity, the phenotypic variation within a strain is reduced. Thus, reduced animal numbers are needed to show significant phenotypic changes induced by environmental effects or in response to genetic modifications, such as transgenes, recombined knockout or knockin genes and natural or induced mutations.[10, 11] Furthermore, the identification of spontaneous or induced mutations on the genetic background of sequenced inbred strains is enhanced by the availability of the genomic sequence.

1.4. *Limitations of Inbreeding*

Wild populations and outbred stocks may contain recessive deleterious and lethal mutant alleles. As a result of inbreeding, lethal and deleterious recessive alleles are fixed, causing death or reduced viability and fecundity of offspring. In addition, inbreeding causes loss of alleles. During evolution, coadaption of alleles occurs so that specific combinations of allele variants are more favored than every allele separately. Thus, the loss of alleles during inbreeding might interrupt interallelic networks contributing to a trait.

The result of inbreeding depends on the environmental conditions under which inbreeding occurs and whether organisms that inbreed have mechanisms to dampen the anticipated problems of reduced genetic variation. Research animals live in controlled environments where much attention is paid to parameters such as pathogen-free environment, airflow, ammonia levels, humidity, and temperature. Most inbred mouse strains will not be viable in the wild.

In the case of overdominance, when heterozygous animals have a more extreme phenotype than either parent, homozygosity may cause a loss of the phenotype observed in the outbred population. Overdominance plays a role in reproduction traits, immune response, learning ability and, in general, in many characteristics that are essential for high viability in the wild.

2. Generation of Mouse Inbred Strains

2.1. *General Inbreeding Procedure*

The fastest way to attain inbred strains is the continued mating of close relatives To create a mouse inbred strain, a number of full-sib pairs of outbred mice are initially selected. Subsequently, repeated brother-sister matings are made for several generations. At least 20 generations are necessary to reach homozygosity for nearly all genetic loci.[12] Inbred strains can also be developed by consequent backcrosses to an inbred strain over 10 generations or by half-sib mating over more than 20 generations (Table 1).

Table 1. Changes of inbreeding coefficients under various systems of close inbreeding (adopted from Falconer and Mackay[13])

Generation (t)	Inbreeding coefficient (F) Repeated backcrosses to an inbred strain	Brother × sister mating	Half sib mating (females half sisters)
0	0	0	0
1	0.500	0.250	0.125
2	0.750	0.375	0.219
3	0.875	0.500	0.305
4	0.938	0.594	0.381
5	0.969	0.672	0.449
10	0.999	0.886	0.691
15	~1.00	0.961	0.827
20	~1.00	0.986	0.903
Recurrence equation	$1/2(1+F_{t-1})$	$1/4(1+2F_{t-1}+F_{t-2})$	$1/8(1+6F_{t-1}+F_{t-2})$

A relatively high number of full-sib pairs of outbred mice have to be selected initially, and in every generation as many brother-sister matings should be made as there are males and females available. These matings are necessary because 80–90% of the attempts to create inbred strains fail, typically becoming extinct between 5 and 7 generations. The unwanted extinction results from the reduction of heterozygosity and increase of homozygosity, which leads to the fixation of recessive genes with deleterious effects. Furthermore, the physical separation of interacting allele variants that ensures high fitness in the outbred population might additionally reduce fecundity and lead to a lesser degree of inbreeding depression at generations 14 to 20.[11]

2.2. *Inbreeding of Wild–Derived Mice*

Wild-derived inbred mice are descendants of mice originally caught in the wild. Often they originate from only one pair or trio of captured animals. Most wild-derived inbred strains have their origin in *Mus musculus*. Other wild-derived mouse inbred strains descend from the *Mus* species *caroli*, *castaneus*, *hortulames*, *molossinus*, *praetextus*, *pahari*, and *spretus* (Table 2). The large number of genetic differences between wild–derived and common inbred laboratory mice makes wild–derived inbred mice valuable tools for evolution and systematic research and makes progeny from interspecific crosses especially useful for genetic mapping. The introduction of wild mice into research led to the discovery of many new and interesting phenotypes and genotypes. The major histocompatibility complex is the best example of an extraordinary degree of polymorphism uncovered by wild mice.[18]

Table 2. Origin of wild-derived inbred strains (data from Jax Mice Database – Wild-derived Inbred Web Site)[19]

Species[a]	Geographic origin	Strain
M. caroli (Ricefied mouse)	Thailand	Mus caroli/EiJ
M. musculus castaneus (Southeast Asian house mouse)	Thailand	CASA/RkJ, CAST/EiJ
M. musculus domesticus (Western European house mouse)	California	CALB/RkJ
	Lewes, Delaware	LEWES/EiJ
	Ohio, USA	MOR/RkJ
	Tirano, Italy	TIRANO/Ei
	Monastir, Tunisia	WMP/PasDnJ
	Centreville, Queen Anne City, Maryland	WSB/EiJ
	Zalende, Switzerland	ZALENDE/Ei
M. hortulanus (Garden mouse)	Pancevo, Serbia	PANCEVO/EiJ
M. musculus (House mouse)	Studenec, Morovia, Czechoslovakia	CZECHI/EiJ, CZECHII/EiJ
	Rimac valley, Peru	PERA/EiJ, PERC/EiJ
	Marin County, California	SF/CamEiJ
M. musculus musculus (Central European house mouse)	Kunratice, Czech Republic	PWD/PhJ
	Lhotka, Czech Republic	PWK/PhJ
M. musculus molossinus (Laboratory mouse)	Japan	JF1/Ms
	Fukuoka, Kyushu, Japan	MOLC/RkJ, MOLD/RkJ, MOLF/EiJ, MOLG/DnJ
	Mishima, Shizuoka, Japan	MSM/Ms
M. musculus / domesticus [b]	Skive, Denmark	SKIVE/EiJ
M. musculus praetextus / musculus [c]	Israel	IS/CamRkJ
M. pahari (Indochina shrew-mouse)	Thailand	Mus pahari/EiJ
M. spretus (Orange mouse)	Puerto Real, Cadiz Province, Spain	SPRET/EiJ

Abbreviation: *M.* - *Mus*,

[a]Common names are accessible online at the the MURINAE Web Site.[20]

[b]Nuclear DNA was identified as being of *M. musculus* origin, while mitochondrial DNA is of *M. domesticus* origin.[21–23]

[c]A *M. musculus praetextus* male caught in an Israeli port was bred with a *M. musculus musculus* female from laboratory stock.

2.3. *Inbreeding of Selected Strains*

Repeated selection for divergent characters from heterogeneous outbred populations leads to selected strains differing extremely in the selected phenotype. Breeding for traits such as growth, behavior or fecundity has generated selected mouse lines, which are often polygenic models for complex traits. The process of continual selection over many generations starting from a highly heterogeneous base populations is expected to fix gene variants contributing to the selection response.[14] But if there are multiple genes that can contribute to the phenotype, several or many genes might have a higher performance in the heterozygous state. Therefore, subsequent inbreeding ensures fixation of genes which are homozygous as a result of selection and will fix many, but not necessarily all, alleles contributing to the selection response. Such selected strains are available for different complex traits, among them growth and obesity[15, 16] and behavior.[9, 17]

3. Widely Used Mouse Inbred Strains

3.1. *Genetic Diversity*

About 2800 inbred, recombinant inbred, congenic, consomic and genetically modified inbred strains are currently available at the Jackson Laboratory, the world's largest holder and distributor of mouse strains (The Jackson Laboratory Web Site).[24] Rules for the nomenclature of strains indicating the original inbred strain(s) and modifications are given by the International Committee on Standardized Genetic Nomenclature for Mice.[25] About 450 mouse strains are common inbred strains.[26] A comprehensive list with a detailed description of their origins, known mutations and phenotypes is given by Festing.[27, 28]

Phylogenetic analyses show the genealogical relationships between inbred strains. According to their genetic origin, the strains may be divided into seven categories: Swiss mice, Castle's mice, strains from Asia, unknown sources, C57-related mice, wild mice, and mixtures of inbred strains.[26] The use of different types of genetic data, such as protein loci, immune loci, or endogenous viral loci may give qualitatively different estimates of phylogenetic relationships.[29] Molecular studies have shown that although the base for most laboratory inbred strains is the wild-derived *Mus musculus* type, there is increasing evidence that laboratory mice have been developed with contributions from species or subspecies of wild mice. For example, some strains carry the *Mus musculus domesticus* Y-chromosome, while others have the *Mus musculus musculus* type. The following classification was found: Mice of the strains A/J, AEJ/Gn, AU/SsJ, BALB/cJ, BDP/J, BXSB/MpJ, CBA/J, CE/J, C3H/HeJ, DA/HuSn, DBA/2J, HRS/J, HTG/Go, I/Ln, LP/J, NZB/B1N, NZW/Lac, P/J, RIIIS/J, SB/Le, SEA/Gn, SEC/1ReJ, SF/Cam, SK/Cem, SM/J, WB/ReJ, WC/ReJ, YBR/Ei, 129/J carry the *Mus musculus musculus* Y chromosome type, while mice of the strains AKR/J, BUB/J, MA/MyJ, PL/J, RF/J, SJL/J, ST/bJ, SWR/J, SWV contain the *Mus musculus domesticus* type.[30] In contrast to 36 other standard inbred strains, C57BL/6 carries a Y chromosome of Asian origin[31] and a LINE-1 element derived from *Mus spretus*.[32]

Genetic analyses require variation that can be followed by generations. Therefore, many mouse inbred strains have been genotyped for microsatellite markers and single nucleotide polymorphisms. Information on genetic marker alleles can be found for several mouse inbred strains at the Mouse Genome Informatics Web Site.[7, 33]

3.2. *Phenotypic Diversity*

Both the development of a new knockout or transgenic mouse and the mapping of genes in crossbred populations require extensive knowledge of the endogenous traits of inbred strains to design the most efficient experiments. Furthermore, in recent years, an increasing number of scientific articles reported that the genetic background genes from a parental inbred strain might interact with the mutated gene in a manner, which could modulate the phenotype and severely compromise the interpretation of the observations.[34] During inbreeding of heterozygous mutant mice on a specific genetic background of an inbred strain, the effect of the mutation can even vanish, as was found for the mutant allele of the retinoblastoma tumor-suppressor gene (*Rb-1*) on pituitary tumorgenesis, for example.[35]

Therefore, strain distributions have been analyzed extensively for many distinct phenotypes. For behavior, for example, strain distributions are described for open field activity, learning and memory tasks, aggression, sexual and parental behaviors, acoustic startle and prepulse inhibition, and the behavioral actions of ethanol, nicotine, cocaine, opiates, antipsychotics, and anxiolytics.[36] Moreover, the strains C57BL/6J, C57BL/6NTac, 129P3/J, 129S6/SvEvTac and FVB/NTac, which are commonly used in transgenic and knockout production, were compared with regard to genetic background and behavioral habituation to the open field.[37] As a conclusion, the authors suggested transferring any transgene on the FVB/NTac background to C57BL/6.

The Mouse Phenome Database[38] is an international effort to collect phenotypic data on commonly used and genetically diverse inbred mouse strains. Among the strains that are intensively phenotyped are the mouse inbred strains 129S1/SvImJ, A/J, BALB/cByJ, C3H/HeJ, C57BL/6J, CAST/EiJ, DBA/2J, FVB/NJ, SJL/J, and SPRET/EiJ, which belong to a group of strains that is widely used. Present in this strain set are strains with available genomic sequence, strains that are progenitors in transgenesis or studies, and strains that are progenitors of recombinant inbred, consomic, or congenic strains. The strain set is genetically diverse with the inclusion of the wild-derived inbred strains CAST/EiJ and SPRET/EiJ. These strains are generally easy to maintain with good reproductive performance. Thus, they are recommended for all projects.

Using the information on strain characteristics, molecular geneticists can choose parental background strains for crossbred experiments or for targeted mutations, which best fit the traits they want to evaluate. Table 3 shows exemplary profiles of diversity between strains for selected phenotypes.

4. Deviations from Common Inbred Strains

4.1. *Substrains of Inbred Strains*

According to the definition, substrains are branches of an inbred strain that have shown probable genetic differences (Figure 2). These genetic differences among branches may result from genetic drift of residual unfixed alleles, from mutations occurring spontaneously, or from genetic contamination with foreign mice.[44] Substrains are formed when branches of a strain are separated before the 40th generation of inbreeding, when branches of a strain have been maintained separately from other branches for more than ten generations of inbreeding or when genetic differences from other branches are discovered (The Jackson Laboratory Web Site).[24] The degree of genetic diversity is often not well characterized.

For example, there are several substrains for C57BL and 129, which are widely used for genetic research. At least nine substrains exist for C57BL (MGI2.98 Inbred Strains of Mice: C57BL Web Site).[45] In extensive studies, a large degree of genetic diversity among 129 substrains was identified.[46, 47] The

Table 3. Phenotypic diversity between inbred strains for selected traits (data from the Mouse Phenome Database)[384]

Trait/Phenotype (unit)	Sex	n[a]	Mean	Median	s.d.	Min	Max
Anxiety[39]							
Latency to enter open quadrant (s)	f	8	55.6	30.7	54.4	6.34	140
	m	8	47.8	37.1	41.4	8.34	127
Total time in open quadrants (s)	f	8	54.1	42.3	31.2	16.3	96.5
	m	8	50.1	34.1	31.6	18.8	103
Hematology[40]							
Red blood cell count (units per	f	43	9.11	9.1	0.54	7.96	10.6
volume × 10^6) (n/μl)	m	43	9.23	9.25	0.63	7.79	10.4
White blood cell count (units	f	43	7.31	6.89	2.37	2.6	12.6
per volume × 10^3) (n/μl)	m	43	7.66	7.16	2.51	2.76	13.3
Anxiety[41]							
Time in dark box (s)	f	11	443	451	35.9	382	487
Number of transitions between light and dark compartments (n)	f	11	20.4	16.8	9.89	10.6	40.6
Nociception[42]							
Hot plate, latency to respond (s)	m	11	26.1	23.8	9.47	14.7	44.2
Tail clip, latency to respond (s)	m	12	17.3	12.8	14.5	4.7	56.1
Body composition[43]							
Body mass index after 8 wks on	f	23	32.2	31.6	4.52	24.5	41.2
atherogenic diet (g/cm^2)	m	20	36.4	37.4	5.61	26.8	47.4
Weight of fat portion of tissue mass	f	32	5.41	5.01	2.86	1.71	14.4
after 8 wks on atherogenic diet (g)	m	36	6.82	6.06	3.41	1.57	15.8
Weight of lean portion of tissue mass	f	32	16.5	16.6	3.5	9.17	23
after 8 wks on atherogenic diet (g)	m	36	20.7	21.6	4.19	11	27.3

[a]Number of strains analyzed.

overall result of scrutinizing the genealogy of strain 129 is a change in nomenclature that specifies groups of substrains related by their common parental lineage.[48] The major parental lineages include the letter P, S, T or X indicating whether it is from one of the 129 strains derived from the original "parental" strain, from a congenic strain made by outcrossing to introduce the "Steel" mutation, from the 129 congenic that originally carried the "Teratoma" mutation or a genetically-contaminated "X" substrain, respectively. The numbers following the letters (e.g., P3) distinguish the different 129 parent strains within each lineage (MGI2.98 Inbred Strains of Mice: 129 Web Site).[49]

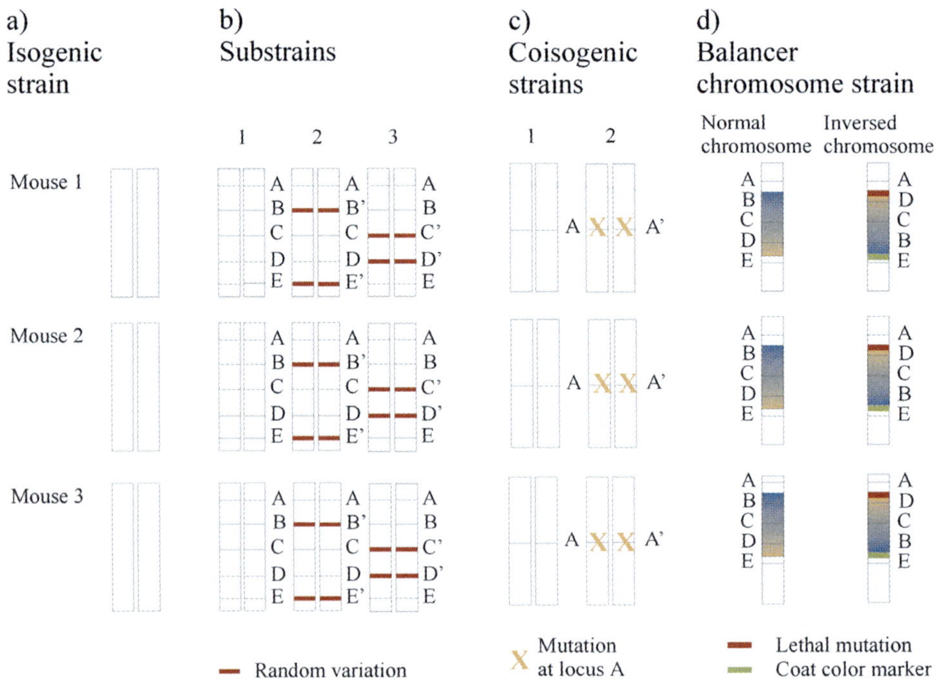

Figure 2. Different types of inbred strains. (a) Inbred strains are isogenic strains; every individual carries the same gene variants. (b) Substrains of inbred strains are branches of an inbred strain, which might result from genetic drift, mutations, or contamination with foreign DNA. (c) Coisogenic strains differ from one another by a single mutation. (d) A balancer chromosome strain is an inbred strain that carries an inversion of a chromosomal region. Recombination is suppressed within the heterozygous region of inverted and wild-type chromosome. A lethal mutation and a coat color marker allow the selection of heterozygous balancer chromosome carriers.

4.2. *Coisogenic Strains*

Several spontaneous single gene mutations occurred in inbred mice and have been selected from the base inbred strain (isogenic strain). Such inbred strains that differ from one another only by a single gene as the result of a mutation are designated as coisogenic (Figure 2). For example, severe mutations in the genes encoding leptin and leptin receptor, which cause the extreme obese and diabetic phenotype, respectively, were selected from the inbred strain C57BL/6J. These strains are coisogenic with the inbred strain C57BL/6J.

Other types of coisogenic strains have been created by induced random or targeted mutations on the background of inbred strains. The production of transgenic mice with overexpression or targeted deletion of genes[50] and the random generation of mutant mice[51–53] play a major role as model systems for functional genome analysis. To test the effect of the genetic modification, the modified mice have to be compared with the base strain. Therefore, these experiments ideally require inbred mouse strains that ensure the identical reproduction of the base and the genetically modified individuals. Several mouse inbred strains have been identified as being well suited for the different purposes.

4.2.1. *Gene Transfer*

Transgenesis is the stable integration of foreign DNA into a host genome. The most direct way to generate transgenic mice is by microinjection of a solution of foreign DNA into the pronucleus of fertilized oocytes. In general it is advisable to generate mice on a defined genetic background of inbred strains. This need is particularly important when subtle phenotypes are expected or genetic background effects might influence the effect of the transgene. Mouse inbred strains which produce a large number of fertilized oocytes with large and clearly visible pronuclei and which are resistant to the mechanical injury of the cell by the glass micro needle during injection are particularly suitable for the production of transgenics. Mouse strains that are well suited for gene transfer into oocytes include the inbred strains C57BL/6J and FVB/N, but also the F_1 hybrids of CBA × C57BL/6J and the NMRI outbred strain.[54] If the transgene is integrated into the genome of an inbred strain, heterozygous offspring are mated to produce a homozygous transgenic strain. The transgenic and the base strains are coisogenic.

4.2.2. *Gene Targeting*

Using transgenic technology, researchers can target specific mouse genes for inactivation (knockout) or activation (knockin). A knockout mouse is a mutant mouse containing a null allele for a gene under study, generated by homologous recombination in embryonic stem (ES) cells. Progenitors of the 129 strains have been used extensively to establish ES cell lines for the production of knockout mice. ES cell lines have been established also for the inbred strains BALB/c and C57BL/6.[55] Genetically modified ES cells are injected into blastocysts or are aggregated with morulae to form chimeras. For the selection of host embryos, preferentially inbred strains are chosen which differ from the ES cell line by coat color so that chimeras can be readily identified. The most common choice of host strain is C57BL/6, followed by BALB/c.[56]

4.2.3. *Randomly Induced Mutagenesis*

At present, artificial mutagenesis of the mouse is used as a means of obtaining new single-gene mutants. The goals of international mouse mutagenesis projects are directed towards the production of at least one heritable mutation, in either ES cells or mice, in every gene of the genome to identify every gene that affects key traits of biomedical interest.[51–53, 57] Mutations are induced either chemically or by irradiation. As every base can be potentially mutated, different subtle mutations may be induced. Several rounds of backcrossing are necessary to show that the phenotypic change is a genetic effect and caused by a single mutation. These mutations will be found only if the phenotypic effect is high enough to be significantly different from the base strain. Inbred strains are essential for finding phenotypic effects and for identifying the induced DNA variation.

N-ethyl-N-nitrosourea (ENU) has become the mutagen of choice for many mouse researchers. A number of inbred strains have been tested for their tolerance to ENU. Inbred strains vary widely in their tolerance of ENU. Therefore, strain-specific dose protocols should be determined.[53] The inbred strain BTBR/N is particularly resistant to the toxic effects of ENU and yields a very high mutation rate after exposure.[58] Other inbred strains that have been used successfully for ENU mutagenesis are C57BL/6J, BALB/cJ and several C3H substrains. The strain FVB/N, which is commonly used to create transgenic mice, does not tolerate ENU well, but can recover from low dose of ENU to produce mutant offspring.[59] F_1 hybrid animals generally tolerate higher doses, regain fertility faster and live longer after treatment.

Mutations in laboratory derived stocks can also be induced by radiation. For making nested deletion complexes, ES cells were irradiated. The key to this technology is that F_1 hybrid ES cells of the genotypes resulting from the crosses BALB/cTa × 129/SvJae and C57BL/6J × 129/SvJae retain germline colonizing ability after exposure to levels of ionizing radiation that induce chromosomal deletions.[60]

4.3. *Balancer Chromosome Strains*

Balancer chromosome strains are carriers of inversion mutations embracing entire chromosomal regions (Figure 2). Balancer chromosomes have the characteristic that recombination between chromatids is suppressed as a result of failure of exact pairing of homologous sequences within the inverted chromosomal region. In combination with genetic markers ensuring the visualization of heterozygous carriers of the inversion and markers for lethality in the case of homozygosity of the inversion, balancer chromosome strains can be used to identify mutations in the inverted chromosomal segment. Such chromosomal inversions occur rarely as natural mutations. Since they have been discovered in mice, they have been used for the analysis of recessive mutations. Recent progress in recombinant genome technology enables researchers to produce "designer" balancer chromosomes for specific chromosomal regions that combine all requirements for targeted genetic research.[61]

5. Crossbreeding

5.1. *Crossbreds*

Crossing is the mating of individuals from different populations; in mice these are matings between different outcross populations, between inbred strains, or between an inbred and an outbred strain. Animals that result from crossing are called crossbreds or first filial generation, symbolized by F_1. All F_1 animals that result from a mating between two inbred strains are genetically identical and heterozygous for all autosomes and the pseudoautosomal part of the X chromosome, carrying one chromosome from the paternal and one chromosome from the maternal strain. Reciprocal F_1 offspring are not genetically identical. They differ in the origin of the Y chromosome and mitochondrial DNA.

As soon as different inbred strains are crossed, the inbreeding completely disappears. The heterozygous crossbred progeny usually show an increase in the mean of those traits that previously suffered a reduction from inbreeding.[13] Heterosis (dominance or overdominance) occurs when the average performance of a crossbred progeny is superior to the average performance of the parental strains. In general, heterosis is greatest in traits associated with reproduction and viability. The greater the genetic diversity between the two strains, the greater is the heterosis in crosses between them.

The completely heterozygous F_1 population from a cross between two inbred strains is an ideal basis for the subsequent generation of segregating populations by either intercrosses between F_1 offspring to produce a second filial generation, designated as F_2, or by backcrosses of F_1 offspring to animals of one of the parental strains, to generate a first backcross generation, called BC_1. In F_2 or BC_1 progeny the inheritance of each of the parental chromosomes can be scrutinized, and the offspring can be grouped for the inherited parental alleles.

In linkage analyses, usually two mouse strains, which differ in a phenotypic trait (e.g. body weight as juveniles or adults) and in molecular markers (e.g. microsatellites, RFLPs, or SNPs) are crossed. The resulting F_1 population is then sib mated or backcrossed to produce an F_2 or BC_1 population. This population is phenotyped and genotyped to associate a genetic marker with the phenotype. The chromosomal locus that contributes to the variation of a phenotypic trait has been called a quantitative trait locus (QTL).[62]

The application of different breeding strategies beginning with F_1 offspring generates structured populations which are suitable for mapping loci that affect segregating traits of Mendelian or polygenic characters and estimating their additive, dominance and epistatic effects (Figure 3).

5.2. *Mapping Populations*

To decide whether an F_2 or backcross design is more powerful for QTL detection, the purpose of the research must be defined. Soller *et al.* demonstrated that in most cases of genetic inheritance, the F_2 design requires fewer offspring for equivalent power than the backcross design so that in planning linkage experiments, the F_2 design is recommended.[63] Falconer and Mackay stated that for additive effects, four times as many individuals are required for backcross populations as compared to F_2 pedigrees to obtain the same power of QTL detection.[13] In contrast, a backcross requires only about half the progeny of an F_2 population for the detection of recessive effects.[64] The backcross makes linkage analysis more powerful due to the lower significance thresholds required for QTL detection,[65] and the estimated QTL effects might be higher because of the reduction in residual genetic variance, which might be caused by genetic interaction.[66] The conclusion from the different findings is that the F_2 design is preferred for the simultaneous detection of additive and dominance QTL effects, and it is more efficient for the analysis of genetic interaction between QTL.

5.3. *Fine Mapping Populations*

As a consequence of the restricted number of recombinations, QTL mapping by linkage analysis in F_2 and BC_1 populations is limited due to the large size of confidence intervals of mapped loci and in its power to distinguish closely linked loci. If we assume 40,000 genes in the mouse genome of a genetic length of about 1300 cM, we expect on average 30 genes per cM or 15 genes per mega-

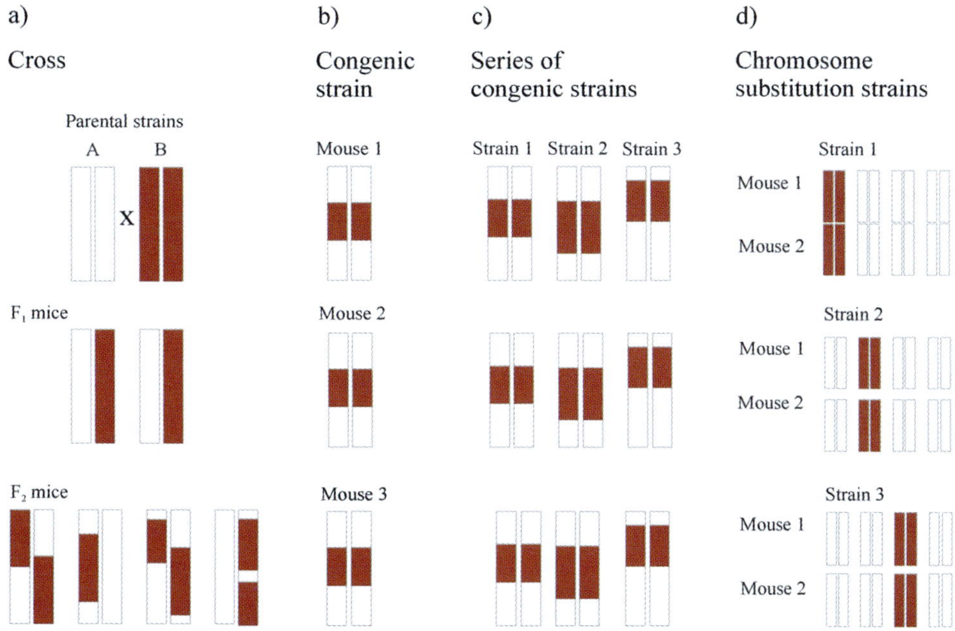

Figure 3. Crosses between inbred strains are the basis for the generation of specified genetic models for mapping genes, estimating their effects and analyzing their interaction with modifier loci. (a) An intercross population from the cross between two mouse lines that differ in phenotypic traits and in molecular markers is a genetic resource for gene mapping by linkage analyses and for the construction of congenic strains. (b) Congenic strains harbor a chromosomal region from a donor mouse strain on the background of an inbred recipient mouse strain. (c) Series of congenic strains can be used to fine map gene effects in a small chromosomal interval. (d) Chromosome substitution strains have one chromosome of a recipient strain substituted by the homologous chromosome of a donor strain.

base. Thus, the reduction of the confidence interval for the most likely position of a QTL by fine mapping is necessary to support the search for genes underlying a QTL effect.

There are several different strategies for QTL fine mapping in mouse experiments. Darvasi compared selective genotyping, recombinant progeny testing, interval specific congenic strains and recombinant inbred segregation tests as alternative designs for QTL fine mapping.[64]

5.3.1. *Advanced Intercross Populations*

A promising approach for fine mapping is the construction of advanced intercross populations.[66] These populations are generated from an initial intercross between different inbred strains followed by at least ten consecutive generations of random mating. In this way, the number of recombinations on every chromosome can be considerably increased, which makes the dissection of single QTL into small chromosomal fragments feasible in advanced generations. So far, this strategy has been successfully applied for the fine-mapping of QTL for trypanosomiasis resistance where the initial QTL region could be reduced by the factor 2.5 to more than tenfold.[67] Recently, an advanced intercross population has been generated from a cross between two inbred strains, which were initially selected for high and low heat loss.[68]

5.3.2. *Heterogeneous Stocks*

In principle, the detection of QTL is also possible in population studies and in outbred populations of initial crosses between many inbred mouse strains.[69, 70] However, usually many more individuals have to be phenotyped and genotyped to identify QTL effects.

6. Inbred Strains Generated from Crossbred Populations

6.1. *Congenic Strains*

The strains are denoted as congenic if a chromosomal region has been transferred from one strain to another (Figure 3).[71] Congenic strains are generated by recurrent backcrossing of a founder individual to a recipient strain. A minimum of ten backcross generations is necessary to recover homozygosity of the background genome (Table 1). Traditionally, this process was controlled by the phenotype of the recipient strain. For example, Snell discovered the key genetic locus of histocompatibility in mice via the development of congenic strains that are resistant to tissue transplants.[72] With the availability of genome-wide chromosomal markers, the transfer of a genomic region of interest may be scrutinized by the inheritance of genetic marker alleles characterizing the donor chromosomal region of interest. In the same way, the reduction of genetic background may be controlled. The marker-assisted breeding of congenic mouse strains (speed congenics) is theoretically possible over five generations.[73]

Congenic strains are generated to fine map QTL and to transfer natural or artificial mutations from one strain to another. The latter strategy is used to identify modifier genes influencing the effect of a mutation. A set of interval-specific congenic strains consists of strains that differ from one another only with respect to the substituted chromosomal segment, which covers a whole chromosomal interval, e.g. a QTL region (Figure 3). Such strains allow the precise location of the chromosomal segment influencing a trait and the estimation of the gene effects. For example, an X chromosome-linked body weight QTL could be mapped to a 2 cM region by a fine mapping approach that combines the use of congenic strains and progeny testing of recombinants. The latter approach is especially useful if the number of animals of the desired genotype is small.[74] Genome-wide overlapping congenic strains can be used for mapping and fine mapping of QTL across the whole genome.[75] The cross between strains carrying different isolated chromosomal regions of a donor strain on the same genetic background strain allows the systematic test for epistatic interaction between these chromosomal regions.

In some cases, mostly unreported, the expected phenotypic effect of a QTL in a small isolated chromosomal segment gets lost. For complex traits, the correlation between genotype and phenotype at a single locus is imperfect. Obviously, the organism has the ability to compensate for mutations to maintain the homeostasis of the physiology by a network of gene actions. Genes having overlapping functions and interactions between genes with unrelated functions are two major reasons for the maintenance of homeostasis.[76] The process of repeated backcrossing during the generation of a congenic strain can lead to the disruption of complexes of coadapted gene variants from the parental strains. Therefore, the loss of the phenotype in small isolated chromosomal parts of congenic strains could result from incorrect mapping of a QTL but also from malfunction of the network of interacting genes.

6.2. *Recombinant Congenic Strains*

Recombinant congenic strains (RCS) are formed by crossing two inbred strains, followed by a few backcrosses to one of the parental strains, and subsequent inbreeding without selection.[77] Such inbred strains consist of randomly fixed chromosomal segments of the donor strain on the background of the recipient strain to which it was backcrossed. The proportion of the genome of the donor strain depends on the number of backcrosses to the recipient strain before inbreeding was carried out. Two backcross generations reduce the proportion of the donor genome to an average of 12.5%.

6.3. *Recombinant Inbred Strains*

Recombinant inbred strains (RIS) are derived from an intercross population of two inbred strains by mating a pair of randomly chosen F_2 offspring and subsequent full sib mating over 20 generations. RIS are composed of 50% each of the two parental genomes. No selection is applied. Therefore, the fixation of alleles occurs in a random fashion.

Multiple RIS that were independently derived from the same original cross of parental strains constitute a very valuable genetic resource for fine mapping of chromosomal regions, as recombinations have been accumulated in every strain. Each of the recombinant inbred strains must be typed genome-wide by genetic markers to determine which of the founder alleles is fixed at each locus. The comparison of the phenotypes with the recombinant genotypes of the different RIS allows the mapping of chromosomal regions affecting the trait. Many different traits can be analyzed in a set of RIS, as identical individuals of every RIS can be produced and phenotyped.

RIS for the identification of obesity related QTL were derived from the systematic inbreeding of randomly selected pairs of F_2 animals from the initial cross SM/J × A/J.[78] Analyzing 10 to 28 mice in each of 21 strains detected several QTL for body weight, insulin, triglycerides, and cholesterol. It was important to notice that the detected effects in the different RIS showed notable differences from the base strains, indicating neutralizing effects of different acting genes and interaction between genes controlling the phenotype in the base strains. The same set of RIS has also been used to search for loci influencing mRNA amounts.[79]

The increase in genetic variance between RIS correlates to the variance of gene effects influencing the trait in the founder population and results from the tendency of different lines to become homozygous for different genes. The accuracy of the mapping position of a gene effect depends on the number of accumulated recombination events before alleles are being fixed. Therefore, RIS from an advanced intercross population of multiple inbred strains as founders are powerful genetic resources.[80]

6.4. *Chromosome Substitution Strains*

Chromosome substitution strains (CSS) have one chromosome of a recipient strain substituted by the homologous chromosome of a donor strain (Figure 3). A mouse chromosome substitution panel consists of 22 mouse strains, each substituted for one of the 19 autosomes, the X and Y chromosome, and mitochondrial DNA. Typically, such strains are produced by an initial cross between two inbred strains, the donor and the recipient strain, and subsequent repeated backcrossing to the recipient strain while maintaining one whole chromosome of the donor strain. Finally, heterozygous carriers of the donor chromosome, being homozygous at the background chromosomes, are crossed to fix the substituted chromosome in the recipient strain. Marker assisted introgression is used to follow the inheritance of the target chromosome and to ensure the transfer of a non-recombinant donor chromosome to the next generation. The genotyping of the background chromosomes can support the selection against foreign donor DNA on all chromosomes except the substituted one and thus might accelerate the generation of inbred strains.

CSS are particularly useful for the analysis of quantitative traits that are influenced by many genes, which have only small effects.[81] The chromosome-wise partitioning of the genome in combination with the generation of identical individuals of every strain allows the detection of chromosome-wise direct additive and dominance effects of genes located on the substituted chromosome. Furthermore, targeted crosses between selected CSS promise to be a powerful tool for discovering modifier loci contributing to phenotypic variance, and for a better understanding of the effects of interaction as compared to single gene effects.

Recently, the construction of CSS has been reported for the mapping of a susceptibility locus for testicular cancer on mouse chromosome *19*.[82] The first complete mouse chromosome substitution panel was derived from the genome of the mouse strain A/J which was transferred in chromosome-wise portions and as mitochondrial DNA to the genetic background of C57BL/6J mice.[83] This panel was used to characterize chromosomal effects on 53 complex traits like serum levels of sterols and amino acids, diet induced obesity and anxiety. The study revealed evidence for 150 QTL; one or more CSS showed significant effects for nearly all traits analyzed. The comparison of results with other published mapping studies[84] showed that usually significantly more QTL are detected with CSS than with crosses of roughly comparable size.

As a CSS is inbred on the genetic background of the recipient inbred strain, a CSS is an excellent base for the subsequent construction of congenic strains to further fine map and isolate specific QTL. Finally, the CSS are valuable resources for combined studies of genetic mapping and functional analyses of quantitative trait loci.

6.5. *Conplastic Strains*

Conplastic strains are identical in the nuclear DNA, but differ in their cytoplasm. Such strains are developed by crossing male mice of one strain with females of another strain. To recover the cytoplasm of the maternal founder strain, female offspring are crossed back repeatedly to males of the paternal founder strain. Conplastic strains are resources to analyze the interaction between nuclear and mitochondrial genome.[83]

7. Future Prospects for Inbreeding and Crossbreeding

Without question, the elucidation of single gene functions by mutant and genetically modified mice on the background of inbred strains and crosses between well characterized mouse strains supports the analysis of the function and regulation of genes. However, there are two major limitations to these standard genetic techniques. First, natural variation is fundamentally different from mutagenesis in respect to phenotypic changes, and second, the identification of genomic regions that seem to play a role in complex traits requires fine mapping on a heterogeneous background. One of the factors confounding the search for genes contributing to complex traits like diabetes or obesity is that individual mice that carry a susceptibility gene may also carry other protective genes masking the effect of the disease-prone allele. This situation means that the effect of a tested gene, especially if it contributes to a complex trait, is dependent on the zygosity stage of the gene itself, the genetic background and environmental influence, and, thus, might be visible as an expected phenotype, or as a sub-phenotype, or might get lost. Therefore, the identification of quantitative trait loci requires novel models that enable fine mapping on heterogeneous, but reproducible, genetic background. Thus, polygenic animal models that are based on natural genetic variation and defined heterogeneity are unique resources to study the complexity of most traits.

In this context, recombinant inbred strains descending from highly heterogeneous advanced intercross populations[85] and chromosome substitution strains[83] are valuable genetic resources for both fine mapping of QTL and the search for gene-gene and gene-environment interaction effects. Novel model populations designed to mimic natural variation will help to find modifier loci contributing to many complex traits including growth regulation, cancer, and behavior under different environmental conditions. They will give insight into the effects of interacting genes as compared to single gene effects. The novel substrains will likely provide sub-phenotypes, which beside main effects on one trait, might display pleiotropic effects on other traits like fecundity or disease resistance.

References

1. von Nathusius, H. 1872. *Vorträge über Viehzucht und Rassenkenntnis. Allgemeines.* p. 161. Verlag von Wiegandt und Hempel, Berlin.
2. Jones, D.F. 1925. Inbreeding. In: *Genetics in Plant and Animal Improvement.* ed J.G. Lipman. p. 292. John Wiley and Sons Inc., New York.
3. Weismann, A. 1893. The Germ-plasm. Cited by Jones, D.F. 1925. In: *Genetics in Plant and Animal Improvement.* ed. J.G. Lipman. p. 295. John Wiley and Sons Inc., New York.
4. Silver, L.M. 1995. *Mouse Genetics: Concepts and Applications.* pp. 9–17. Oxford University Press, Oxford.
5. Haldane, J.B.S. 1915. Cited by Malakoff, D. 2000. The rise of the mouse, biomedicine's model mammal. *Science* **288**:248–253.
6. Botstein, D., R.L. White, M. Skolnick and R.W. Davis. 1980. Construction of a genetic linkage map in man using restriction fragment length polymorphisms. *Am. J. Hum. Genet.* **32**:314–331.
7. Rhodes, M., R. Straw, S. Fernando, A. Evans, T. Lacey, A. Dearlove, J. Greystrong, J. Walker, P. Watson, P. Weston, M. Kelly, D. Taylor, K. Gibson, C. Mundy, F. Bourgade, C. Poirier, D. Simon, A.L. Brunialti, X. Montagutelli, J.L. Gu'enet, A. Haynes, S.D. Brown. 1998. A high-resolution microsatellite map of the mouse genome. *Genome Res.* **8**:531–542.
8. Ensembl Web Site, World Wide Web (URL: *http://www.ensembl.org/Mus_musculus*).
9. Crabbe, J.C., D. Wahlstein and B.C. Dudek. 1999. *Science* **284**:1670–1672.
10. Festing, M.F.W. 2003. Principles: The need for better experimental design. *Trends Pharmacol. Sciences* **24**:341–345.
11. Gill, J.T. 1980. The use of randomly bred and genetically defined animals in biomedical research. *Am. J. Pathol.* **101**:S21–32.
12. Green, E.L. 1981. *Genetics and Probability in Animal Breeding Experiments.* Oxford University Press, Oxford.
13. Falconer, D.S. and T.F.C. Mackay. 1996. *Introduction to Quantitative Genetics.* p. 90. Pearson Education Limited, Essex.
14. Robertson, A. 1960. A theory of limits in artificial selection. *Proc. Royal Soc. London, Series B, Biol. Sciences* **153**:234–249.
15. Eisen, E.J. 1975. Population size and selection intensity effects on long-term selection response in mice. *Genetics* **79**:305–323.
16. Bünger, L., A. Laidlaw, G. Bulfield, E.J. Eisen, J.F. Medrano, G.E. Bradford, F. Pirchner, U. Renne, W. Schlote and W.G. Hill. 2001. Inbred lines of mice derived from long-term growth selected lines: unique resources for mapping growth genes. *Mamm. Genome* **12**:678–686.
17. Grahame, N.J. 2000. Selected lines and inbred strains. Tools in the hunt for the genes involved in alcoholism. *Alcohol Res. Health* **24**:159–163.

18. Klein, J. 1986. *Natural History of the Major Histocompatibility Complex*. John Wiley and Sons, New York.
19. Jax Mice Database – Wild-derived Inbred Web Site, The Jackson Laboratory, Bar Harbor, Maine. World Wide Web (URL: *http://jaxmice.jax.org/jaxmicedb/html/wild.shtml*). May 2004.
20. MURINAE Web Site, World Wide Web (URL: *http://www.il-st-acad-sci.org/mammals/murids002.html*). May 2004.
21. Ferris S.D., R.D. Sage and A.C. Wilson, 1982. Evidence from mtDNA sequences that common laboratory strains of inbred mice are descended from a single female. *Nature* **295**:163–165.
22. Ferris S.D., R.D. Sage, C.M. Huang, J.T. Nielsen, U. Ritte and A.C. Wilson. 1983. Flow of mitochondrial DNA across a species boundary. *Proc. Natl. Acad. Sci. USA* **80**:2290–2294.
23. Ferris S.D., R.D. Sage, E.M. Prager, U. Ritte and A.C. Wilson. 1983. Mitochondrial DNA evolution in mice. *Genetics* **105**:681–721.
24. The Jackson Laboratory Web Site. The Jackson Laboratory, Bar Harbor, Maine. World Wide Web (URL: *http://www.jax.org*).
25. International Committee on Standardized Genetic Nomenclature for Mice Web Site. The Jackson Laboratory, Bar Harbor, Maine. World Wide Web (URL: *http://www.informatics.jax.org/mgihome/nomen/strains.shtml*). May 2004.
26. Beck, J.A., S. Lloyd, M. Hafezparast, M. Lennon-Pierce, J.T. Eppig, M.F.W. Festing and E.M.C. Fisher. 2000. Genealogies of mouse inbred strains. *Nature Genet.* **24**:23–25.
27. Festing, M.F.W. 1996. Origins and characteristics of inbred strains of mice. In: *Genetic Variants and Strains of the Laboratory Mouse*. eds. M.F. Lyon, S. Rastan and S.D.M. Brown. pp. 1537–1576. Oxford University Press, Oxford.
28. Mouse Genome Informatics Database Web Site, The Jackson Laboratory, Bar Harbor, Maine. World Wide Web (URL: *http://www.informatics.jax.org/external/festing/mouse/*). May 2004.)
29. Atchley, W.R. and W. Fitch. 1991. Gene trees and origins of inbred strains of mice. *Science* **254**:554–558.
30. Nishioka, Y. 1987. Y-chromosomal DNA polymorphism in mouse inbred strains. *Genet. Res.* **50**:69–72.
31. Tucker, P.K., B.K. Lee, B.L. Lundrigan and E.M. Eicher. 1992. Geographic origin of the Y chromosomes in "old" inbred strains of mice. *Mamm. Genome* **3**:254–261.
32. Rikke, B.A., Y. Zhao, L.P. Daggett, R. Reyes and S.C. Hardies. 1995. Mus spretus LINE-1 sequences detected in the Mus musculus inbred strain C57BL/6J using LINE-1 DNA probes. *Genetics* **139**:901–906.
33. Mouse Genome Informatics Web Site, The Jackson Laboratory, Bar Harbor, Maine. World Wide Web (URL: *http://www.informatics.jax.org/*).
34. Montagutelli, X. 2000. Effect of the genetic background on the phenotype of mouse mutations. *J. Am. Nephrol.* **11**:S101–S105.
35. Armstrong, J.F. and M.L. Hooper. 1998. Inbreeding abolishes the effect of parental origin of a mutant Rb-1 allele on pituitary tumorgenesis in mice. *British J. Cancer* **78**:484–485.
36. Crawley, J.N., J.K. Belknap, A. Collins, J.C. Crabbe, W. Frankel, N. Henderson, R.J. Hitzemann, S.C. Maxson, L.L. Miner, A.J. Silva, J.M. Wehner, A.Wynshaw-Boris and R. Paylor. 1997. Behavioral phenotypes of inbred mouse strains: implications and recommendations for molecular studies. *Psychopharmacology* **132**:107–124.

37. Bothe, G.W.M., V.J. Bolivar, M.J. Vedder and J.G. Geistfeld. 2004. Genetic and behavioral differences among five inbred mouse strains commonly used in the production of transgenic and knockout mice. *Genes, Brain & Behavior* **3**:149–157.
38. Mouse Phenome Database Web Site, The Jackson Laboratory, Bar Harbor, Maine. World Wide Web (URL: *http://www.jax.org/phenome*).
39. Flaherty, L., M.N. Cook and R.W. Williams. 2001. Flaherty1 - Anxiety-related behaviors in the elevated zero-maze, MPD:118. Mouse Phenome Database Web Site, The Jackson Laboratory, Bar Harbor, Maine. World Wide Web (URL: *http://www.jax.org/phenome*). May 2004.
40. Justice, M. 2002. Justice2 - Clinical hematology parameters. MPD:132. Mouse Phenome Database Web Site, The Jackson Laboratory, Bar Harbor, Maine. World Wide Web (URL: *http://www.jax.org/phenome*). May 2004.
41. Koide, T., K. Ikeda, K. Moriwaki, H. Niki and T. Shiroishi. 2000. Koide1 - Behavioral studies on wild-derived inbred strains. MPD:18. Mouse Phenome Database Web Site, The Jackson Laboratory, Bar Harbor, Maine. World Wide Web (URL: *http://www.jax.org/phenome*). May 2004.
42. Mogil, J.S., M. Devor and W.R. Lariviere. 1999–2002. Mogil1 - Heritability of nociception. MPD:22. Mouse Phenome Database Web Site, The Jackson Laboratory, Bar Harbor, Maine. World Wide Web (URL: *http://www.jax.org/phenome*). May 2004.
43. Naggert, J.K., B. Paigen and L.L. Svenson, 2003. Naggert1 - Diet effects on bone mineral density and content, body composition, and plasma glucose, leptin, and insulin levels. MPD:143. Mouse Phenome Database Web Site, The Jackson Laboratory, Bar Harbor, Maine. World Wide Web (URL: *http://www.jax.org/phenome*). May 2004.
44. McLaren, A. and A. Tait. 1969. Cytoplasmic isocitrate dehydrogenase variation within the C3H inbred strain. *Genet. Res.* **14**:93.
45. MGI2.98 Inbred Strains of Mice: C57BL Web Site, Mouse Genome Informatics. The Jackson Laboratory, Bar Harbor, Maine. World Wide Web (URL: *http://www.informatics.jax.org/external/festing/mouse/docs/C57BL.shtml*). May 2004.
46. Simpson, E.M., C.C. Linder, E.E. Sargent, M.T. Davisson, L.E. Mobraaten and J.J. Sharp. 1997. Genetic variation among 129 substrains and its importance for targeted mutagenesis in mice. *Nat. Genet.* **16**:19–27.
47. Threadgill, D.W., D. Yee, A. Matin, J.H. Nadeau and T. Magnuson. 1997. Genealogy of the 129 inbred strains: 129/SvJ is a contaminated inbred strain. *Mamm. Genome* **8**:390–393.
48. Festing, M.F., E.M. Simpson, M.T. Davisson and L.E. Mobraaten. 1999. Revised nomenclature for strain 129 mice. *Mamm. Genome* **10**:836.
49. MGI2.98 Inbred Strains of Mice: 129 Web Site, Mouse Genome Informatics. The Jackson Laboratory, Bar Harbor, Maine. World Wide Web (URL: *http://www.informatics.jax.org/external/festing/mouse/docs/C57BL.shtml*). May 2004.
50. Zambrowcz, B.P., G.A. Friedrich, E.C. Buxton, S.L. Lilleberg, C. Person and A.T. Sands. 1998. Disruption and sequence information of 2000 genes in mouse embryonic stem cells. *Nature* **392**:608–611.
51. Hrabé de Angelis, M., H. Flaskwinkel, H. Fuchs, B. Rathkolb, D. Soewarto, S. Marshall, S. Heffner, W. Pargent, K. Wuensch, M. Jung et al. 2000. Genome-wide, large-scale production of mutant mice by NEU mutagenesis. *Nat. Genet.* **25**:444–447.

52. Nolan, P.M., J. Peters, M. Strivens, D. Rogers, J. Hagan, N. Spurr, I.C. Gray, L. Vizor, D. Brooker, E. Whitehill et al. 2000. A systematic, genome-wide, phenotype-driven mutagenesis programme for gene function studies in the mouse. *Nat. Genet.* **25**:440–443.
53. Justice, M.J. 2000. *Mutagenesis of the mouse germline. In Mouse Genetics and Transgenics.* eds. I.J. Jackson and C.M. Abbott. pp. 185–216. Oxford University Press, Oxford.
54. Hammes, A. and A. Schedl. 2000. *Generation of transgenic mice from plasmids, BACs and YACs. In Mouse Genetics and Transgenics.* eds. I.J. Jackson and C.M. Abbott. p. 221. Oxford University Press, Oxford.
55. Kawase, E., H. Suemori, N. Takahashi, K. Okazaki, K. Hashimoto and N. Nakatsuji. 1994. Strain difference in establishment of mouse embryonic stem (ES) cell lines. *Intern. J. Devel. Biol.* **38**:385–390.
56. Plagge, A., G. Kelsey and N.D. Allen. 2000. Directed mutagenesis in embryonic stem cells. In *Mouse Genetics and Transgenics.* eds. I.J. Jackson and C.M. Abbott. p. 271. Oxford University Press, Oxford.
57. The International Mouse Mutagenesis Consortium: Nadeau J.H., R. Balling, G. Barsh, D. Beier, S.D.M. Brown, M. Bucan, S. Camper , G. Carlson, N. Copeland, J. Eppig, C. Fletcher, W.N. Frankel, D. Ganten, D. Goldowitz, C. Goodnow, J.-L. Guenet, G. Hicks, M. Hrabe de Angelis, I. Jackson, H.J. Jacob, N. Jenkins, D. Johnson, M. Justice, S. Kay, D. Kingsley, H. Lehrach, T. Magnuson, M. Meisler, A.M. Poustka, E.M. Rinchik, J. Rossant, L.B. Russell, J. Schimenti, T. Shiroishi, W.C. Skarnes, P. Soriano, W. Stanford, J.S. Takahashi, W. Wurst and A. Zimmer. 2001. Functional annotation of mouse genome sequences. *Sc*ience **291**:1251–1255.
58. Shedlovsky, A., J.D. McDonald, D. Symula and W.F. Dove. 1993. Mouse models of human phenylketonuria. *Genetics* **134**:1205–1210.
59. Davis, A.P., R.P. Woychik and M.J. Justice. 1999. Effective chemical mutagenesis in FVB/N mice requires low doses of ethylnitrosourea. *Mamm. Genome* **10**:308–310.
60. You, Y.R. Bersgtram, M. Klemm, H. Nelson, R. Jaenisch and J. Schimenti. 1998. Utility of C57BL/6J x 129/SvJae embryonic stem cells for generating chromosomal deletions: tolerance to gamma radiation and microsatellite polymorphism. *Mamm. Genome* **9**:232–244.
61. Hentge, K.E. and M.J. Justice. 2004. Checks and balancers: balancer chromosomes to facilitate genome annotation. *Trends Genet.* **20**:252–259.
62. Geldermann, H. 1975. Investigations on inheritance of quantitative characters in animals by gene markers. I. Methods. *Theor. Appl. Genet.* **46**:319–330.
63. Soller, M., T. Brody and A. Genizi. 1976. On the power of experimental design for the detection of linkage between marker loci and quantitative loci in crosses between inbred lines. *Theor. Appl. Genet.* **47**:35–39.
64. Darvasi, A. 1998. Experimental strategies for the genetic dissection of complex traits in animal models. *Nat. Genet.* **18**:19–24.
65. Lander, E.S. and L. Kruglyak. 1995. Genetic dissection of complex traits: guidelines for interpreting and reporting linkage results. *Nature Genet.* **11**:241–247.
66. Darvasi, A. and M. Soller. 1995. Advanced intercross lines, an experimental population for fine genetic mapping. *Genetics* **141**:1199–1207.
67. Iraqi, F., S.J. Clapcott, P. Kumari, C.S. Haley, S.J. Kemp and A.J. Teale. 2000. Fine mapping of trypanosomiasis resistance loci in murine advanced intercross lines. *Mamm. Genome* **11**:645–648.

68. Leamy, L.J., K. Elo, M.K. Nielsen, L.D. Van Vleck and D. Pomp. 2004. Genetic variance and covariance patterns for body weight and energy balance characters in an advanced intercross population of mice (submitted).
69. Wright, A.F., A.D. Carothers and M. Pirastu. 1999. Population choice in mapping genes for complex diseases. *Nature Genet.* **23**:397–404.
70. Talbot, C.J., A. Nicod, S.S. Cherny, D.W. Fulkner, A.C. Collins and J. Flint. 1999. High-resolution mapping of quantitative trait loci in outbred mice. *Nat. Genet.* **21**:305–308.
71. Flaherty, L. 1981. Congenic strains. In *The Mouse in Biomedical Research*, Vol. 1. eds. H.L. Foster, J.D. Small and J.G. Fox. pp. 215–222. Academic Press, New York.
72. Snell, G.D. 1978. Congenic resistant strains of mice. In *Origins of Inbred Mice*. ed. H.C. Morse. pp. 1–31. Academic Press, New York.
73. Markel, P., P. Shu, C. Ebeling, G.A. Carlson, D.L. Nagle, J.S. Smutko and K. Moore. 1997. Theoretical and empirical issues for marker assisted breeding of congenic mouse strains. *Nat. Genet.* **17**:280–284.
74. Liu, X., F. Oliver, S.D. Brown, P. Denny and P.D. Keightley. 2001. High-resolution quantitative trait locus mapping for body weight in mice by recombinant progeny testing. *Genet. Res.* **77**:191–197.
75. Iakoubova, O.A., C.L. Olsson, K.M. Dains, D.A. Ross, A. Andalibi et al. 2001. Genome–tagged mice (GTM): two sets of genome-wide congenic strains. *Genomics* **15**:89–104.
76. Wagner, A. 2000. Robustness against mutations in genetic networks of yeast. *Nature Genet.* **24**:355–361.
77. Demant, P. and A.A.M. Hart. 1986. Recombinant congenic strains - a new tool for analyzing genetic traits determined by more than one gene. *Immunogenetics* **24**:416–422.
78. Anunciado, R.V., T. Ohno, M. Mori, A. Ishikawa, S. Tanaka, F. Horio, M. Nishimura and T. Namikawa. 2000. Distribution of body weight, blood insulin and lipid levels in the SMXA recombinant inbred strains and the QTL analysis. *Exp. Anim.* **49**:217–224.
79. Koza, R.A., S.M. Hohmann, C. Guerra, M. Rossmeisl and L.P. Kozak. 2000. Synergistic gene interactions control the induction of the mitochondrial uncoupling protein (Ucp1) gene in white fat tissue. *J. Biol. Chem.* **275**:34486–34492.
80. Threadgill, D.W., K.W. Hunter and R.W. Williams. 2002. Genetic dissection of complex and quantitative traits: from fantasy to reality via a community effort. *Mamm. Genome* **13**:175–178.
81. Nadeau, J.H., J.B. Singer, A. Matin and E.S. Lander. 2000. Analysing complex genetic traits with chromosome substitution strains. *Nat. Genet.* **24**:221–225.
82. Matin, A., G.B. Collin, Y. Assada, D. Varnum and J.H. Nadeau. 1999. Susceptibility to testicular germ cell tumors in 129.MOLF-Chr19 mice. *Nature Genet.* **23**:237–240.
83. Singer, J.B., A.E. Hill, L.C. Burrage, K.R. Olszens, J. Song, M. Justice, W.E. O'Brien. D.V. Conti, J.S. Witte, E.S. Lander and J.H. Nadeau. 2004. Genetic Dissection of Complex Traits with Chromosome Substitution Strains of Mice. *Science* **304**:445–448.
84. Brockmann, G.A. and M. Bevova. 2002. Using mouse models to dissect the genetics of obesity. *Trends Genet.* **18**:367–376.
85. Peirce, J.L., L. Lu, L.M. Silver and R.W. Williams. 2004. A new set of BXD recombinant inbred lines from advanced intercross populations in mice. *BMC Genet.* **5**:7.

CHAPTER 5

GENOTYPE BY ENVIRONMENT INTERACTION: LESSONS FROM THE MOUSE

William D. Hohenboken

Animal and Poultry Sciences Department
Virginia Polytechnic Institute and State University, Blacksburg, VA, USA
whohenbo@vt.edu

1. Introduction

Despite ongoing controversy on the importance of nature versus nurture, most people agree that biological characteristics generally are affected by inheritance *and* the environment. Quantitative geneticists express this belief in the formula, P = G + E. That is, an individual's phenotypic merit for a variable trait (P) is dependent upon genetic potential for that trait (G) and the net effect of relevant environmental circumstances (E). Such geneticists expend considerable effort determining the relative importance of G versus E for specific traits in specific populations. This has enabled them to predict the outcome of selective breeding in domestic plants and animals and partially to understand speciation, adaptation and evolution in natural populations.

P = G + E is not, however, the whole story. Perceptive observers realize that genetic influences are not expressed in an environmental vacuum. For example, an individual's genetic potential may be more fully expressed in some environments than in others. Equivalently, a change in some aspect of the environment may affect individuals of different genotypes to different degrees. Quantitative geneticists define these possibilities as genotype by environment interaction (GEI) and account for them by amending the previous formula to P = G + E + GE. Phenotype is determined not only by genetics and the environment but, in some cases, by their interaction as well. Furthermore, $V_P = V_G + V_E + V_{GE}$. Statistical analysis of suitably designed experiments can partition total phenotypic variance (V_P) into variance attributable directly to genetics (V_G), directly to environment (V_E) and to their joint or interaction effect (V_{GE}).

This chapter begins with definitions of genotype, environment and genotype by environment interaction, followed by a description of procedures for the investigation of GEI. Interactions will then be illustrated between various categories of genotype and various kinds of environments. It next will be shown that effects of mating systems can be modulated by the environment, another type of GEI. Finally, it is shown that selection response can be influenced by differences among the environments in which the selection occurs.

My objectives are two-fold, to illustrate the contributions of laboratory and wild mice to our understanding of the phenomenon of GEI and, no less important, to illustrate what investigations of GEI have taught us about the biology of the mouse.

2. Genotypes, Environments and Genotype by Environment Interaction Defined

In the context of GEI, there are many categories of *genotypes*. Important to biomedical research are inbred lines, of which some 400 have been developed.[1] Each is a sample of the *Mus musculus* genome, its uniqueness from all other lines dependent upon the evolutionary history of the population from which its founders were sampled and by genetic drift and natural and artificial selection during its formation. Other categories of *genotypes* are congenic inbred lines, outbred laboratory strains, populations artificially or naturally selected for one or more quantitative traits, non-selected control populations bred concurrently with such lines, transgenic strains, single-locus genotypes and naturally occurring or knockout mutants versus their wild type counterparts. *Genotypes* could also represent differences within an outbred population, as illustrated by families within lines or individuals within a family that differ in genetic value for a trait. As a special case, genetic interventions such as selection or the application of a mating system can be thought of as *genotypes* because the impact of such interventions may depend upon the environment in which they occur.

Environments in the context of GEI also can represent a range of effects. External environments include variable temperatures, habitats, diets, feeding programs, housing types, management conditions, social contexts and experimental manipulations. Internal environments include an animal's sex, age and physiological state (pregnant versus non-pregnant, for example). Contemporary animals treated alike in the same habitat or laboratory also experience micro-environmental differences, but these are not amenable to investigation in a GEI format.

Genotype by environment interaction exists when phenotypic differences between groups of individuals of two or more genotypes depend upon the environment in which those genotypes are compared. Equivalently, GEI exists when environments, management systems or experimental interventions have a greater impact on phenotypic expression of some genotypes than of others. For the special case in which genetic interventions serve as genotypes, GEI occurs when the impact of an intervention such as selection or a mating system is dependent upon the environment in which that intervention occurs.

3. Experimental and Analytical Methods to Study GEI

3.1. *Analysis of Variance*

The most straightforward manner to investigate GEI is to record phenotypes for the same trait on individuals of two or more distinct genotypes in each of two or more distinct environments. The most straightforward analysis of variance of data from such an experiment is to test the statistical significance of differences among genotypes, among environments and among genotype by environment subclasses and to quantify the contribution of main and interaction effects to total variance for the trait. Hahn and Schanz[2] provided an example for rate of ultrasonic calling by mouse pups in response to stress-inducing procedures. Genotypes were two F_1 inbred-line crosses created by mating C57BL/10J females to DBA/2J and to SJL/2J males. (Thus, C57BL/10J females provided the pre- and postnatal maternal environment to all experimental subjects.) Environments were to stress one half of each litter by short term exposure to cold and to stress the remaining half by short term rotation in a tilted container. Measurements were made daily on mice between two and eight days of age.

The analysis of variance for number of alarm calls recorded during each testing procedure on each day included sources of variation for genotype, environment and their interaction. Genotypes differed significantly for frequency of alarm calls across environments; mice sired by DBA/2J males called more frequently than mice sired by SJL/2J males. Averaged across days, this genotypic difference accounted for 15% of variation in calling frequency. Environments differed significantly across genotypes, with rotation eliciting a higher calling frequency than exposure to cold. Averaged across days, testing stimulus accounted for 53% of total variance. Even though GEI was statistically significant on four of seven testing days, it accounted for only 5% of the variation in calling frequency averaged across those four days.

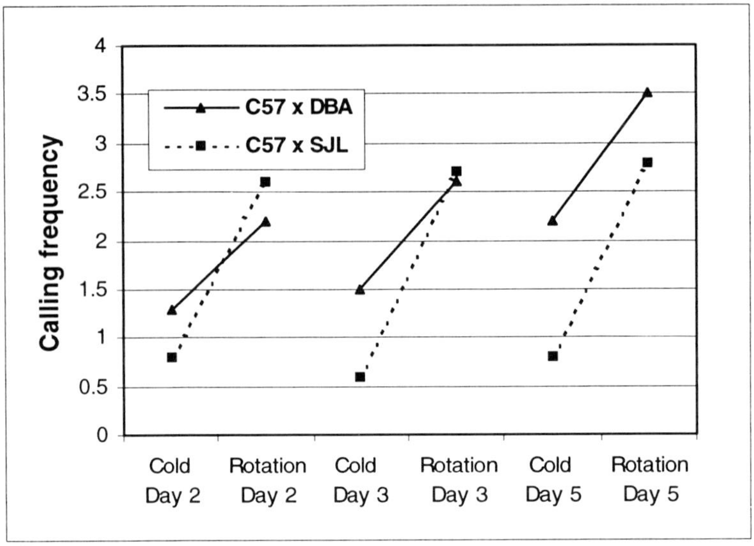

Figure 1. The effect of cold exposure and of rotation in a tilted container on frequency of ultrasonic calls in mouse pups of two linecross groups at two, three and five days of age. Genotypes and environments interact because, on each of three days, the difference between genotype in calling frequency depends upon the aversive stimulus. Prepared from tabular data of Hahn and Schanz.[2]

Three manifestations of GEI are illustrated in Figure 1. At two days of age, C57BL/10J × DBA/2J mice called more frequently than C57BL/10J × SJL/J mice in response to cold; but in response to rotation, C57BL/10J × SJL/J mice called more frequently than C57BL/10J × DBA/2J contemporaries. Thus, lines switched rank for calling frequency when stressed by different procedures. At three days of age, lines differed in calling frequency when stimulated by cold exposure but not when stimulated by rotation. Thus, only one of the two environments elicited a differential line response. At five and six days of age (results at five days of age are shown), C57BL/10J × DBA/2J mice called more frequently than C57BL/10J × SJL/J mice in both environments, but the difference between lines was larger during cold exposure than during rotational stress.

3.2. *Genetic Correlation*

The additive genetic correlation between two polygenic traits quantifies the extent to which both of them are influenced by the same genes (pleiotropy) and/or by distinct but closely linked genes. When a genetic correlation is close to plus or minus one, both traits are influenced by mostly the same genes and thus are essentially equivalent. A zero genetic correlation between traits means that no genes affecting one trait simultaneously affect the other. A genetic correlation between zero and plus or minus one indicates that genes at some loci affect both traits but that each trait also is influenced by genes that do not affect the other.

Conceptually, one trait expressed in two environments can be thought of as two separate traits.[3] Each will have its own genetic architecture (influential loci, heritability, dominance variance, inbreeding depression and heterosis), which might be quite different from one *trait* to another. The degree of difference can be examined by computing the genetic correlation of the same trait as expressed in two environments, and the magnitude of this genetic correlation is a direct reflection of GEI. When it approaches unity, genotypes within the population do not interact with the two environments. When it deviates markedly from unity, they do interact. If the genetic correlation equaled zero, then entirely different sets of genes would control variation in the two environments.

Within the F_4 generation of a population created by crossing four inbred lines, Lynch *et al.*[4] reported a genetic correlation of 0.55 ± 0.09 between nest building score (amount of nesting material used in four days) of male mice maintained at 4 versus 21°C temperatures. Heritability of the trait was 0.35 ± 0.08 in the cold environment and 0.52 ± 0.06 in the warm environment. Thus, the constellation of genes controlling this complex thermoregulatory trait apparently differed between temperatures. Results differed in contemporary females. Again, heritability of nest building score was higher at 21 than at 4°C (0.55 ± 0.08 versus 0.29 ± 0.08 respectively), but the genetic correlation between the two traits did not differ from 1.00. GEI was restricted to males.

3.3. *Reaction Norm*

As discussed by Falconer[5] and Sarkar and Fuller,[6] a procedure known as reaction norms has also been used to visualize the nature of genotype by environment interactions and to assess their statistical significance. Individuals of two or more genotypes are scored for the same trait at three or more levels for an environmental variable affecting the trait in question. In a two dimensional diagram (Figure 2), phenotypic values are represented on the vertical axis, ordered environments are represented on the horizontal axis and points are plotted within the graph to represent each genotype's average phenotypic value at each level of environment. Linear or higher order polynomial regression lines fit for each genotype can be tested for variation in intercept and slope. Any two genotypes whose regression lines are not parallel interact within the range of environmental levels that are represented.

Davis and Lamberson[7] used reaction norm concepts to examine stability of the expression of reproductive traits of outbred lines and linecrosses in benign versus stressful environments. ICR and CF1 females were mated to ICR, CF1 and ICR × CF1 F_1 males to create contemporary pure line, reciprocal F_1 and backcross litters. At 24 days of age, females of each genotype were transferred to environmental chambers maintained at 22, 15 or 8°C and paired with a male of the unrelated ND4 strain. Age and weight at puberty and ovulation rate, implantation rate and number of fetuses 10-days post-mating subsequently were recorded.

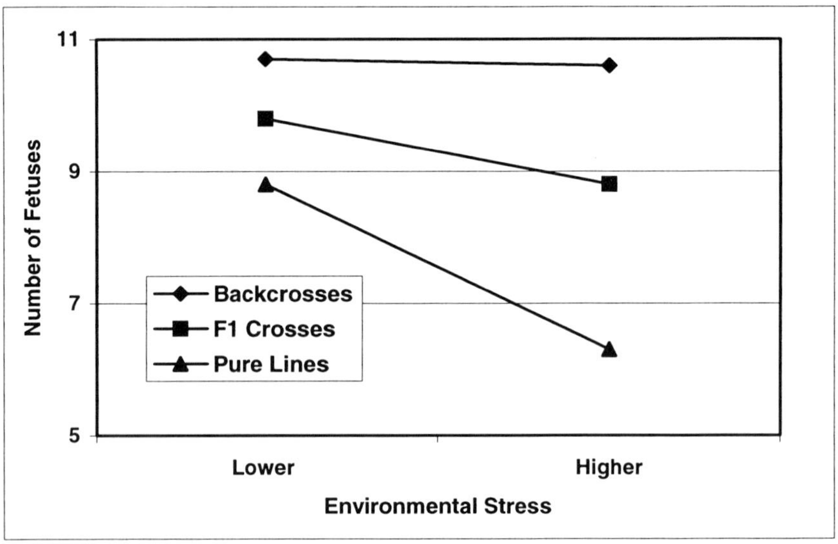

Figure 2. The effect of increasing environmental stress on the number of fetuses present per female 10 days after mating in backcross, F_1-cross and pure line mouse genotypes. Prepared from tabular data in Davis and Lamberson.[7]

Recall that reaction norms may be used to quantify response of a genotype to an environment. To accomplish this, the average impact of each level of the environment on expression of the trait must be quantified. The individual response of each genotype to increasing environmental effects must then be assessed, to determine whether it deviates from other genotypes or from the norm.

Davis and Lamberson quantified environmental impact by computing average merit of all genotypes for each trait at each temperature. For example, the number of fetuses present 10-days post-mating (averaged across genotypes) equaled 10.1 at 22°C, 10.0 at 15°C and 8.6 at 8°C. They assessed responses of each genetic group to temperature stress by regressing trait performance of that group on average environmental impact, computed as just described.

There was a common pattern for each of the traits measuring fecundity. Results for number of fetuses are shown in Figure 2. Backcross genotypes were most stable. Their performance changed little from less through more stressful temperature environments. F_1 linecrosses were intermediate in stability. For these females, number of fetuses decreased moderately under increasing thermal stress. Pure line genotypes were the most sensitive; the reduction in their fecundity under increasing environmental challenge was largest of all.

4. Interactions of Genotype with the External Environment

Animals face diverse opportunities and challenges from the environment. External environments (those over which animals exert little control) may interact with genotypes in many ways. Several are illustrated below.

4.1. *Laboratory Environment and Handling*

Inbred lines, outbred strains, selection lines, transgenic lines and mouse stocks harboring induced mutations play an important part in biomedical research. For research results to be robust and useful, differences among strains for medically important traits and responses should not be dependent upon the laboratory in which an experiment was performed. This may not always be the case. Laboratory management and environment may differ in profound or subtle ways, and standard procedures may be conducted idiosyncratically by different investigators.

Crabbe *et al.*[8] sought to determine whether behavioral differences among inbred and mutant mouse strains were dependent on the laboratory in which the behaviors were recorded. How robust, in other words, were among-strain differences? Investigators evaluated males and females from eight lines (six inbreds, one knockout mutant and its wild type control) at three laboratories for seven behavioral traits. Genetic lines differed significantly for all traits, and the expression of five traits differed significantly among laboratories as well. GEI was statistically significant for four behaviors. Thus, strain differences sometimes differed among laboratories. The authors concluded that large differences among strains were not likely to be affected by site-specific interactions. They warned, however, that "for behaviors with smaller genetic effects (such as those likely to characterize most effects of a gene knockout) there can be important influences of environmental conditions specific to individual laboratories, and specific behavioral effects should not be uncritically attributed to genetic manipulations such as targeted gene deletions."

Reliability of biomedical research also is dependent upon the well-being of experimental subjects, which is partially dependent upon their physical and social environment. Accordingly, research has been conducted to define requirements of laboratory mice for housing, handling and care. One theme has been to provide cage environments that allow expression of species-specific behavior. In one such experiment, Nevison *et al.*[9] studied behavioral and physiological responses to an enriched cage environment of male mice from two outbred strains and four inbred lines. Four adult males of the same strain were housed per cage during the four week experiment. The enriched environment was a standard cage to which bedding material and a transparent tunnel had been added. Time budgets were estimated for mice of each genotype by environment group by repeated recording of aggression, social investigation, non-social investigation, sleep, food or water ingestion, grooming, stereotypic behavior and attentive behavior. Physiological variables included pre- and post-experimental concentrations of serum corticosterone, testosterone and immunoglobulin G. Organ weights were recorded at termination of the experiment.

In both cage types, the two outbred strains and one inbred line were similar in total aggression and had considerably more aggressive encounters than mice in the remaining three inbred lines. Time spent investigating the cage was greater in enriched than in standard cages in all strains except one of the inbred lines. In another inbred line, stereotypic behaviors accounted for a higher proportion of activity in enriched than in standard cages while in other strains, differences in stereotypic behaviors between cage types were small. In the three aggressive lines, cage enrichment resulted in substantial increases in the concentration of serum testosterone over the course of the experiment, but there was virtually no change for this trait in the three less aggressive lines. In one line, corticosterone concentration was lower at the end than at the beginning of the trial; in two lines it did not change; and in the remaining three lines, it increased substantially. There were small but significant interactions between strain and cage environment for weight of heart and spleen. In agreement with Sluyter and Van Oortmerssen,[10] Nevison *et al.* concluded that impacts of enriched environments on laboratory animals are likely to be strain specific and sometimes counter intuitive. Modifications that enhance well-being of (and research reliability from) some genetic stocks may reduce it in others.

Lieblich and Guttman[11] examined whether differences in behavioral responses of mouse strains varied with experimental situations. Genotypes were the C57BL/6J and DBA/1J inbred lines, their reciprocal F_1 crosses and backcrosses to both parental strains. Environments were mild versus severe imposed stress. To create the former, each mouse was placed for five minutes into an unfamiliar cage with internal barriers. To create severe stress, each mouse was placed one day later in an aluminum pail and subjected to the sound of a bell. The behavioral phenotype was emotional defecation, the number of fecal pellets deposited during the imposition of each stress.

GEI did occur. Under mild stress, DBA mice produced more fecal pellets than C57BL mice, whereas under severe stress, C57BL mice had higher defecation scores than DBA individuals. Under mild stress, gene action was additive in that the average score for F_1 hybrids was intermediate to those of the parental lines. Under severe stress, F_1 hybrids closely resembled the more reactive parent (C57BL), so the trait expressed a dominant mode of inheritance. The estimated degree of genetic determination for defecation score was 25 ± 7 % under mild stress and 6 ± 14% under severe stress. From these results, the authors concluded that emotional defecation is not a single genetic trait but should be considered a distinct emotional response to defined conditions of environmental stress.

4.2. *Thermal Environment*

Temperature and humidity are important aspects of any animal's environment, and much research has been devoted to physiology and inheritance of thermoregulatory mechanisms. Genotype by thermal environment interactions have been investigated in the mouse.

Barnett and Dickson[12] maintained closed lines of mice for 10 generations in cold (3°C) versus warm (23°C) laboratory environments. No artificial selection was exerted, but populations had the opportunity to adapt to cold versus warm environments by natural selection. Mice born in generation 5 were grandparents of experimental subjects in the cross-fostering and environmental transfer experiment discussed here. In generation 6, half of 40 litters born in each environment were transferred at birth to the opposite environment, while the remaining half of each group remained in the environment of their ancestors.

Thus, biological parents of generation 7 experimental subjects were classified into four categories: cold-adapted versus warm-adapted genotypes maintained in cold versus warm ambient environments. In generation 7, all litters were standardized to five pups. Half of all litters were then cross-fostered to another dam of their own genotype, the remaining half to a dam of the other genotype. (Litters were not transferred out of the environment into which they were born.) Each generation 7 mouse thus had a genotype (its ancestors were selected for adaptation to a cold or to a warm environment), a postnatal maternal environment (its foster dam's ancestors had been selected for adaptation to a cold or to a warm environment) and an ambient environment (cold or warm). Notice that in this experiment, what is an environmental effect for the offspring (the maternal environment) is a genetic effect for the dam. Some interactions can thus be thought of as GGEI (individual genotype by maternal genotype by environment interaction).

Earlier work (Barnett and Dickson[13]) had established that cold genotype mice grew more rapidly, were fatter and heavier at maturity and had higher fecundity than warm genotype mice in both temperature environments. Direct genetic effects were observed in the cross-fostering/environmental transfer experiment as well. Cold genotype mice generally were heavier than warm genotype contemporaries at 10 days, 21 days, 16 weeks and 30 weeks of age, but the magnitude of the difference varied according to maternal environment and ambient environment. For example, the superiority for preweaning growth rate of cold genotype mice reared by cold genotype foster dams to mice of all other groups was greater in the cold than in the warm rearing environment. Genotype of the foster dam was an important source of variation for many traits. It was generally beneficial for a mouse of either genotype to be reared by a cold genotype mother, although the magnitude of this benefit was dependent upon the temperature of the rearing environment. In the warm environment, for example, having a cold-adapted foster mother increased 16-week weight of warm genotype and cold genotype males, while in the cold environment, having a cold genotype foster mother did not benefit either genotype. Percentage of body fat at 30 weeks of age did not vary among individual or dam genotypes in the warm environment, but in the cold environment, fat percentage was highest for cold genotype mice with cold genotype foster dams.

4.3. *Habitat*

Observation of nest building activities of C57BL/6J and BALB/c females led Van Oortmerssen[14] to speculate about the evolutionary history of the ancestors of those inbred lines. Virgin adult females were individually placed in an enclosure that allowed them to build sleeping nests on a grass surface or in holes that they dug into the earthen substrate. Nearly all the BALB/c females built surface nests, whereas approximately 50% of C57BL/6J females nested in holes while 35% nested on the surface. In this and other experimental situations, Van Oortmerrsen observed that BALB/c mice readily explored open spaces, constructed free-standing, spherical nests from self-processed materials and would investigate and dig into the substrate but were not skillful tunnel builders. In contrast, C57BL/6J mice were more curious toward their substrate than their surface environment. They were compulsive diggers and skillful tunnel builders. Rather than constructing elaborate surface nests, they lined self-constructed tunnels or holes with non-processed nesting material.

From these and other behavioral differences, Van Oortmerssen speculated as follows: The BALB/c line may have descended from *Mus musculus domesticus*, a non-tunneling, commensal species for which surface exploration and well constructed and insulating nests would have been important. The C57BL/6J line may have descended from *Mus musculus musculus*, a non-commensal species adapted to outdoor habitats in which tunneling skills would have been beneficial. Further, co-adapted gene complexes enabling these and other habitat-specific behaviors were maintained during the inbreeding process.

This experiment does not illustrate GEI *per se*, but it does illustrate how natural selection may have created genotypes adapted to specific environments, how remnants of that adaptation survived the passage of time, the relaxation of natural selection and the inbreeding process and how those genotypes, when given the opportunity to do so, can choose an environment appropriate to the adaptations of their ancestors.

4.4. *Nutrition*

Nutrition is an important component in any animal's environment, and differences in the quality, quantity or balance among nutrients affect many traits. When a nutritional change affects different genotypes differently, GEI exists. Many experiments have investigated such phenomena in mice. In addition to examples reviewed here, the issue is discussed in "The Environment for Selection" later in the chapter.

Mercer and Trayhurn[15] examined the effects of diets differing in fatty acid composition (corn-oil, high in polyunsaturated fatty acids, versus beef tallow, high in saturated and mono-unsaturated fatty acids) on energy balance and brown fat thermogenesis of obese (ob/ob) mice and phenotypically normal (+/+ or ob/+) littermates. Compared to a low-fat control diet, feeding either high fat diet for two weeks caused increased metabolizable energy intake, body energy storage and energy expenditure (presumably from induced brown fat thermogenesis). These traits were increased in both genotypes, but obese mice were more sensitive to the effect. For example, in comparison to the low energy diet, metabolizable energy intake from the high fat diets was increased by 41% and 21% in obese mice and normal littermates, while body energy gain increased 86% for obese and 25% for normal individuals. Compared to the low fat diet, the corn oil diet caused increases in apparent energy expenditure (the difference between the metabolizable energy intake and energy gain in body tissue) of 36% versus 22% in obese and normal mice, respectively. The beef tallow diet had little impact on energy expenditure in lean mice (+4%) and a somewhat larger impact (+22%) in obese littermates. Thus, animals differing in genotype at a locus with major effects on obesity reacted differently to these substantial changes in their diet.

In work of West *et al.*,[16] inbred lines differing in propensity for fat deposition (AKR/J was prone to obesity and SWR/J was not) were provided diets varying in fat and protein content from five to 17 weeks of age. In the experimental design, three levels of fat (45, 30 and 15 kcal%) were cross-classified with three levels of protein (30, 20 and 10 kcal%). All diets were equivalent in vitamin and mineral content and in the ratio of saturated to unsaturated fatty acids. Phenotypes were growth rate during the 12 week trial, various physiological variables and body composition at termination of the trial.

Increased fat content in the diet caused increased growth rate in the obesity-prone AKR/J line but did not affect growth rate of SWR/J mice at all. Increased dietary protein content decreased growth rate in AKR/J mice but had no effect in the SWR/J strain. Although lipid percentage in the carcass was higher in AKR/J mice than SWR/J mice on all diet combinations, increasing fat composition of the diet had a greater impact on carcass lipid content in the AKR/J than in the SWR/J line. Similarly, AKR/J mice had higher insulin concentrations than SWR/J mice on all diets, but only in the AKR/J line did insulin concentration increase with increased fat content in the diet. The highest level of dietary protein led to reduced carcass lipid in AKR/J mice on all fat levels, but only on mice at the highest level of dietary fat in the SWR/J line. All of these results suggest that growth, body composition and metabolic traits of a line prone to obesity were more sensitive to changes in the diet than were those same traits in a line not prone to obesity.

Also working with genotypes differing in genetic propensity for fat deposition, Bertram *et al.*[17] report a situation in which genotypes and nutritional treatments did not interact. These authors examined the effects of dietary protein percentage on growth, food conversion efficiency and body composition in mouse lines divergently selected for fat deposition and in their unselected controls. Mice from each line were randomly allotted to 12, 16, 20 or 24% protein diets from weaning to nine weeks of age, after which they were killed and body composition was assessed. On each protein level, mice selected for high fat content grew more rapidly and had higher food conversion efficiency than mice selected for low fat content, with control mice similar to the low fat line. All three lines had maximum efficiency and growth on the 16% protein diet. Because GEI were neither significant nor important for growth, food conversion efficiency or body composition traits, Bertram *et al.* concluded that divergent selection for fat composition had not altered requirement for or utilization of protein in either of the selected lines.

4.5. *Toxins and Drugs*

Tall fescue plants (*Festuca arundinacea)* infected with a particular endophytic fungus produce ergovaline and other toxins.[18] Ingestion of seed from toxic fescue depresses food intake, growth rate and reproduction in the laboratory mouse.[19] Hohenboken and Blodgett[20] conducted eight generations of divergent selection for resistance or susceptibility to fescue toxicosis using depression in postweaning gain caused by a toxin-containing diet as the selection criterion. In a subsequent experiment, Wagner *et al.*[21] examined effects of the toxin-containing diet on reproduction in resistant and susceptible line mated pairs. Their results illustrate how artificial selection, as it creates distinct genotypes, simultaneously may create GEI. In comparison to a control diet containing non-endophyte infected fescue, the toxin-containing diet reduced total pups born during 36 weeks of cohabitation by 13 and 28% in resistant and susceptible lines, respectively, with reductions of 10 versus 25% in total number of pups weaned. Total weight of offspring weaned was reduced by 30 and 42% in resistant versus susceptible line mated pairs, respectively. Mature weight also was more severely reduced in males and females of the susceptible than the resistant line.

Belknap *et al.*[22] studied responses of body temperature and physical activity to morphine dose (control versus 4, 8, 16 or 32 mg/kg) in 15 inbred lines of mice. Lines, doses and their interaction contributed significantly to variation for both traits. In the *average* strain, activity was decreased from control at the lowest morphine dose and increased in a dose response fashion thereafter. However, two strains were particularly sensitive and three were totally refractory to the effect. Body temperature decreased with increasing morphine dose in the *average* strain, but there was considerable variation in the pattern. One line was refractory, the same line whose activity level did not respond to morphine dose. Several lines were particularly sensitive, but not those lines whose physical activity responded most acutely to morphine dose. Analyses suggested that both response traits were heritable but genetically independent. A line's response for physical activity did not predict its thermal response to morphine.

5. Interactions of Genotype with the Internal Environment

5.1. *Sex*

As stated by Lynch *et al.*,[4] "Sex differences in gene expression may be viewed as a special case of genotype by environment interaction, since the complement of autosomal genes controlling a quantitative character may be expressed differently depending on the sex of the individual." Although males and females within a population have similar complements of genes at autosomal and mitochondrial loci, many traits are expressed differently in females and in males. Examples of sexually dimorphic traits in mice include morphology, growth and exploratory behavior.

Eisen and Legates[23] examined genotype by sex interaction in a randomly mating ICR mouse line. Phenotypes of interest were weaning weight, six week and eight week weight and three to six week gain. Genotypes were paternal half-sib and full sib families produced from a hierarchical mating design. The environment was mouse sex. Sexual dimorphism for each trait was quantified as the average difference in phenotypic value between male and female offspring within each litter. Additive genetic variance for sexual dimorphism thus defined was shown to be linearly related to phenotypic variance attributable to genotype by sex interaction. Thus, the existence of one implies existence of the other.

Heritability for the *difference* between male and female sibs in phenotypic value was near zero for weaning weight, 0.08 for six week weight and approximately 0.29 for eight week weight and three to six week gain. Thus, genes responsible for growth differences between males and females became active following weaning. Furthermore the average difference between males and females for such traits could be modified by selection, as was shown experimentally by Korkman[24]. Additive genetic variance and the genotype by sex interaction variance component differed significantly from zero for eight week weight and three to six week gain.

All of the postweaning traits were more highly heritable in males than females, suggesting that genetic influence on growth was not precisely the same trait in the two sexes. Estimated genetic correlations between the same trait expressed in males and females were 0.64 for weaning weight, 0.90 for six week weight, 0.80 for eight week weight and 0.68 for three to six week gain. If these correlations had all equaled 1.00, within-family differences between male and females would not have been heritable, sexual dimorphism would not have existed and there would have been no interaction between genotype and sex for these traits.

The existence of sexual dimorphism for a trait implies a past selective advantage for that difference and natural selection to create and sustain it. Such knowledge facilitates understanding of the genetic architecture of traits and evolutionary adaptation of populations.

5.2. *Age*

During a five week post-weaning trial, Timon and Eisen[25] compared growth, food intake and food conversion efficiency of a non-selected control mouse line to that of a line selected for nine generations for increased three to six week gain. As expected, mice from the selected line grew more rapidly, consumed more food and had higher food conversion efficiency than control line mice. Regressions of each of these traits on age were non-linear and heterogeneous between lines, so the investigators re-examined the relationships during sequential 12 day intervals. Among those short but dynamic segments of the life cycle, genotype and age did interact. Voluntary food intake was higher in selection line than control mice throughout the period, but the magnitude of the difference decreased with advancing age. Males from the selected line had higher food conversion efficiency than control line males during the first two 12-day periods but not during the third. Thus, line differences were dependent on (and interacted with) age at measurement.

5.3. *Physiological State*

Physiological state can be illustrated by the presence or absence in contemporary and otherwise comparable individuals of a reproductive event, pregnant versus non-pregnant or lactating versus non-lactating, for example. If the impact of such conditions affects different genotypes differently, then GEI exists.

Wallinga and Bakker[26] report such a case. Their genotypes were a control line of laboratory mice compared to a line originating from the same base population but selected for 25 generations for increased first-parity litter size. Females from both lines were randomly allotted to two mating management systems. Reproductive performance was recorded from 10 through 54 weeks of age. Males and females were continuously cohabited in the more intensive system, and females frequently experienced concurrent pregnancy and lactation. In the discrete mating system, males were removed from cages before each litter was born and were returned as that litter was weaned.

GEI was striking (Figure 3) in that genotypes changed rank according to the mating system in which they were compared. Under the discrete mating regime, females from the high prolificacy line bore and reared more total offspring than control line females. In contrast, the total number of offspring born and reared under continuous reproduction was higher for control line than for selection line females. Selection response for prolificacy in the first litter could not be maintained under the challenge of frequent reproduction and concurrent pregnancy and lactation.

Eisen and Saxton[27] also compared a mouse line selected for increased first-parity litter size (23 generations) to its unselected control. The trait of interest was reproduction in a female's second litter. In the first environment, males were removed before a female's first litter was born and were reintroduced after the first litter was weaned. Half of the females in this environment reared first litters standardized to eight and half raised first litters standardized to 16 pups. In the second environment, males were housed continuously with females until post-parturition mating was detected or, failing that, seven days after the first parturition. Thus, the second environment allowed concurrent pregnancy and lactation. Females in this environment reared first litters standardized to four, eight, 12 or 16 pups. In both lines and in both environments, second litters were standardized to 10 pups.

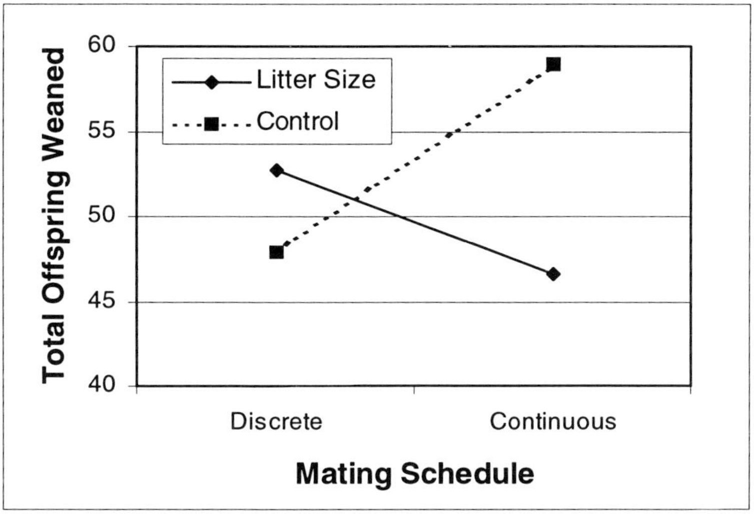

Figure 3. The effect of discrete versus continuous cohabitation of males and females on the number of pups weaned in 34 weeks by control line females versus females of a line selected for increased first parity litter size. Prepared from tabular data in Wallinga and Bakker.[26]

GEI were detected, all suggesting that the high prolificacy line was more sensitive to increased intensity of reproduction than was the unselected control. Pregnancy rates of second matings were lower under continuous mating than post-partum mating, and the magnitude of the reduction was greater in the selected line than in the control line. For females mated after their first litter was weaned, pregnancy rate decreased as number of offspring reared in the first litter increased. This decrease was larger in the line selected for prolificacy than in the control line. The two lines had equal second-litter pup mortality when subjected to post-partum mating, but pup mortality was higher in the selected than in the control line following concurrent pregnancy and lactation. Higher reproductive potential seems to have come at the cost of higher sensitivity of the components of successful reproduction to environmental stress. Females of the selected strain expressed superiority to their unselected control only under mating conditions characteristic of those experienced by their ancestors during selection.

6. Mating System by Environment Interactions

Selection determines who reproduces, and for how long. Mating systems are defined by the description of who mates with whom. In random mating, the chances are equal of a female mating with any male from among those that are available. This mating system (when several other conditions are met) leads to an equilibrium of genotypic frequencies in offspring in which the frequency of each homozygous genotype is equal to the frequency of the contributing allele squared and the frequency of each heterozygous genotype equals twice the product of the frequencies of the two contributing alleles.

In non-random mating, some criterion other than chance determines allocation of mates. In inbreeding, mates are more closely related than if they had been chosen at random, leading to an increase in homozygous genotypic frequencies and a concomitant decrease in frequency of heterozygotes. In outbreeding, mates are less closely related than if they had been mated at random, leading to increased heterozygote and decreased homozygote frequencies. Mating systems and environments interact when the effect of a mating system on the mean and/or variance of a trait depends upon the environment in which that mating system is applied.

6.1. *Inbreeding Depression and the Environment*

Inbreeding increases homozygosity, leading to increased expression of deleterious recessive genes, leading to inbreeding depression, particularly for traits affecting reproductive fitness and health. If the severity of inbreeding depression varies across environments, this would constitute a mating system by environment interaction.

Such a case was clearly shown in work of Meagher *et al.*[28] Ancestors of experimental subjects were from two populations of house mice trapped from the wild. Genotypes were created by random mating (males were not related to their mates) or inbreeding (each mated pair were full sibs, so the inbreeding coefficient of the offspring was 25%). The *standard* environment involved mating single males and females in typical laboratory cages under management and environmental conditions commonly practiced in contemporary research settings. In the *semi-natural* environment, six replicate groups of eight male and 16 female mice each were housed in indoor enclosures divided into inter-connected rooms. Space allocated to these group-housed mice created densities typical of wild house mouse populations. In contrast to the standard environment, this environment allowed the expression of territoriality, aggression, mate selection and cooperative nesting and maternal care. Phenotypes were reproduction and survival during 40 weeks of cohabitation and quantification of male to male encounters in the semi-natural environment.

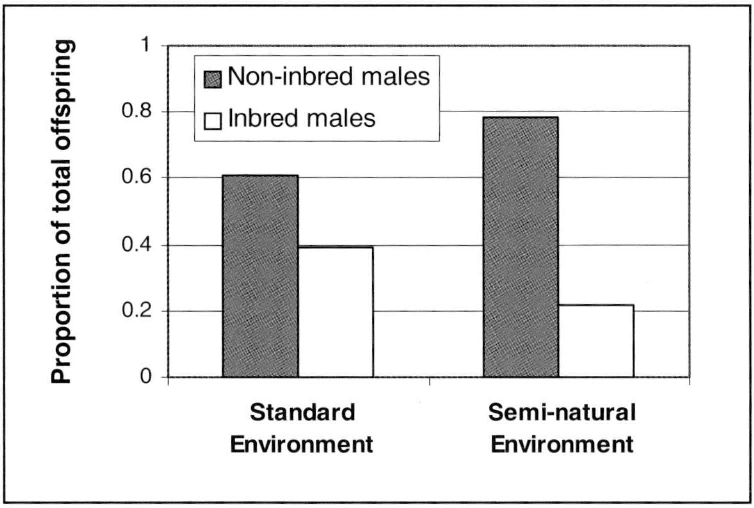

Figure 4. Proportions of total offspring produced by inbred versus non-inbred males during 40 weeks cohabitation of single mated pairs in standard laboratory cages versus group-mated males and females in a semi-natural environment. Inbreeding depression was greater in the environment in which male to male competition could be expressed. Prepared from tabular data in Meagher *et al.*[28]

Proportions of total offspring to reach weaning age that were fathered by inbred versus non-inbred sires in each environment are shown in Figure 4. In the standard environment, approximately 40 and 60% of pups had inbred versus non-inbred fathers, respectively, so inbreeding did have a negative impact on male fertility. In the semi-natural environment, however, non-inbred sires produced approximately 80% of all pups. The penalty to a male of being 25% inbred was, therefore, substantially greater in the environment in which males had to compete for territory and mates. Several factors contributed to the poorer reproductive fitness of inbred than non-inbred males in the semi-natural environment. Inbred males were less successful than non-inbred males in claiming and defending territories, they lost more aggressive encounters and they died at younger ages and at higher frequencies than their non-inbred contemporaries.

Comparable results for females are shown in Figure 5. As was the case for males, inbred females produced a smaller proportion of total offspring in both environments. Unlike the case for males, however, the penalty to a female of being 25% inbred was not larger in the semi-natural than in the standard environment. Meagher *et al.* propose that mating system and environment did not interact for females because they did not compete with other females for food, space or other resources in either of the two environments. It was, in other words, competition among males that magnified inbreeding depression in that sex.

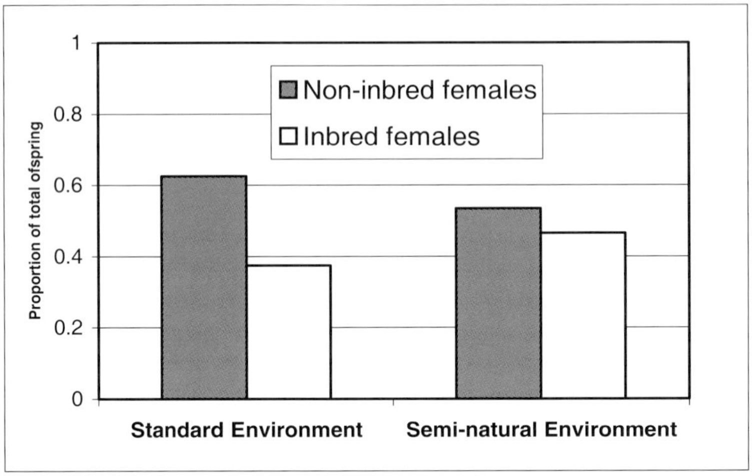

Figure 5. Proportions of total offspring produced by inbred versus non-inbred females during 40 weeks cohabitation of single mated pairs in standard laboratory cages versus group-mated males and females in a semi-natural environment. Inbreeding depression was not larger in the semi-natural than in the standard environment, possibly because females were not competing for territory or food in either case. Prepared from tabular data in Meagher *et al.*[28]

6.2. *Response of Inbred versus Outbred Individuals to the Environment*

In mice, inbred lines commonly are formed by mating a brother and sister from the same outbred or linecross litter, then mating a brother and sister from the resultant litter and continuing this process for a minimum of 20 generations.[29] Each such inbred line will be genetically different from all other inbred lines and from all outbred populations. Individuals within such a line will be homozygous at nearly all loci and will be essentially identical, genetically, to one another.

If, as a group, inbred lines do not respond to some environmental variable in the same manner as outbred populations, this constitutes a mating system by environment interaction. Nevison *et al.*[30] examined one such case. There were two behavioral phenotypes. The first was scent marking, in which a male deposits urine at locations throughout a territory, in order to make his presence known to potential adversaries. The second was counter-marking, in which a male detects the scent mark of another male and then, to establish his dominance over the individual making the original scent, deposits fresh urine to mask the original scent. (Urine provides information on individual identity because odors are genetically determined and individually unique, at least within outbred populations of animals. Because animals within an inbred line are genetically identical, however, so are their scent markings, unless they have been modified by environmental circumstances unique to each individual.)

Genotypes in the experiment were two outbred laboratory mouse lines (ICR and TO) versus one inbred line (BALB/c). Environments were testing the counter marking behaviour of each genotype against scent marking of individuals of its own line versus testing against scent markings of individuals of the other two lines. Dominant individuals from both outbred lines identified and counter marked urine spots of potential competitors from their own line, from the other outbred line and from the BALB/c inbred line with equal facility. Dominant BALB/c males readily identified urine spots from males of all three lines. They counter marked those from males of both outbred lines, but they did not counter mark urine spots from males of their own line. This suggests that, as suggested above, they were unable to differentiate urine from males of their own line from the scent of their own urine. The experiment illustrates that for some behavioral traits, inbred lines differ from outbred populations in the effect of contrasting environments on such traits. Generalizations from the experiment are constrained, however, by the fact that only one inbred line was evaluated.

6.3. *Impact of the Environment on Heterosis*

As discussed earlier, Davis and Lamberson[7] studied reproductive traits of female mice from two inbred lines, their reciprocal F_1 crosses and the backcross to each parent line in 8, 15 or 22°C ambient environments. Heterosis for reproductive traits was quantified across environments, but the authors did not report heterosis within temperature environments. For presentation here, those statistics were computed from their published results.

Four traits were significantly affected by heterosis: weight of the female at time of vaginal opening (an indicator of sexual maturity) and number of corpora lutea, number of implantation sites and number of fetuses present when females were killed 10 days after evidence of mating. Heterosis for all four of the traits was smallest in the 15°C environment. It was largest in the 22°C environment for weight at vaginal opening and largest in the 8°C environment for all three traits reflecting fecundity. Laboratory mice usually are maintained at temperatures around 22°C, as was true for mice in this experiment prior to their assignment to the three temperature environments. Wild ancestors of the mice may, however, have evolved under average temperatures closer to 15°C than to 22°C. If that is so, a possible explanation for the above observations is that the proportional expression of heterosis increases as a population undergoes greater stress. Equivalently, inbreeding depression may be greater at super- or sub-optimum than at optimum temperatures, creating a larger linecross advantage and numerically greater heterosis in sub-optimum environments.

7. The Environment for Selection

Rapid response to selection is an important goal in plant and animal breeding. Success requires an optimum balance between selection differential, generation interval and heritability of the selected traits, each of which can be affected by the environment in which selection occurs. It also is important that selection response achieved in the environment of selection be expressed by descendents in other environments. For both of these reasons, the appropriate environment in which to conduct selection has long been an important concern for applied breeding programs.

In an influential 1947 publication, John Hammond[31] concluded that an optimum environment (abundant resources and minimum stress) would allow fuller expression of heritable differences among individuals than a sub-optimum environment, leading to higher heritability for the trait in question and more rapid response to selection. He further concluded that response achieved from selection in an optimum environment should be expressed in other, less favorable environments as well, "provided that other characters, specially required by the new environment, are present in the animals." Hammond's conclusions were a justification, if not an endorsement, for the common practice of elite livestock breeders providing the best possible nutritional and environmental conditions to their young stock, a practice with merchandising as well as purported genetic advantages. Frequently, such treatment is not economically feasible for commercial livestock farmers purchasing breeding stock from individuals following Hammond's advice.

Falconer and Latyszewski[32] subsequently examined "whether environmental conditions that enhance the expression of the desired character will render selection for that character more successful than (selection under) unfavorable conditions." They identified two criteria that must be met for Hammond's conclusion to be correct. First, most of the same genes would have to affect expression of the character in both optimum and sub-optimum environments, and relatively few genes could affect the trait in one but not the other environment.

Second, a sub-optimum environment would have to affect superior genotypes more severely than it affected average or inferior genotypes, thus preventing the expression of genetic merit of superior individuals such that they could not be differentiated from mediocre contemporaries.

Experimental evidence on the question was scant, so Falconer and Latyszewski initiated a selection experiment to test the Hammond hypothesis. Although practical interest was in livestock, they chose in the interest of economy and time to work with laboratory mice. Lines from the same base population were selected for increased six week weight in two nutritional environments, *ad libitum* access to a balanced diet versus provision of the same diet at 75% of the amount consumed by contemporaries under *ad libitum* conditions.

To determine whether selection was more effective under optimum than sub-optimum conditions, the heritability of six week weight and cumulative response to eight generations of selection were compared in the *ad libitum* versus the restricted feeding lines. In contradiction to Hammond's hypothesis, heritability was non-significantly lower in the *ad libitum* line than in the restricted line (20 ± 6 vs 29 ± 13%, respectively). In agreement with Hammond's hypothesis, selection response was non-significantly higher in the *ad libitum* than in the restricted line (0.33 ± 0.10 vs 0.26 ± 0.11 g/generation, respectively). This was because of a larger cumulative selection differential, despite the lower heritability of weight, in the *ad libitum* versus the restricted line.

To determine whether response to selection achieved under one set of environmental conditions was expressed in the other environment as well, mice from generation 5 and 7 of each line were transferred to the other line's environment. That is, from three to six weeks of age, *ad libitum* line mice were raised on the restricted diet and restricted line mice were raised on the *ad libitum* diet. In contradiction to Hammond's hypothesis, selection response achieved under *ad libitum* feeding was not expressed under restricted feeding conditions. In fact, when fed the restricted diet, mice from the *ad libitum* line were smaller at six weeks of age than contemporary restricted line mice. In contrast, restricted line mice were nearly equal to *ad libitum* line contemporaries for six week weight when both were fed the *ad libitum* diet. Thus, selection response under restricted feeding conditions was robust. It was expressed both in the environment of selection and in the environment more favorable to the expression of increased growth rate. Selection response for increased growth under optimum conditions was not robust. It was hardly expressed in the restricted feeding environment at all.

To determine whether six week weight was influenced by the same genes in both the optimum and the sub-optimum environment, Falconer[3] estimated the genetic correlation between the trait expressed in the two environments to equal 0.65. Thus, although some genes did affect growth in both environments, others did not. At some loci, selection for growth on *ad libitum* feeding favored different genotypes than selection for growth when food access was restricted.

In a follow-up experiment, Falconer[33] selected lines of mice for increased (+) and for decreased (–) three to six week gain on high (H) and low (L) energy density diets. The first litter of each mated pair was raised on its diet of selection, and the second was raised on the other diet. Both direct and correlated selection responses were monitored throughout 13 generations in all four lines. As in the earlier experiment, selection for increased growth was effective on both the H and the L diet, selection on the L diet achieved increased growth on the H diet, but selection on the H diet did not achieve increased gain on the L diet. Response to selection was robust when conducted in the environment less beneficial (L) to expression of the selected trait (+).

Selection for decreased growth also was effective on both diets. Selection for decreased growth on the H diet resulted in decreased growth on the L diet, but selection for decreased growth on the L diet had little impact on growth on the H diet, a mirror image of results from the lines selected for increased growth. In both cases, selection response was robust for lines selected under environmental conditions detrimental to expression of the selected trait. Falconer concluded that "if good performance under a variety of conditions is desired, then selection should be made under the conditions *least* favorable to the desired expression of the character." Furthermore, the same measured trait, three to six week gain in this instance, was not the same biological and genetic trait when expressed under sufficiently divergent environmental conditions.

From a genetically heterogeneous mouse population, Nielsen and Andersen[34] selected replicate lines for increased, decreased and no change (control) in three to nine week growth on normal and low protein diets. After six generations, response was achieved in the predicted direction in all of the selected lines on both diets. Realized heritabilities for divergence between growth of upward and downward lines were 0.33 on the normal diet and 0.26 on the low protein diet. In generation 7, direct response was greater than correlated response for each direction of selection and diet combination. In agreement with the experiments of Falconer and colleagues, correlated responses generally were more robust when selection was conducted in the environment least favorable to expression of the selected trait. In other words, to increase growth on both adequate and inadequate diets, selection should be conducted on the inadequate diet.

8. Conclusions

(a) The laboratory mouse has contributed significantly to verification of GEI theory and development of methods to investigate such interactions. Experiments in a GEI format have contributed markedly to understanding the biology of the mouse.

(b) The various forms that GEI can assume are well illustrated in experiments with mice. Genotypic ranking for a trait might change from one environment to another. Differences among genotypes might be expressed in some environments but not in others. The magnitude of differences among genotypes might differ across environments.

(c) Another form of GEI is manifested when genotypes differ in the magnitude of their response to an environmental challenge. Stable genotypes maintain performance in spite of increasing challenge; sensitive genotypes do not.

(d) The genetic architecture of a trait may vary, depending upon the environment in which it is expressed. The heritability of nest building behavior, for example, differed between warm and cold environments, and the genetic correlation between the trait expressed in different thermal environments was 0.65. Thermoregulatory behavior apparently is controlled by partially different sets of genes in warm versus cold conditions.

(e) Genotype by environment interactions are created as populations adapt to different environmental challenges. Rankings of and differences among selected genotypes are likely to be quite different in their adapted versus in an alien environment.

(f) The severity of inbreeding depression for fitness traits may increase with greater environmental challenge. Equivalently, fitness superiority of linecross mice to that of their inbred parents may be higher in stressful than in optimum environments.

(g) Response to selection to change a quantitative trait may vary, depending upon the environment in which the selection occurs. Response to selection may be most robust across environments when conducted under environmental conditions detrimental to expression of the selected trait.

(h) This chapter emphasizes the importance of GEI, but they do not always occur. That is, genotypic rankings for many traits frequently are robust across divergent environments. In general, GEI are most likely to be significant and important when differences among genotypes and environments both are substantial.

References

1. Van Zutphen, L. F. M., V. Baumans and A. C. Beynen. 2001. *Principles of Laboratory Animal Science. A Contribution to the Humane Use and Care of Animals to the Quality of Results.* Elsevier, Amsterdam
2. Hahn, M. E. and N. Schanz. 2002. The effects of cold, rotation, and genotype on the production of ultrasonic calls in infant mice. *Behav. Genet.* **32**:267–273.
3. Falconer, D. S. 1952. The problem of environment and selection. *Am. Nat.* **86**:293–298.
4. Lynch, C. B., D. S. Sulzbach and M. S. Connelly. 1988. Quantitative genetic analysis of temperature regulation in *Mus domesticus.* IV. Pleiotropy and genotype-by-environment interactions. *Am. Nat.* **132**:521–537.
5. Falconer, D. S. 1990. Selection in different environments: effects on environmental sensitivity (reaction norm) and on mean performance. *Genet. Res.* **56**:57–70.
6. Sarkar, S. and T. Fuller. 2003. Generalized norms of reaction for ecological developmental biology. *Evol. Dev.* **5**:106–115.
7. Davis, J. A. and W. R. Lamberson. 1991. Effects of heterosis on performance of mice across three environments. *J. Anim. Sci.* **69**:543–550.
8. Crabbe, J. C., D. Wahlsten and B. C. Dudek. 1999. Genetics of mouse behavior: Interactions with laboratory environment. *Science* **284**:1670–1672.
9. Nevison, C. M., J. L. Hurst and C. J. Barnard. 1999. Strain-specific effects of cage enrichment in male laboratory mice (*Mus musculus*). *Anim. Welf.* **8**:361–379.
10. Sluyter, F. and G. A. Van Oortmerssen. 2000. A mouse is not just a mouse. *Anim. Welf.* **9**:193–205.
11. Lieblich, I. and R. Guttman. 1968. Analysis of emotional defecation under severe and mild stress- Evidence for genotype-situation interaction. *Life Sciences* **7**:301–309.
12. Barnett, S. A. and R. G. Dickson. 1986. Interaction of genotype and parental environment in the adaptation of wild House mice *Mus musculus* to cold. *J. Zool., Lond (A)* **208**:531–539.
13. Barnett, S. A. and R. G. Dickson. 1984. Changes among wild House mice (*Mus musculus*) bred for ten generations in a cold environment, and their evolutionary implications. *J. Zool., Lond.(A)* **203**:163–180.
14. Van Oortmerssen, G. A. 1971. Biological significance, genetics and evolution of variability in behavior within and between strains of mice: a behavior genetic study. *Behav.* **38**:1–92.

15. Mercer, S. W. and P. Trayhurn. 1987. Effects of high fat diets on energy balance and thermogenesis in brown adipose tissue of lean and genetically obese ob/ob mice. *J. Nutr.* **117**:2147–2153.
16. West, B. B., J. Waguespack and S. McCollister. 1995. Dietary obesity in the mouse: interaction of strain with diet composition. *Am. J. Physiol.* **268**:R658–R665.
17. Bertram, M. J., W. D. Schoenherr, E. J. Eisen and M. T. Coffey. 1996. Effects of dietary protein level on growth and body composition in mice divergently selected for fat content. *Can. J. Anim. Sci.* **76**:613–620.
18. Thompson, F. N. and J. A. Stuedemann. 1993. Pathophysiology of fescue toxicosis *Agric.Ecosyst. Environ.* **44**:263–281.
19. Zavos, P. M., D. R. Varney, M. R. Siegel, R. W. Hemken, J. A. Jackson and L. P. Bush. 1987. Effect of feeding endophyte-infected tall fescue seed on the reproductive performance in male and female CD-1 line mice via combination crosses. *Theriogenology* **27**:541–548.
20. Hohenboken, W. D. and D. J. Blodgett. 1997. Growth and physiological responses to toxicosis in lines of mice selected for resistance or susceptibility to endophyte-infected tall fescue in the diet. *J. Anim. Sci.* **75**:2165–2173.
21. Wagner, C. R., T. M. Howell, W. D. Hohenboken and D. J. Blodgett. 2000. Impacts of an endophyte-infected fescue seed diet on traits of mouse lines divergently selected for response to that same diet. *J. Anim. Sci.* **78**:1191–1198.
22. Belknap, J. K., J. Riggan, S. Cross, E. R. Young, E. J. Gallagher and J. C. Crabbe. 1998. Genetic determinants of morphine activity and thermal response in 15 inbred mouse strains. *Pharmacol. Biochem. Behav.* **59**:353–360.
23. Eisen, E. J. and J. E. Legates. 1966. Genotype-sex interaction and the genetic correlation between the sexes for body weight in *Mus musculus*. *Genetics* **54**:611–623.
24. Korkman, N. 1957. Selection with regard to the sex difference in body weight in mice. *Hereditas* **43**:665–678.
25. Timon, V. M. and E. J. Eisen. 1970. Comparisons of *ad libitum* and restricted feeding of mice selected and unselected for postweaning gain. I. Growth, feed consumption and feed efficiency. *Genetics* **64**:41–57.
26. Wallinga, J. H. and H. Bakker. 1978. Effect of long-term selection for litter size in mice on lifetime reproductive rate. *J. Anim. Sci.* **46**:1563–1571.
27. Eisen, E. J. and A. M. Saxton. 1984. Effects of concurrent lactation and postpartum mating on reproductive performance in mice selected for large litter size. *J. Anim. Sci.* **59**:1224–1238.
28. Meagher, S., D. J. Penn and W. K. Potts. 2000. Male-male competition magnifies inbreeding depression in wild house mice. *Proc. Natl. Acad. Sci.USA* **97**:3324–3329.
29. Green, E. L. 1981. *Genetics and Probability in Animal Breeding Experiments*. pp. 124–151. Oxford University Press, New York.
30. Nevison, C. M., C. J. Barnard, R. J. Beynon and J. L. Hurst. 2000. The consequences of inbreeding for recognized competitors. *Proc. R. Soc. Lond. B* **267**:687–694.
31. Hammond, J. 1947. Animal breeding in relation to nutrition and environmental conditions. *Biol. Rev.* **22**:195–213.
32. Falconer, D. S. and Latyszewski. 1952. The environment in relation to selection for size in mice. *J. Genet.* **51**:67–80.

33. Falconer, D. S. 1960. Selection of mice for growth on high and low planes of nutrition. *Genet. Res.* **1**:91–113.

34. Nielsen, V. H. and Andersen, S. 1987. Selection for growth on normal and reduced protein diets in mice. *Genet. Res.* **50**:7–15.

CHAPTER 6

GENETICS OF GROWTH IN THE MOUSE

James M. Cheverud

Department of Anatomy & Neurobiology
Washington University School of Medicine, St. Louis, MO, USA
Cheverud@pcq.wustl.edu

1. Introduction

The study of the genetics of murine growth has a long history and has often been taken as a model for mammalian growth in general. Some of the early experiments on the evolutionary response to selection used mouse growth and weight as key characters.[1-4] These early studies were followed by more detailed quantitative genetic analyses of growth in outbred lines to determine the genetic architecture underlying the growth process in some detail. Many of these studies were carried out by E. J. Eisen and his collaborators[5-13] and students or researchers influenced by their work.[14-30] These later experiments dissected out the genetic architecture of growth in different growth periods and looked more closely at the relationships among factors affecting growth, such as maternal effects and body composition. We will review the general pattern of mouse growth and its physiological basis and then examine the genetic architecture of mouse growth as revealed by quantitative genetic and quantitative trait loci studies.

2. Mouse Growth Pattern

Mammalian growth is a complex physiological process involving the time and tissue specific effects of genes acting in a coordinated fashion to produce a functional adult. However, there is structure in this complexity that has been revealed over the past twenty years of research. The typical mouse growth curve has a sigmoidal shape with the rate of growth increasing for the first few weeks

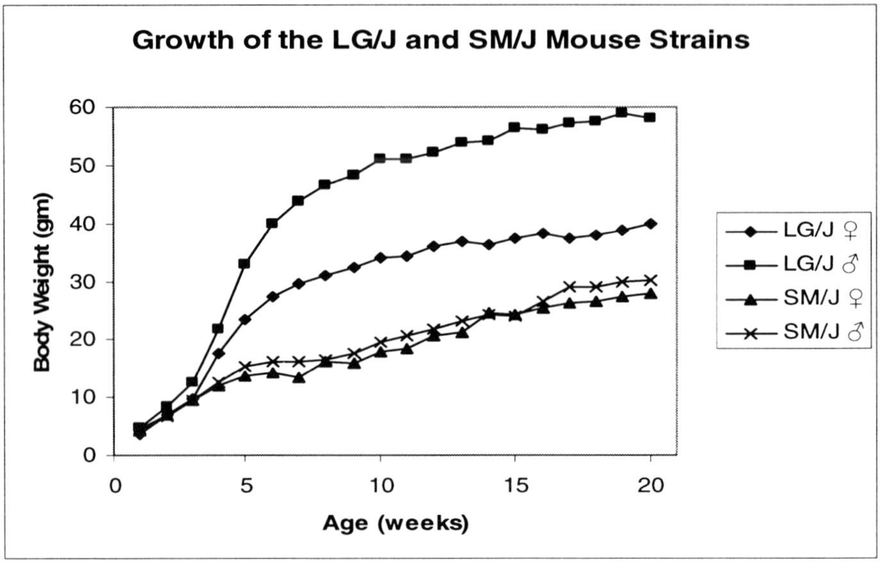

Figure 1. Growth curves for male and female LG/J and SM/J mice.[31]

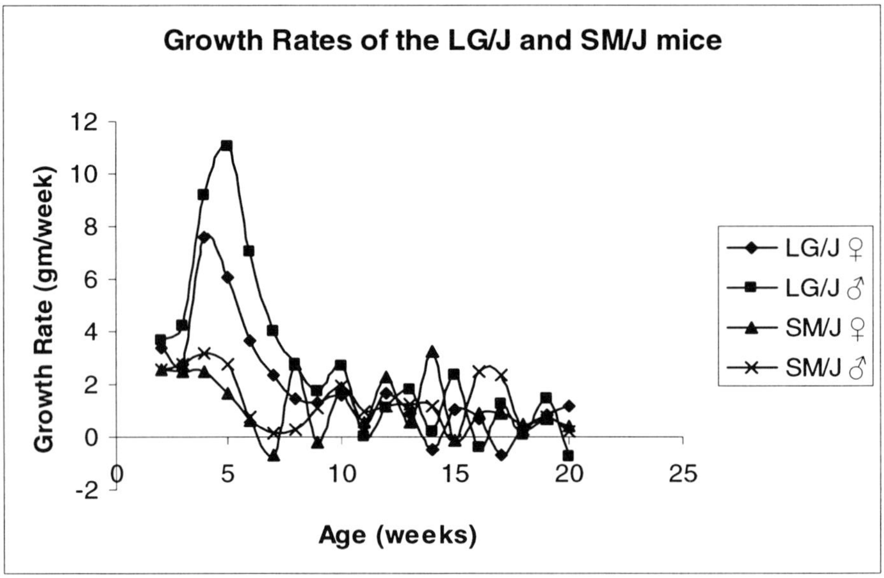

Figure 2. Growth rate (velocity) curves for male and female LG/J and SM/J mice.[31]

after birth, reaching a maximum rate at the inflection point (point of highest growth rate), followed by a decrease in growth rate. Figures 1 and 2 display the growth and velocity curves, respectively, for two of the most distinctive mouse strains with respect to their growth, Large (LG/J) and Small (SM/J)[31] LG/J males are twice as heavy as SM/J males (60 versus 30 grams), while LG/J females are about 50% larger than SM/J females (40 versus 28 grams) at 20 weeks of age. Growth rate peaks between 3 and 5 weeks of age and then rapidly declines until a slow steady-state level of growth (about 1 gm/week) is reached by 10 weeks of age. This later slow growth period continues to about 20 weeks of age. Skeletal growth seems complete by about 10 weeks of age, but soft tissue continues to be added over the next ten weeks. Growth in body weight is usually completed by about 20 weeks of age although this varies from strain to strain. Sexual dimorphism in body size usually appears at about 3 weeks of age, with males growing at a faster rate than females after this period.

3. Physiology of Growth

There is a striking difference in the physiological basis for growth pre- and perinatally relative to later postnatal growth. One obvious difference is that the effects of the environment provided by a mother for her growing offspring are strongest pre- and perinatally, fading with time after weaning.[27-30] Even so, pre- and perinatal maternal effects can persist into adulthood.[21-24] The physiology of the early growth period is dominated by the actions of Insulin-like growth factor II (IGF-II).[32] While both IGF-I and IGF-II are expressed during prenatal life, IGF-II is responsible for the bulk of prenatal growth through its effects on the placenta and its direct paracrine effects on fetal tissues. Over-expression of IGF-II leads to fetal overgrowth.[33-35] While prenatal suppression of IGF-I leads to reduced size,[36] it seems that IGF-I's primary prenatal role is in modulating the response to nutrients and nutrient-sensitive hormones.[32] Nevertheless, short term overexpression of IGF-I during fetal life does not seem to lead to overgrowth in body weight.[37-38] The expression of IGF-II in most tissues disappears at about weaning in rodents, and it is not usually detectable in adult animals.[39,40] Thus, IGF-II's growth-promoting effects are prenatal and perinatal in mice, not directly affecting later postnatal growth.

After the loss of IGF-II's growth-promoting effects, the primary factors affecting growth operate through the growth hormone (GH)-IGF-I axis. Growth hormone is produced by the pituitary gland and passes through the circulatory system to cells throughout the body that express the growth hormone receptor (GHR). Therefore, GH can have major direct effects on growth and metabolism over a wide array of tissues. One major effect of GH is that it induces production of IGF-I by the liver. IGF-I then acts to enhance growth in the skeleton, muscle and other tissues. While linked in this way, IGF-I and GH also act independently to promote growth by modulating the effects of multiple inputs. Double knockout mouse mutants lacking growth hormone signaling (*Ghr–/–*) and IGF-I (*Igf1–/–*) production reach only 17% of normal body weight, having an adult body size of only 5 grams.[41] Clearly, genes affecting postnatal growth are likely to do so in the context of these two important factors. Unlike IGF-I, growth hormone seems to have no effect on prenatal growth.[42,43] The effects of GH first become evident after about 2 weeks of age and then persist through the postnatal growth phase. The major effects of the growth hormone axis on growth appear to be concentrated in the period between 15 and 40 days of age when growth rate is at its peak.[41] After 40 days, growth differences are retained but only increase slightly. The relative importance of IGF-I and growth hormone for murine growth can be examined using single and double knockout mutants. Mechanisms independent of these two factors, such as IGF-II mediated effects, account for about 17% of adult size and are inactive after the first few weeks of life. Adding growth hormone but not IGF-I function results in an animal 30% of normal adult size while adding IGF-I but not growth hormone results in animals 52% of normal adult size. These results indicate an ordering of direct effects on adult size from largest to smallest of IGF-I, overlapping effects of IGF-I and growth hormone, effects independent of the GH-IGF-I axis, and finally, growth hormone (Figure 3).[41]

Male and female mice typically grow at different rates, with males becoming substantially larger than females. Growth differences between the sexes first appear during the later postnatal growth period after about 3 weeks of age (Figure 1). These differences are the result of distinctive gonadal hormone activity in the two sexes. The gonadal hormones result in sexually dimorphic patterns of pulsatile growth hormone secretion from the pituitary.[44] IGF-I and double GH-IGF-I knockout animals lack sexual dimorphism while GH-signaling deficient knockouts reverse dimorphism, with females being slightly larger than males.[41] This observation indicates that the interaction of both factors is important in producing normal levels of sexual dimorphism. Prenatal and perinatal factors do not produce sexual dimorphism.

The organs that compose the body grow at different rates at different ages. One of the major differences among organs in their growth is that the brain completes its growth relatively early,[16–19,26] before the time of growth hormone activity.[45] Thus, variation in brain size is unlikely to be influenced by the GH-IGF-I axis. Instead, it is under the control of the earlier acting IGF-II based system. This factor results in a relatively large brain size in mice with postnatal growth deficiencies because the brain has virtually reached its full size before the

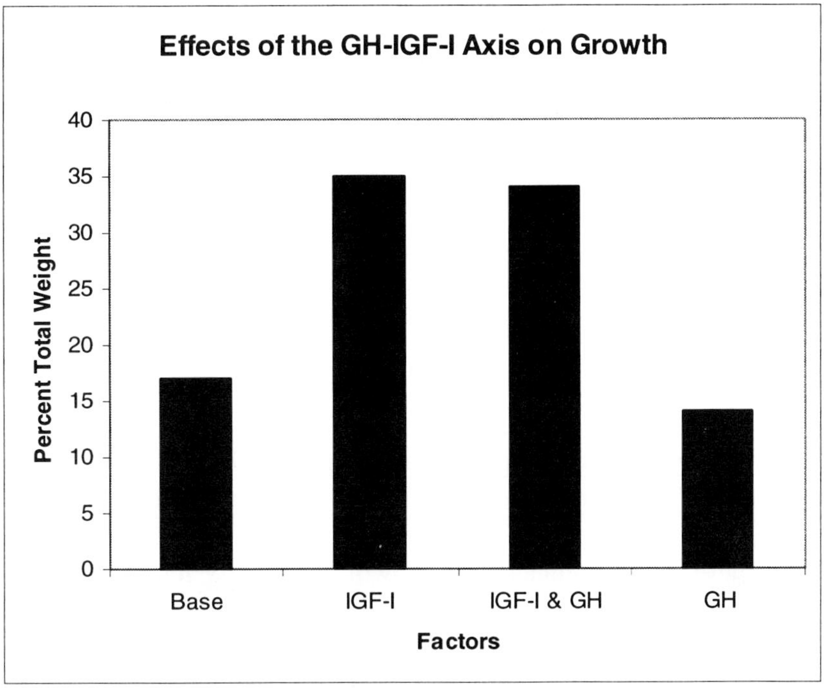

Figure 3. Percent of total adult weight contributed by activity of various hormones, including Insulin-like Growth Factor I (IGF-I) and Growth Hormone (GH).[41]

growth deficiency takes effect. In contrast to these results for the brain, other organs, such as the heart, lungs, liver, and testes scale equally with body size while the kidneys and spleen are smaller in relative size. The earlier pre- and perinatal growth period typically involves cell multiplication while the later postnatal period is primarily achieved by increases in cell size.[26] Thus, mouse growth has two distinct physiological stages, first the pre- and perinatal period dominated by IGF-II and then the later postnatal period dominated by the GH-IGF-I axis. The changeover between these physiological systems occurs between the third and fourth weeks after birth in mice, near the time of maximum growth rate.

4. Longitudinal Study of Growth

Growth is a continuous process over time. While growth acts in a pulsatile fashion on an hourly or daily scale in mice, it can be modeled continuously at the level of weekly growth. Traditionally, there are two different ways to approach the analysis of the growth process. One is to consider age-specific periodic weights measured over time as separate but correlated characters. This is the way that the data are presented in Figures 1 and 2, with time successive weights being linked by straight edges. The second is to fit a smooth curve of a specific functional form to the age-specific weight data. In this second case growth is measured by the parameters of a function fit to the observations. The most commonly used growth function and the one which best follows the patterns of growth described in Figures 1 and 2 is the logistic function

$$y_t = A/(1 + e^{K(B-t)}) \tag{1}$$

where A is the final adult size, K is the growth rate, B is the age at point of inflection in the growth rate (age at maximum growth rate) and t is age. This function was used to estimate these parameters for the growth of male and female LG/J and SM/J mice presented in Figure 1.[31] The values obtained are given in Table 1. These values indicate that growth is virtually identical in males and females. The only parameter estimate significantly different between the sexes is the final size of the LG/J males and females. However, the strains are significantly different in all three parameter estimates, with the final size and growth rate higher in the LG/J than in the SM/J and the age of inflection being three weeks later in SM/J than in LG/J. While these differences in parameter estimates are significant, it can be difficult to compare them one at a time among strains because there are strong error correlations among the parameter estimates. For example, the correlation between the final adult size estimate and the growth rate estimate is –0.90 while the correlation between final adult size and age of inflection is 0.96. These high correlations indicate that the parameters are not independently estimated, making independent interpretation difficult. For this

Table 1. Logistic growth curve parameter estimates for male and female LG/J and SM/J mice[31]

	Growth Parameters*		
	A	K	B
LG/J ♀	37.3	0.53	4.47
LG/J ♂	55.6	0.57	4.76
SM/J ♀	31.0	0.18	7.96
SM/J ♂	33.0	0.18	7.90

*A is the asymptotic or final size (g), K is the rate of change in growth rate (1/wk), and B is the age at point of inflection (wk) when the maximum growth rate is reached.

reason, some researchers prefer to work with the age-specific weights directly rather than doing so indirectly through fitted functional parameters. Even so, the functional method has been used to good effect by some researchers.[46–48]

5. Quantitative Genetics of Growth

A series of studies on the quantitative genetic architecture of growth was carried out by W. R. Atchley, J. M. Cheverud and colleagues through the 1980s.[14–30] These studies have produced relatively consistent results concerning the patterns of heritability, genetic variance, and genetic correlation for age-specific weights. A seminal paper in this program was that by Riska *et al.*[17] concerning the dynamics of genetic variances and covariances for age-specific weights through the growth process. They modeled age-specific weights (W_t) as a linear combination of a series of growth rates (ΔW_i) starting from a baseline value at time zero (ΔW_0),

$$W_t = \sum_{i=0}^{t} \Delta W_i \tag{2}$$

Then the variance in weight is given by

$$\text{var}(W_t) = \sum_{i=0}^{t} \text{var}(\Delta W_i) + 2 \sum_{i=1}^{t} \sum_{j=0}^{i-1} \text{cov}(\Delta W_i, \Delta W_j) \tag{3}$$

Thus, the genetic variance of an age-specific weight is composed of the sum of the genetic variances of the growth rates preceding it plus twice the genetic covariances among the growth rates. Given this relationship, we would expect the genetic variance to increase with age as the variance in each growth rate accumulates due to the first term in the equation. However, the variance can be either greatly enhanced or reduced by genetic covariances between growth rates. If these covariances are positive, genetic variance is greatly enhanced, but if the covariances are negative, genetic variance can be eroded over the growth process.

A negative covariance between growth rates is a sign of targeted growth. Targeted growth occurs when there is a preferred final weight that will be reached regardless of the underlying growth dynamics producing it. A negative genetic covariance between weight gains indicates that individuals that are larger than average at age X grow slower than average to age Y, and individuals that are smaller than average at age X grow faster than average to age Y. This relationship results in a narrow population distribution of genetic values (breeding values) and, hence, results in a reduction in genetic variance with age. Riska *et al.*[17] found just this sort of pattern in their population of ICR randombred mice and suggested that there was antagonistic pleiotropy between the early pre- and perinatal growth and later postnatal growth periods. This result was consistent with their findings of negative genetic correlations between early and later growth rates and the subsequent reduction in genetic variance seen at the later age-specific weights (Figure 4). Riska and Atchley[19] also found a strong negative genetic correlation between the early growing brain and later, postnatal body size growth (r = –0.56). However, subsequent studies on other mouse populations have not always found evidence for this negative correlation between early and later growth rates.[22,23,49]

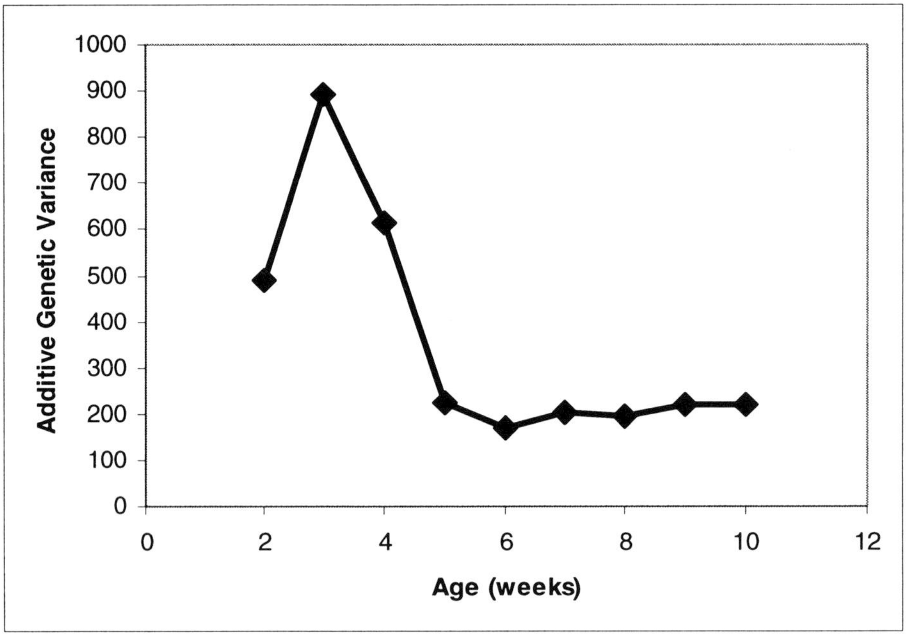

Figure 4. Changes in additive genetic variance for body weight, transformed to a logarithmic scale, with age in ICR random-bred mice.[17]

In contrast to the genetic variances, heritabilities tend to remain fairly constant throughout growth.[17,27-30] The percentage of phenotypic variance due to maternal effects declines with age but was still significant even in 10-week old mice. The decline in the proportional contribution of maternal effects to phenotypic variance with age is matched by a corresponding increase in the proportional effects of non-heritable, or environmental, factors. While the heritability of age-specific weights is relatively constant, the heritability of weekly growth rates first increases from about 30% to 60% from 2 to 5 weeks of age and then decreases, reaching low levels from 7 to 10 weeks (Figure 5). Note that the decline of growth rate heritability after 49 days corresponds to the age when the effect of the GH-IGF-I growth axis declines.[41]

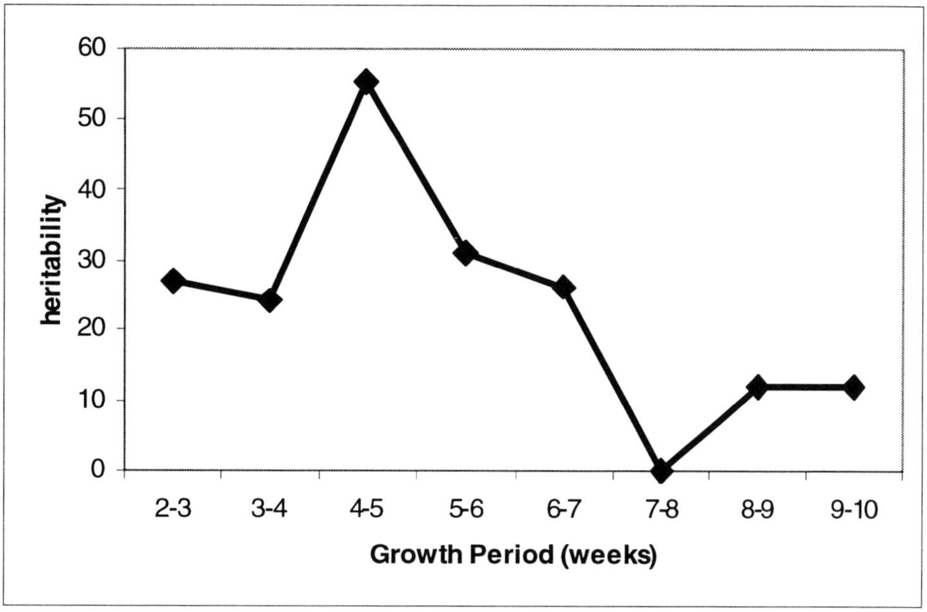

Figure 5. Heritability of weekly growth rates in ICR randombred mice.

Genetic correlations among age-specific weights follow a predictable pattern that can be considered using the Riska *et al.*[17] linear combination model. The genetic correlation between age-specific weights declines as the time between the ages increases and as new heritable variance in body weight is generated by growth (see Table 2).[17,28,29,49,50] The decline in correlation will be even more precipitous if there are negative correlations between early and later growth rates. Correlations between week 1 weight and later weights decline after week 4 when the percent shared genetic variation is reduced below 50%.[50] Week 1 weight shares only a minor portion of its genetic variation with adult weight ($r^2 = 0.26$). This low amount of shared variation is consistent with the observed changes in growth physiology during the third and fourth week of life. The GH-IGF-I axis plays little role in weight gain until after weaning (3 weeks) at which time much

Table 2. Genetic correlations among age-specific weights for the intercross of LG/J and SM/J.[28]

Week	1	2	3	4	5	6	7	8	9	10
1	1.00									
2	0.76	1.00								
3	0.81	0.93	1.00							
4	0.79	0.85	0.94	1.00						
5	0.68	0.78	0.82	0.91	1.00					
6	0.61	0.74	0.77	0.86	0.97	1.00				
7	0.56	0.69	0.71	0.79	0.93	0.98	1.00			
8	0.53	0.67	0.69	0.77	0.91	0.96	0.99	1.00		
9	0.52	0.67	0.68	0.75	0.89	0.95	0.98	0.99	1.00	
10	0.51	0.66	0.67	0.74	0.87	0.93	0.96	0.97	0.99	1.00

new, uncorrelated genetic variation is produced. As expected, given the lack of heritable growth differences at later growth periods, the correlations among age-specific weights from 42 to 70 days are all above 0.90.

Cheverud *et al.*[28] interpreted the principal components of the genetic correlation matrix for age-specific weights as representing three kinds of gene effect (Table 3). Most genetic variance (78%) was accounted for by a general size factor that resulted in an individual being larger or smaller than average throughout its life but not affecting growth rates. The second principal

Table 3. First two principal components (PC1 and PC2) of the genetic correlation matrix for age-specific weights in the F_3 intercross of LG/J and SM/J[28]

Trait	PC1	PC2
Week 1	0.62	−0.77
Week 2	0.78	−0.56
Week 3	0.80	−0.57
Week 4	0.92	−0.23
Week 5	0.95	0.17
Week 6	0.95	0.27
Week 7	0.94	0.31
Week 8	0.95	0.30
Week 9	0.95	0.29
Week 10	0.92	0.33
% Variance	78.3%	17.6%

component accounts for 17% of the variation and contrasts the early with the later ages; animals larger than average at young ages would tend to be smaller than average at later ages and vice versa. These kinds of gene effects indicate variation in the shape of the growth curve and were interpreted as evidence for genes with antagonistic pleiotropic effects on early and later growth. Growth produced by the pre- and perinatal IGF-II axis and the later postnatal GH-IGF-I axis would actually be negatively associated at some loci rather than being independent. Further principal components, explaining only 4% of the variance, indicate age-specific effects independent of effects at other ages.

Unlike genetic correlations, phenotypic correlations among age-specific maternal effects remain high throughout life.[28,29] Age-specific maternal effects have a correlation of 1.0 from age to age. Unlike the genetic correlations, maternal correlations do not decline as the time between ages increases. This lack of decline occurs because after weaning, maternal environments no longer produce new variation among offspring either uncorrelated or negatively correlated with earlier growth.

6. Quantitative Trait Loci (QTL) Affecting Growth

During the 1990s the advent of abundant variable molecular markers in mice[51] and the new statistical methods of interval mapping[52] allowed the individual genomic regions affecting growth to be mapped in the genome and their phenotypic effects to be measured. A quantitative trait locus (QTL) is a genomic region that has a quantitative effect on a phenotype of interest. QTL analyses allow for measurement of individual gene effects on traits. Cheverud *et al.*[49] used the QTL interval mapping approach to identify genomic regions affecting growth and test the hypothesis that some genes would show antagonistic pleiotropy for growth. It was hypothesized that if a variant at the locus produced faster than average early growth, it would also produce slower than average later growth as seen in the second principal component of the genetic correlation matrix for age-specific weights. Given that 17% of the genetic variance is associated with this antagonistic pleiotropy component, we might expect one such locus among every five QTL discovered. They tested the hypothesis in a population of 535 F_2 intercross mice formed by crossing females from the LG/J strain with males of the SM/J strain. In the F_2 generation of an intercross, both the molecular markers and the phenotypes segregate. Interval mapping identifies the genomic regions whose segregation correlates with the segregation of specific molecular markers.

Cheverud *et al.*[49] surprisingly, did not find any antagonistic pleiotropy QTL. Instead, they discovered that early and later growth QTL mapped to different genomic positions (Figure 6). There were 13 QTL affecting weight at 1 week of age or growth from 1 to 3 weeks and another 15 QTL affecting later growth. Early and later growth QTL mapped to different genomic locations. These results were repeated by Vaughn *et al.*[53] in a second, replicate F_2 intercross of the LG/J and SM/J strains. This same pattern, a lack of antagonistic pleiotropy and separate QTL for developmentally or physiologically separate traits, has been a common finding in our studies of growth, body composition, and skeletal morphology in the LG/J by SM/J intercross.[54] Indeed, the braincase, which grows in response to the early growing brain, is affected by a largely different set of QTL than the later growing facial skeleton.[55]

These results appear to conflict with the earlier quantitative genetic studies of early and later growth and brain size and body size reviewed above. They also conflict with the interpretation provided by Cheverud *et al.*[28,29] for the principal components of age-specific traits. However, having separate loci affect early and later growth produces the same pattern of principal components as antagonistic pleiotropy would. It is a mistake to interpret as a physiological process the results of a purely mathematical manipulation.

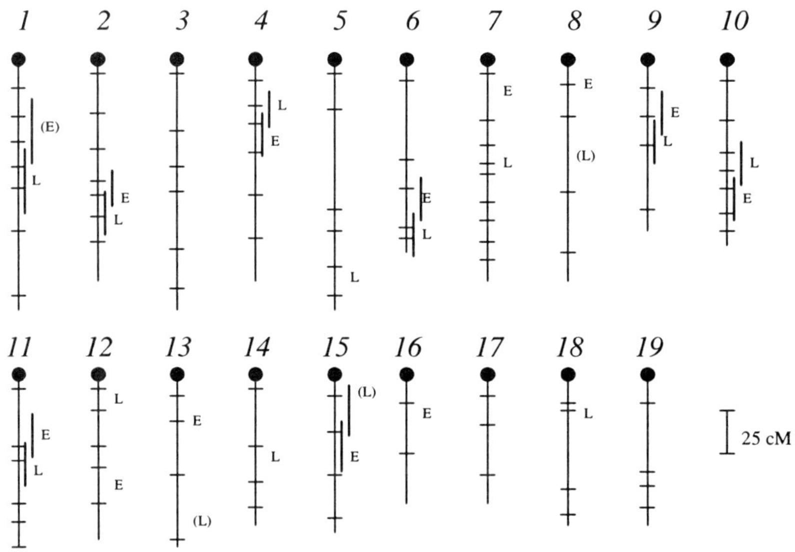

Figure 6. Locations of early and late growth QTL in the replicated LG/J by SM/J F_2 intercross.[49]

The main finding that variation in early and later growth is due to discrete sets of genes corresponds to the differences in growth physiology for these two periods. They were also confirmed by studies of other mouse populations[12,13,56–58] so that the finding is likely to be a general pattern for murine growth. It is important to keep in mind that these QTL studies only analyzed the direct effects of genes in the neonate and growing mouse, neglecting potential genetic effects that operate through the mother on the environment she provides for her growing offspring. Wolf *et al.*[59] mapped these maternal effect loci in a three generation cross between the LG/J and SM/J strains using a cross-fostering design. They found a series of four maternal effect QTL accounting for 31% of the variance in early growth. Direct effect QTL for early growth only accounted for 11% of the variance, indicating that maternal genetic factors are responsible for more genetic variation in early growth than direct factors.

7. Spontaneous Mutants and Transgenic Models

Growth in mice has also been studied extensively using either spontaneous mutations or transgenic manipulations. While study of these alleles indicates little about naturally occurring variation in growth, the alleles can provide important windows into the growth process. Keightley and Hill[60] estimated that nearly 25% of major mutations in mice have a pleiotropic effect on growth. A search of the Mouse Genome Database (http://www.informatics.jax.org/) uncovered 105 spontaneous mutations affecting postnatal growth, including such well known alleles as non-agouti, pygmy, Ames dwarf, Snell dwarf, and little. Most of the recorded spontaneous mutations for growth are diminishing mutations, possibly because these are more obvious at an earlier age than overgrowth mutants. Indeed, the Ames dwarf mutant arose in the stock selected for large body size by Goodale (LG/J) and was particularly obvious amongst its larger littermates.[2] Many of these growth reduction mutants affect important growth factors, such as failure of anterior pituitary development and consequent loss of growth hormone secretion in the Ames dwarf and the null mutation in the growth hormone releasing hormone receptor in the little allele.[61] Most oversized growth mutants, such as yellow agouti, fat, obese, diabetes, and tubby, are obese rather than being especially large skeletally. These mutants typically interfere with metabolism or appetite control rather than growth factors *per se*.

Targeted gene mutations have also been used to study the genetic basis for variation in growth. A large number of gene knockouts have indirect effects on growth through their effects on the intact system of hormonal growth factors.[61]

However, direct study of genes directly involved in growth control has been very important in delineating the physiological control of growth. The effects of knockouts and combinations of knockouts were discussed in section 3.[41] Conditional knockouts have also proven important in determining the means by which these factors affect growth. For example, liver-specific IGF-I knockouts show that when IGF-I is knocked out in the liver, so that the liver does not respond to GH by producing IGF-I, circulating levels of IGF-I decrease by 80%. However, postnatal growth was unaffected, presumably because autocrine and paracrine production of IGF-I compensated for the loss of liver function.[62]

8. Conclusions

Genetic variation in murine growth can involve many different genes, usually having small effects. The physiological genetic basis for growth and its variation involves two largely distinct systems. An early growth system, involved in prenatal and perinatal growth, centers on IGF-II and its direct effects on fetal growth and indirect effects mediated through the placenta. At 2 to 4 weeks of age, IGF-II disappears and a later postnatal growth system dominates. This system is mediated by the GH-IGF-I growth axis and has its strongest effects between 2 and 6 weeks of age, after which growth slows substantially. The separate nature of these two growth axes is indicated in the lack of genetic correlation between them and the separate mapping of QTL affecting early and later growth. Some studies even indicated a negative genetic correlation between early and later growth rates, although this result was not replicated and may have been due to linkage disequilibrium between separate genes affecting early and later growth in the study population. Perhaps a greater effect of the GH-IGF-I axis on prenatal and perinatal growth would be through its effects on the size of the mother and on the environment she provides for her offspring before weaning. These maternal effects are critical for the early phase of growth and can cause more variation in growth than the direct effects of genes expressed in the offspring themselves.

References

1. MacArthur, J. 1944. Genetics of body size and related characters. I. Selection of small and large races of the laboratory mouse. *Am. Nat.* **78**:142–157.
2. Goodale, H.D. 1938. A study of the inheritance of body weight in the albino mouse by selection. *J. Heredity* **29**:101–112.

3. Falconer, D.S. 1953. Selection for large and small size in mice. *Genetics* **51**:470–501.
4. Falconer, D.S. and J.W.B. King. 1953. A study of selection limits in the mouse. *J. Genetics.* **51**:561–581.
5. Rutledge J.J., O.W. Robison, E. J. Eisen, J. E. Legates. 1972. Dynamics of genetic and maternal effects in mice. *J. Anim. Sci.* **35**:911–918.
6. Eisen, E.J. 1974. The laboratory mouse as a mammalian model for the genetics of growth. *Proc. First World Congr. Genet. Appl. Livest. Prod.* **1**:467–492.
7. Eisen, E.J., J.E. Legates and O.W. Robison. 1970. Selection for 12-day litter weight in mice. *Genetics* **64**:511–532.
8. Eisen, E.J. 1987. Selection for components related to body composition in mice: Direct responses. *Theor. Appl. Genet.* **74**:793–801.
9. Eisen, E.J. 1992. Restricted index selection in mice designed to change body fat without changing body weight: Direct responses. *Theor. Appl. Genet.* **83**:973–980.
10. Eisen, E.J. 1993. Multitrait restricted and desired gains selection indices designed to change growth and body composition in mice. *J. Anim. Breed. Genet.* **110**:13–29.
11. Eisen, E.J., L.S. Benyon and J. A. Douglas. 1995. Long-term restricted index selection in mice designed to change fat-content without changing body size. *Theor. Appl. Genet.* **91**:340–345.
12. Rocha, J.L., E. J. Eisen., L.D. Van Vleck and D. Pomp. 2004. A large-sample QTL study in mice: I. Growth. *Mamm. Genome* **15**:83–99.
13. Rocha J.L., E.J. Eisen, L.D. Van Vleck and D. Pomp. 2004. A large-sample QTL study in mice: II. Body composition. *Mamm. Genome* **15**:100–13.
14. Atchley, W. R. and J. J. Rutledge. 1980. Genetic components of size and shape variation. I. Dynamics of components of phenotypic variability and covariability during ontogeny in the laboratory rat. *Evolution* **34**:1161–1174.
15. Atchley, W.R. 1984. Ontogeny, timing of development, and genetic variance-covariance structure. *Am. Nat.* **123**:519–540.
16. Atchley, W.R. 1984. The effect of selection on brain and body size associations. *Genet. Res.* **43**:289–298.
17. Riska, B., W. R. Atchley and J. J. Rutledge. 1984. A genetic analysis of targeted growth in mice. *Genetics* **107**:79–101.
18. Atchley, W. R., B. Riska, L. A. P. Kohn, A. A. Plummer and J. J. Rutledge. 1984. A quantitative genetic analysis of brain and body associations, their origin and ontogeny: Data from mice. *Evolution* **38**:1165–1179.
19. Riska, B. and W. R. Atchley. 1985. Genetics of growth predicts patterns of brain-size evolution. *Science* **229**:668–671.
20. Atchley, W. R. 1987. Developmental quantitative genetics and the evolution of ontogenies. *Evolution* **41**:316–330.
21. Cowley, D. E., D. Pomp, W. R. Atchley, E. J. Eisen and D. Hawkins-Brown. 1989. The impact of materal uterine genotype on postnatal growth and adult body size in mice. *Genetics* **122**:193–203.
22. Atchley, W. R., D. E. Cowley and S. Xu. 1997. Restricted index selection for altering developmental trajectories in mice. *Genetics* **146**:629–640.
23. Atchley, W. R. and J. Zhu. 1997. Developmental quantitative genetics, conditional epigenetic variability and growth in mice. *Genetics* **147**:765–776.

24. Rhees, B. K., C. A. Ernst, C. H. Maio and W. R. Atchley. 1999. Uterine and postnatal maternal effects in mice selected for differential rate of early development. *Genetics* **153**:905–917.
25. Ernst, C. A., P. D. Crenshaw and W. R. Atchley. 1999. Effect of selection for development rate on reproductive onset in female mice. *Genet. Res.* **74**:55–64.
26. Atchley, W. R., Wei, R., and P. D. Crenshaw. 2000. Cellular consequences in the brain and liver to age-specific selection for rate of development in mice. *Genetics* **155**:1347–1357.
27. Cheverud J., L. Leamy, J. Rutledge and W. Atchley. 1983. Quantitative genetics and the evolution of ontogeny. I. Ontogenetic changes in quantitative genetic variance components in randombred mice. *Genet. Res.* **42**: 65–75.
28. Cheverud, J., J. Rutledge, and W. Atchley. 1983. Quantitative genetics of development: Genetic correlations among age-specific trait values and the evolution of ontogeny. *Evolution* **37**:895–905.
29. Leamy, L., and J. Cheverud. 1984. Quantitative genetics and the evolution of ontogeny. II. Genetic and environmental correlations among age-specific characters in randombred house mice. *Growth,* **48**: 339–353.
30. Cheverud, J., and L. Leamy. 1985. Quantitative genetics and the evolution of ontogeny. III. Ontogenetic changes in correlation structure among live-body traits in randombred mice. *Genet. Res.* **46**: 325–335.
31. Ehrich, T., J. Kenney, T. Vaughn, L. S. Pletscher, and J. Cheverud. 2003. Diet, obesity, and hyperglycemia in LG/J and SM/J Mice. *Obesity Res.* **11**: 1400–1410.
32. Fowden, A.L. 2003. The insulin-like growth factors and feto-placental growth. *Placenta* **24**:803–812.
33. Lau M.M.H., C.E.H. Stewart, Z. Liu, H. Bhatt, P. Rotwein, and C.L. Stewart. 1994. Loss of the imprinted IGF2/cation-independent mannose-6-phosphate receptor results in fetal overgrowth and perinatal lethality. *Genes Dev.* **8**:2953–2963.
34. Ludwig, T., J. Eggenschwiler, P. Fisher, A.J. D'Ercole, M.L. Davenport, and A. Efstratiadis. 1996. Mouse mutants lacking the type 2 IGF receptor (IGF2R) are rescued from perinatal lethality in *Igf2* and *Igf1r* null backgrounds. *Dev. Biol.* **177**:517–535.
35. Eggenschwiler, J., T. Ludwig, P. Fisher, P. Leighton, S.M. Tilghman, and A. Efstratiadis. 1997. Mouse mutant embryos overexpressing IGF-II exhibit phenotypic features of the Beckwith-Wiedemann and Simpson-Golabi-Behmel syndromes. *Genes Dev.* **11**:3128–3142.
36. Baker, J., J.-P. Liu, E.J. Robertson, and A. Efstratiadis. 1993. Role of insulin-like growth factors in embryonic and postnatal growth. *Cell* **75**:73–82.
37. Lok, F., J.A. Owens, L. Mundy, J.S. Robinson, and P.C. Owens. 1996. Insulin-like growth factor I promotes growth selectivity in fetal sheep in late gestation. *Am. J. Physiol.* **270**:R1148–1155.
38. Tarantal, A.F., M.K. Hunter, and S.E. Gorsky. 1997. Direct administration of insulin-like growth factor I to fetal rhesus monkeys (*Macaca mulatta*). *Endocrinology* **138**:3349–3358.
39. Lee, J.E., J. Lintar, and A. Efstratiadis. 1990. Pattern of the insulin-like growth factor II gene expression during early mouse embryogenesis. *Development* **110**:151–159.
40. Singh, J.S., L.B. Rall, and D.E. Styne. 1991. Insulin-like growth factors I and II gene expression in Balb/C mouse line during postnatal development. *Biol. Neonate* **60**:7–18.
41. Lupu F., J. D. Terwilliger, K. Lee, G.V. Segre, and A. Efstatiadis. 2001. Roles of growth hormone and insulin-like growth factor 1 in mouse postnatal growth. *Dev. Biol.* **229**:141–162.

42. Fowden, A.L. 1995. Endocrine regulation of fetal growth. *Repro. Fertil. Devel.* **7**:351–363.
43. Pantaleon, M., E.J. Whiteside, M.B. Harvey, R. T. Barnard, M.J. Waters, and P.L. Kaye. 1997. Functional growth hormone (GH) receptors and GH are expressed by preimplantation mouse embryos: A role for GH in early embryogenesis? *Proc. Natl. Acad. Sci. USA* **94**:5125–5130.
44. Davey, H.W., R.J. Wilkins, and D. J. Waxman. 1999. STAT5 signaling in sexually dimorphic gene expression and growth patterns. *Am. J. Hum. Genet.* **65**:959–965.
45. Bishop, K.M. and D. Wahlsten. 1999. Sex and species differences in mouse and rat forebrain commissures depend on the method of adjusting for brain size. *Brain Res.* **81**:358–366.
46. Wu R.L., C.X. Ma, M. Lin, Z.H. Wang and G. Casella. 2002. A general framework for analyzing the genetic architecture of developmental characteristics. *Genetics* **166**: 1541–1551.
47. Kirkpatrick, M. and N. Heckman. 1989. A quantitative genetic model for growth, shape, reaction norms, and other infinite-dimensional characters. *J. Math. Biol.* **27**:429–450.
48. Kirkpatrick, M, D. Lofsvold, and M. Bulmer. 1990. Analysis of the inheritance, selection and evolution of growth trajectories. *Genetics* **124**: 979–993.
49. Cheverud, J., E. Routman, F. M. Duarte, B. van Swinderen, K. Cothran and C. Perel. 1996. Quantitative trait loci for murine growth, *Genetics* **142**: 1305–1319.
50. Kramer, M. G., T. T. Vaughn, L. S. Pletscher, K. King-Ellison, E. Adams, C. Erickson and J. M. Cheverud. 1998. Genetic variation in body weight growth and composition in the intercross of Large (LG/J) and Small (SM/J) inbred strains of mice. *Genetics and Molecular Biology* **21**: 211–218.
51. Dietrich, W., H. Katz, S. Lincoln, H.-S. Shin, J. Friedman, N. Dracopoli and E. S. Lander. 1992. A genetic map of the mouse suitable for typing intraspecific crosses. *Genetics* **131**:423–447.
52. Lander, E. S. and D. Botstein. 1989. Mapping Mendelian factors underlying quantitative traits using RFLP linkage maps. *Genetics* **121**: 185–199.
53. Vaughn, T. T., L. S. Pletscher, A. Peripato, K. King-Ellison, E. Adams, C. Erikson and J. M. Cheverud. 1999. Mapping quantitative trait loci for murine growth – A closer look at genetic architecture. *Genet. Res.* **74**: 313–322.
54. Cheverud, J. M. 2004. Modular pleiotropic effects of quantitative trait loci on morphological traits. In: *Modularity in Development and Evolution.* ed. G. Schlosser and G. Wagner. pp. 132–153. University of Chicago Press, Chicago.
55. Leamy, L., E. Routman and J. Cheverud. 1999. Quantitative trait loci for early and late developing skull characters in mice: A test of the genetic independence model of morphological integration. *Am. Nat.* **153**: 201–214.
56. Morris, K.H., A. Ishikawa, and P.D. Keightley. 1999. Quantitative trait loci for growth traits in C57BL/6J x DBA/2J mice. *Mamm. Genome* **10**: 225–228.
57. Brockmann G.A., E. Karatayli, C.S.Haley, U. Renne, O.J. Rottman and S. Karle. 2004. QTLs for pre- and postweaning body weight and body composition in selected mice. *Mamm. Genome* **15**:593–609.
58. Ishikawa, A. and T. Namikawa. 2004. Mapping major quantitative trait loci for postnatal growth in an intersubspecific backcross between C57BL/6J and Philipine wild mice by using principal component analysis. *Genes Genet. Syst.* **79**:27–39.
59. Wolf, J. B., T. T. Vaughn, L. S. Pletscher and J. M. Cheverud. 2002. Contribution of maternal effect QTL to the genetic architecture of early growth in mice. *Heredity*, **89**: 300–310.

60. Keightley, P.D. and W.G. Hill. 1992. Quantitative genetic variation in body size of mice from new mutations. *Genetics* **131**:693–700.
61. Efstratiadis, A. 1998. Genetics of mouse growth. *Int. J. Dev. Biol.* **42**:955–976.
62. Butler, A.A. and D. LeRoith. 2001. Minireview: Tissue-specific versus generalized gene targeting of the igf1 and igf1r genes and their roles in insulin-like growth factor physiology. *Endocrinology* **142**:1685–1688.

CHAPTER 7

GENETICS OF BODY COMPOSITION AND METABOLIC RATE

Lutz Bünger[1] and William G. Hill[2]

[1]*SAC, Growth Genetics Section, SLS Group, Bush Estate, Edinburgh, UK*
L.Bunger@ed.sac.ac.uk
[2]*School of Biological Sciences, University of Edinburgh,*
West Mains Road, Edinburgh, UK
W.G.Hill@ed.ac.uk

1. Introduction

To advance knowledge of the genetic basis of fatness and its regulation and to consider its interconnection with feed intake and metabolic rate, the mouse has been widely used as a model, both in the analysis of individual genes and in artificial selection. Mice as model animals have great advantages; their short generation interval, high reproductive rate, low cost of animal management, the ability to control and standardize their environment, and the extensive knowledge of their genetics have led to mice becoming an important model for studies of the inheritance of body composition and energetics. The quantitative genetics of body composition in mice was reviewed ca 15 years ago by Eisen.[1] Subsequently, there has been great progress in mapping quantitative trait loci (QTL) and identifying specific genes, for example leptin.[2] Whilst the main focus previously was on the mouse as a model for livestock breeding, research is now mainly using it as a model for the genetics of human obesity and other human diseases.

1.1. *Obesity in Humans*

Today's most visible public health problem is obesity. Individuals having a body mass index [BMI = weight in kg/(height in m)^2] greater than 25 are classified as overweight and those over 30 as obese.[3] Obesity is a complex multifactorial disease where the aggregation of fat and the anatomical site of its deposition is determined by the individual's genes and environment. Environmental factors contributing to obesity include a lack of physical activity and abundant high fat, energy rich food. Obesity is a major risk factor for type 2 diabetes, coronary heart disease, sleep apnea, hypertension and some forms of cancer.[4] In 1997 the WHO declared obesity to be the most serious health threat facing the western world, but obesity is no longer restricted to industrialised societies; as of 2000, it is estimated that the number of obese adults was over 300 million worldwide and that over 115 million people suffer from obesity-related problems in developing countries.[5]

1.2. *Fatness in Farm Animals*

There is a ubiquitous excess of fat in livestock and poultry carcasses. The overproduction of fat has serious consequences for the animal industry: (a) consumer concern about unhealthy food coupled with a higher demand for lean meat, (b) loss of product and labour cost for the meat processor to trim the waste fat off, (c) relative inefficiency of fat growth and so higher production cost, and (d) negative side effects of increased fatness in breeding animals (for detailed discussion see[1]). An understanding of the genetic basis of fatness is therefore important in livestock where selection and potentially genetic manipulation can be practised. Muscle content and distribution between high and low value joints, and intra-muscular fat, as related to meat tenderness, are also significant factors in animal production.

1.3. *Relationships Between Body Composition and Metabolic Rate*

Increased fatness results from a chronic positive energy balance, where metabolisable energy intake exceeds energy expenditure. However, experimental evidence is limited on the relative contribution to weight gain of differences in energy intake and expenditure (resting or due to physical activity) between individuals.[6] Reduced rates of total energy expenditure and low resting metabolic rates have been identified as risk factors for subsequent weight gain in a human population prone to obesity.[7] In Pima Indians various factors including metabolic rate, physical activity, rate of fat oxidation, insulin sensitivity, sympathetic nervous system activity, and plasma leptin concentrations were all found to be important,[8] but the role of energy expenditure in energy regulation remains controversial.[9] Energy expenditure has been shown to be a critical factor in energy regulation, for reduced rates appear to facilitate weight gain and to reduce the extent of body energy loss during dieting in individuals susceptible to obesity. There is evidence that physical activity can influence the underlying determinants of energy balance.[10] Thus, exercise intervention studies have shown that total energy expended after physical activity is greater than would be expected from the calories used for the exercise alone. It is therefore necessary to consider both the caloric intake and expenditure sides of the energy balance equation. Genetic studies will help to identify new pathways involved in the pathophysiology of obesity.[8]

There have been far fewer quantitative genetic studies of the main components of energy balance than on obesity *per se*. This is due to the difficulty in obtaining accurate measurements in large pedigreed populations, and perhaps because of simplistic assumptions that metabolic rate does not show genetic variation, such that feed input of animals varies solely due to differences in tissue accretion and in output of product such as milk. There is, however, clear evidence of genetic variability in metabolic rate and appetite control.

1.4. *Body Weight and Body Composition in Normal Mice*

How body weight and body composition change over age in an unselected and non-inbred mouse population is shown in Figure 1. Although there is variation between mouse strains, the general pattern is that the mouse weighs about 1.5 g at birth. Its pre-weaning growth is strongly affected by maternal effects and number of pups nursing. Typically, there is somewhat restricted growth at weaning (ca 21 d), followed by very rapid growth till about 40 d, around the age of sexual maturity. Growth rate then falls towards a slow period of body weight gain accompanied by increased fat deposition. At maturity, the strain in Figure 1 weighs approximately 40 g and has ca 10–11% fat. The growth pattern of this strain closely resembles that of other unselected strains,[12–14] but mature weight and growth pattern can be changed by selection for growth.[15]

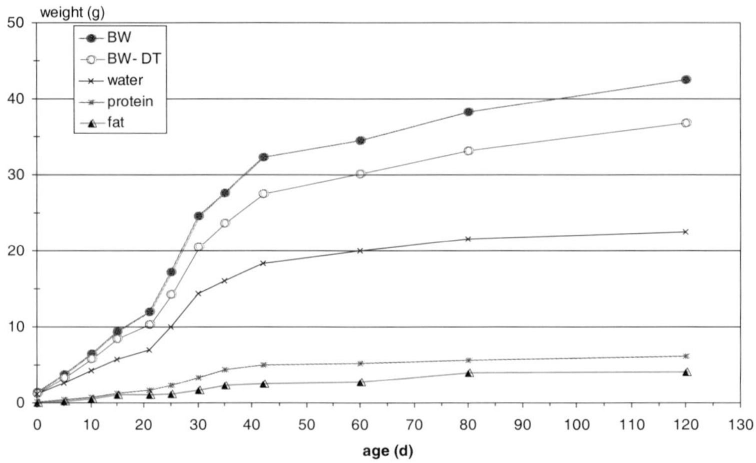

Figure 1. Body weight (BW) of males from an unselected control line, weights of the body without digestive tract (BW-DT), and of water, protein and fat (modified from [11])

2. Measurement of Body Composition and Metabolic Rate in Mice

Whole body content of fat and protein (from nitrogen content) can be estimated at any age by chemical analysis of the dried and ground animal, involving procedures that are time-consuming, destructive and preclude the removal of organs/tissues for others assays or histological examination.[16] Other quicker and cheaper methods of estimating fatness have been evaluated to determine if they are sufficiently sensitive to assess animals that do not differ widely in fat content.

Correlations (Table 1) show that body weight (BW) is usually a poor predictor of fat% unless mice are very fat, but can predict the total protein/lean amount quite well. Some fat depots are highly correlated to total body fat, so the weight of dissected fat depots are useful indicators; for example the gonadal fat pad is relatively easy to dissect out and comprises about 1/8 of the total fat mass at 10 weeks.[17] Fat depots develop at different rates with age. Although the genetic correlations between depots are high, their relative proportions are under some degree of genetic control as shown by selection experiments (see below). Caution is therefore needed when comparing total fat contents estimated from different depots. Particularly high correlations were found for indicators of fatness based on the water content of the body (Table 1). These indicators include dry matter content, estimated as freeze dried mass/wet mass, from which fat percentage can be predicted by linear regression.[18] Body lean content of the dead animal can be estimated from whole body nitrogen or from water content via dry matter content. The mass of various muscles or muscle groups (Table 1) has also been used to indicate total muscle weight or content.

All the methods in Table 1 have to be used *post mortem*. Recent advances in technology and innovative approaches towards investigating relationships between non-destructive measurements[19] and those from dissections and/or chemical analysis have begun to advance our understanding of body composition and its genetic control without euthanasia of the animals (Table 2).

Table 1. Phenotypic and *genetic* correlation coefficients between indicator and adiposity/lean traits

Indicator trait	Predicted adiposity or lean trait	Normal mice (narrow range of fat%)	Fat mice (wide range of fat% or age)	Ref.
Body weight	Fat%	0.13 (ns)	0.83/0.91 (M/F)	20
BW/body length	Fat%		0.84/0.93 (M/F)	20
Density (water displacement)	Fat%	0.12 (ns)		20
Lee index*	Fat%	0.08 (ns)	0.76/0.94 (M/F)	20
Gonadal fat pad%	Fat%	0.82/0.96 (M/F)	0.97/0.96 (M/F)	20
Gonadal fat pad weight	Fat weight		0.81–0.96 (4 to 30wk)	21
Body water%	Fat%	−0.87/−0.99 (M/F)	−0.996/−0.997 (M/F)	20
Dry matter/body weight	Fat%		0.98/0.998	18, 22
Body weight	Protein amount	0.82; *c1.0*		11
Abdominal fat weight	Total fat weight		0.97	23
Body weight	Lean index	0.94		17
Body weight (21 to 91d)	Protein amount		0.99	24
Body weight (0 to 118d)	Protein amount		0.99	13
Body weight	Hind-leg muscle weight (hlmw)		*0.51–0.77*	25
Hindleg weight	hlmw	*0.98*		
Muscle%	hlmw	*0.35*		
Gonadal fat pad weight	Body weight	*0.93 (0.43–0.78)*		26

*Lee index: BW (g)$^{0.33}$/nasoanal length (mm); M/F: Males/Females; ns: not significant.

These methods also allow longitudinal studies incorporating repeated measurements on the same animals at different ages or phases of life, such as reproduction, to replace cross sectional studies. Some of these new methods provide only a prediction of the total fat content, whereas others provide *in vivo* measurements on various fat depots, with accuracy dependent on the software used. A drawback may be the cost of this equipment.

Table 2. Non-destructive methods for measuring body composition in small animals

Method	Animals	Results/traits	Comments	Ref.
TOBEC EM-Scan SA-2 (EM-Scan, Inc., Springfield, IL)	Mice 45d, ffBW: 8–45g	r (TOBEC-ffBW vs. ffBW chemical) = 0.985	Used in a selection experiment	31, 32
TOBEC (various makes & models)	Several mammalian species	r (TOBEC vs. LM) > r (TOBEC vs. FM)	Review; method requires much calibration	33
PIXImus™ Mouse Densitometer		Traits: BMD (mg), Bone area, LM, FM	Images are shown	34
DXA (Norland pDEXA Sabre (Fort Atkinson, WI) & Sabre Research software	Mice 2–15% fat F, 11–38% fat	r (DXA fat area% & dissected fat) r = 0.92 (M), 0.88 (F) r = 0.97	2-D pictures, fat and lean areas identified	35
DXA	Man to mouse		Review	36, 37
MRI (Bruker AMX300, Ettlingen, Germany)	Mice 6–18w	Total fat volume, volume of 3 individual depots	2-D images from cross-sections, different fat depots	38
MRI & NMRS	Mice		Review, NMRS to evaluate tissue metabolism	39
EchoMRI (QMR)	Mice	BMD, fat, lean, with potential to determine spatial properties of body fat	No need for anesthesia, more accurate than DEXA and MRI	40, 41

DXA: dual-energy X-ray absorptiometry; MRI: magnetic resonance imaging, NMRS: nuclear magnetic resonance spectroscopy; TOBEC: total body electrical conductivity; F,M: females, males; LM, FM, ffBW: lean mass, fat mass, fat free body weight; BMD, BMC: bone mineral density, content; QMR: quantitative magnetic resonance.

Resting metabolic rate can be estimated in a respiration calorimeter, but the procedure is time consuming on individual animals and may be influenced by behavioral differences induced by being alone in an alien environment. Indirect measures that have been used include feed intake, which when corrected for body mass and gain presumably estimates energy loss by metabolism, especially when measured in mature animals.

3. Genetic Basis of Obesity

Animal models of obesity can be broadly classified as genetic or 'induced'. The focus here is on genetic models, but in mice obesity can also be induced by chemical agents such as gold thioglucose[27] and bipiperidyl mustard[28] or by providing highly palatable, fat or carbohydrate rich food.[29,30] Evidence for and information on the genetic basis of obesity has emerged from different approaches and different genetic models including monogenic mutations, knockouts, transgenics, QTL mapping studies, and polygenic analyses utilizing selection experiments and correlations of twins and other relatives in pedigreed populations.

3.1. *Effects of Single Gene Mutations on Fatness*

Eleven single-gene mutations at fatness-related loci with major effects have been identified in mice (Table 3).[42] The resulting obesity differs in extent, age of onset, and progression to obesity, and depends on genetic background. In some

Table 3. Spontaneous single gene mutations in mice affecting fatness

Gene	Mutation (former symbol)	MoG	Chr.	Ref.
A^y	Agouti yellow	D	2	48
Atrn	Attractin- Mahagony (*mg*)	R	2	49
Cckar	Cholecystokinin receptor (*OLETF*)	R	5	50, 51
Cpe	Carboxypeptidase-fat mutant (*fat*)	R	8	52
Lep	Obese- Leptin (*ob*)	R	6	2
Lepr	Diabetes-Leptin receptor (*db*)	R	4	53
lit	Little mutation at the GHRH receptor	R	6	54 55
Lpin1	Fatty liver dystrophy (*fld*)	R	12	56
Mgm1	Mahoganoid (*md*)	R	16	57, 58
$Mstn^{Cmpt\text{-}dl1Abc}$	Mutation in the myostatin gene (*Compact*)	>R	1	59, 60
Tub	Tubby	R	7	43

MoG: mode of gene action, D autosomal dominant, R recessive, Chr.: chromosome number. For more details see.[42,61–63]

cases, excessive deposition of adipose tissue can lead to a twofold increase of body weight[43] Although corresponding cases of monogenic obesity in humans are rare,[44–47] these models have contributed to a substantive increase in understanding the molecular pathways and physiological systems underlying the regulation of energy balance and fatness.

3.2. *Candidate Gene Approaches*

Because these single-gene mutations have large effects on fatness and energy balance, some are suitable for candidate gene approaches to address the question of whether polymorphisms in a certain gene have contributed to responses to artificial selection. Such studies have generally shown that the candidate genes contributed little, if any, of the response. The role of the leptin regulatory feedback loop in the genetic changes produced in a polygenic obesity model, made by divergent selectionfor fatness (Table 4; Item 2),[17] was assessed by introgression by repeated backcrossing of the Lep^{ob} and $Lepr^{db}$ mutations (Table 3), causing leptin production deficiency and leptin receptor deficiency, respectively, separately into these Fat and Lean lines. The fat amount increased significantly above wild type in homozygotes for Lep^{ob} or $Lepr^{db}$ in both lines, in Fat females from 6.4 to 24.1 and 27.4 g, respectively, and in Lean females from 2.4 to 18.7 and 23.4 g, maintaining most of the line differences after introgression. Introgression of leptin production and receptor deficiencies thus had different effects from long-term selection, indicating that the genes responsible for the line divergence must act independently of the leptin regulatory system. Energy budget analysis indicated that the lines differed mainly in the level of energy expended on physical activity, suggesting that multiple pathways regulate fatness, each independently responsive to intervention.[64]

The role of genetic changes in sensitivity to leptin in contributing to responses to selection for fatness was also tested by exogenous administration of leptin to the (non-inbred) Fat line and an unselected Control line.[22] Treated animals in both lines had significantly lower mean fatness at the end of test (15% and 3.0%, respectively) than untreated animals (21%, 7.4%), but the differences in response between the lines were all non-significant. There was a much wider range of fatness among treated (3–29%) than untreated Fat line (15–31%) or treated Control (0.7–6.4%) animals. While sensitivity to leptin remained in the Fat line, response appears to vary among animals at the dose level used.

The involvement of growth hormone (GH) in the genetic change produced by long-term selection in growth and fatness was also undertaken using these selected Fat and Lean lines.[65] GH deficiency was achieved by repeated backcrossing into each line the recessive *lit* gene, which has a defective GH releasing factor receptor (Table 3). GH-deficiency increased fat percentage in all lines, especially in males (e.g., males at 98 d, *lit/lit*: Fat 26%, Lean 6.9%: +/+: Fat 22%, Lean 4.8%). For these and other traits of growth or composition recorded in the study, which also included lines selected for high and low body weight, the interaction effects of the *lit* gene and the genetic background were small compared to genotype and line differences, and indicated that loci other than *lit* contributed most of the selection response.

3.3. *QTL for Fatness*

Since the first association study[66] showing a genetic link to obesity in humans, there have been hundreds more with humans or mice.[63] In addition to the known natural single-gene mutations in mouse models of obesity (Table 3), 55 knockout and transgenic rodent models relevant to obesity have been described.[63] QTL reported from animal models, mostly mice, currently number 183. Overall, more than 430 genes, markers and chromosomal regions have been associated or linked with human obesity phenotypes. This indicates that fatness is a typical quantitative trait with a polygenic basis, with the "normal" distribution for fatness in human and animal populations comprising variation from many genes with small effect and a few genes with major or visible effect. Mice have proved to be powerful models for understanding both obesity in humans and carcass fatness in farm animals.

Many QTL have been mapped in crosses between selected and/or inbred mouse lines. Recently, 75 QTL for fatness and 85 QTL for body weight in mice were tabulated.[42] These QTL map to every chromosome except the Y, with high densities on chromosomes 1, 7 and 11, where QTL from different studies coincide frequently. Most QTL effects are small and additive, and often they are modified by diet, age and gender. Some QTL showed a bigger effect, for example *Fob2* contributed almost 20% of the variance in fat% in the F_2.[67] Most of the QTL mapped to wide chromosomal regions (> 20 cM).

Some of the QTL associated with adiposity are related to diet-induced obesity and were identified using crosses between strains that do not differ in fatness on a normal diet, but do differ on a high fat diet (HFD). Crosses of strains (e.g., AKR/J, susceptible to HFD; SWR/J, not susceptible) helped to identify such

QTL.[68,69] Similarly, an intercross between CAST/Ei × C57BL/6J has been used.[30,70]

Congenic, recombinant inbred, advanced intercross, and chromosome substitution strains are needed to fine-map QTL, to identify the genes underlying the traits, and to examine interactions between them. Despite progress, a list of the number, location, and effects of all individual genes contributing to variation in fatness is far from complete. Finally, there is substantial support for epistatic interactions between genes affecting body fat, insulin sensitivity and lipoprotein metabolism,[71] posing special challenges for future genetic analysis.[72] More details are given in reviews on identified murine QTL affecting fatness and body weight.[42,73,74]

Since the comprehensive review by Snyder *et al.* [63] of the genetics of obesity, two recent studies illustrate its complex nature. In one, mapping studies were focused on refining the phenotype and partitioning total fat to single fat depots and body weight to organ weights.[75,76] Fifty QTL were detected on 15 chromosomes affecting weights of various organs and adipose tissue components of growth, including subcutaneous and epididymal fat pad depots. Nearly all the aggregate growth QTL could be interpreted in terms of the organ and fat subcomponents measured. The second study shows the complexity an identified QTL might reveal during the process of fine mapping. In the original cross of divergent selection lines (Table 4-2 and 4-7) four QTL were identified.[67] One of those QTL, *Fob3*, which explained 14% of the variance of fat% in the F_2, was mapped to 34cM from the proximal end of chromosome 15, but fine mapping indicates that *Fob3* is a composite of at least two QTL of smaller effect.[77]

3.4. *Polygenic Inheritance of Fatness*

Many of the most common human diseases such as obesity, atherosclerosis, diabetes, and hypertension exhibit continuous phenotypic variation and have a predominantly multifactorial basis. Genes with roles in energy balance, nutrient partitioning, lipid metabolism and glucose/insulin homeostasis, and a variety of behavioral traits, particularly habitual locomotor activity, seem to interact with environmental factors to regulate body composition. As discussed above, with the current abundance of highly polymorphic genetic markers and the refinement of experimental approaches including new imaging technologies, it is now possible to screen thoroughly the genomes of model organisms for the individual genes or QTL that control polygenic traits such as fatness. With the growing information from comparative mapping, it will be possible to predict the location of a homologous gene in humans and livestock species after first mapping and cloning it in the mouse. Selection lines are important resources in mapping experiments.

Most information on the quantitative genetics of body components in the mouse has come from selection experiments, rather than, for example, the analysis of data on collateral relatives. In many experiments selection has been aimed at changing body composition (e.g., fat%) or simply increasing certain body components like lean or fat amount (Table 4). The results show that selection can easily shift the population mean substantially, indicating that there

Table 4. Selection experiments to change body composition

Trait	Selection criterion	Pop^n size (*reps.*)	Select^n age (d)	Duration (G)	Response	h^{2*} ± SE	Ref.
1 Fat%	GFP/BW indicative for total body fat percentage	N_e = 60 15P (2)	84	10	H 2.1% C 1.0% L 0. 6%	0.66 ± 0.05	78
2 Fat%	GFP/BW to G20 then predicted fat from dry matter/BW	16P then 8P (*3* then *1*)	70/98	11/ > 60	G11: H 36% > C > L 44% G 65: H = 22%, L = 4%	0.43 ± 0.03	17
3 Fat%	Index: GFPW holding BW constant	15P (2)	84	8	IU: L 40 > C > H 60 GFP: L = C < H 160	0.42 ± 0.05	79
4 Lean%	Hind carcass weight/BW, indicative for lean tissue content	N_e = 60 15P (2)	84	10	H 13% C 12% L 11%	0.40 ± 0.04	78
5 Lean (g)	Protein in carcass by CA	40–50P (2)	60	13/ 23	G23: H 53% > C > L 6%	0.16 ± 0.01	80, 81
6 Lean (g)	Protein in carcass by CA	N_e = 70 (*1*)	42	70	78% > C (H line only)	0.56	11
7 Lean (g)	BW 70d – (8 GFPW) to G20, then BW 70	16P then 8P (*3* then *1*)	70	11/ > 60	G11: H 27% > C > L 13% G60: BW70 H 51g L 17g	0.54 ± 0.04	17, 65
8 Lean (g)	Weight of hindleg muscle system	22M × 44F (*1*)	84	7	H 11% > C > L 17%	0.41 ± 0.06	25

TBF: total body fat; G: generation; P: pairs; N_e: effective population size; CA: chemical analysis; h^2*: heritability-pooled, based on divergence or as realised h^2; C: control line; GFP, GFPW: gonadal fat pad, GFP weight; BW*X:* body weight at *X* days of age, reps.: replicates.

is large genetic variation for fat and lean traits, reflected in heritabilities of around 0.5 for both. Further details are given elsewhere.[1,26] Divergent selection on fat% has produced mice with ca 22% and 4% of fat (Figure 2). Selection on body weight has generated a 3-fold divergence in body weight between selected lines and has increased the protein amount by nearly 80%, while the body composition of such lines has not changed appreciably (Table 4-7). As expected if many genes influence the traits, the total response also depends on the length of the experiment, and the highest deviations were observed in the longest experiments (Table 4-2; 6, 7).

3.5. *Correlated Effects of Growth Selection on Fatness*

Direct selection on body composition (Table 4) is costly and labour-intensive, requiring rapidly available measurements on selected traits because of the short generation interval in mice. Growth has therefore been the main target of selection in mice, with estimates of correlated responses in body composition traits then being made on the selected lines and providing most of the available information on pleiotropy of growth and composition traits.[82] There have been many studies of the indirect effects of selection for high body weight on fatness,[82–85] some of which are described in Table 5. Selection for low body weight has been practiced less often and, as expected, usually has the opposite effect on fatness compared to selection for increased growth. Mice from small lines are often leaner,[86] but changes are usually smaller and harder to detect.

There is some confusion in the literature about the effect of growth selection on body composition. As body weight is a composite trait of fat and non-fat tissue in a simple two-compartment model, selection might favour animals containing a higher fat amount or percentage if the selection were just on total body weight. However, correlated response varied considerably in magnitude from no change to high degrees of fatness, which usually can be reached only by direct selection for fatness. The latter can be seen in, for example, the M16, the G-line, the BW56, and the DU-6 (Table 5-5; 7, 8; 12, 13; 14; 19, respectively), with animals reaching around 20% fat, close to the direct response in the F-line (Table 4-2). This variation in the magnitude probably indicates mainly strain differences (e.g., Table 5-1, 2, 3-9), but also genetic sampling or drift, as shown by between replicate variation (e.g., Table 5-9). Indeed the first two selection experiments for growth used the same selection trait (body weight at 60 days, BW60) but different base populations and produced a different correlated response (Table 5-1): Goodale's large strain did not become fat, but MacArthur's did. The influence of the selection trait also cannot be ruled out, with increased fatness following selection for BW42 but not selection for post-weaning weight gain (BWG21-42) (Table 5-2).

Fowler[87] suggested that age of selection had an effect, arguing that if selection was on body weight before the mice began to lay down substantial amounts of fat, the selection response in body weight or gain would consist mostly of lean tissue. In a small short-term experiment, however, a correlated effect in fatness at 42 d was found in the line selected on body weight at 21 d but not in lines selected at later ages.[23] Clarke[88] investigated changes in body composition in Falconer's[89] replicated weight selected lines. Clarke's main conclusion, paraphrased by Roberts,[84] was: When you select for growth up to a certain age, there is little effect on fatness up to the age of selection, but at later ages large mice become progressively fatter. These concepts were extended by Hayes and McCarthy,[90] who selected one line on high BW35 and another on high BW70. Both lines increased in fatness at 70 d and later ages, the line selected on BW35 by slightly more. At 35 d however the BW70 line was leaner than BW35 and even the control (Table 5-9). They argued that in young growing animals the variation in growth is mainly due to variation in food intake and that animals selected at early ages simply eat more. At later ages there is an increasing proportion of fat aggregation and more variation in partitioning, but as fat is energetically more dense than muscle, leaner animals are more efficient and grow more rapidly. Therefore, selection of animals at a late age favors leanness, whereas selection at a young age promotes high food intake and fat deposition.

Although these models contribute to our understanding, the overall picture of the correlated effects of growth selection on fatness is still clouded. This is perhaps not too surprising as these experiments differed in many ways: base populations, selection criteria, age of selection, magnitude of direct response, method of assessing fatness together with age of assessment, with or without prior fasting, sex, food composition, and simply in sampling (small effective population size, few or no replicates). Many studies involved assessment of fat at a single or few age(s) only, which is inadequate for a trait that changes with age and development.[90] Selection also had an impact on the pattern of fat aggregation, although not consistently between experiments (Table 5-11 cf 5-14). Although body weight and protein amount are phenotypically and genetically highly correlated (Table 1), direct selection on protein amount does not lead to an increased fatness, whereas purely selecting on BW42 does (Table 5-19). After the age of selection at 42 d, the DU-6P line also becomes increasingly fat.[11] The notion that selection for protein may avoid negative correlated effects in fatness, at least up to a certain age, is supported by the lean index selection (Table 5-18), where the high and low lines do not differ in fat%. Similar results were found for the Berlin high protein line,[81] but as it is fixed for a myostatin mutation known to reduce fatness (Table 3),[60] the leanness might be confounded with the mutation effect. On a fat-free basis, percent protein, ash and water have not been changed by selection for growth,[91,92] suggesting that it might be difficult to change the chemical composition of the fat-free body.

In summary, these results on correlated responses confirm the existence of substantial genetic variation for total fat content and for the distribution of fat.

3.6. *Importance of Fat Distribution*

In human medicine and in livestock breeding the distribution of fat in the body is important, but mechanisms controlling fat distribution are little understood. In humans the body mass index (BMI), which predicts overall fatness, has usually been the primary variable measured and used to link obesity with several of its co-morbidities, including type 2 diabetes, hypertension, osteoarthritis, and cancer. For example, it was found that individuals with a BMI of 30 have a 27-fold higher risk of developing diabetes than those with a BMI of 20.[93] As early as 1947, however, Vague[94,95] noted sex differences in fat distribution and a higher association between certain fat depots and the risk of metabolic disease; and it is now well established that in the risk assessment for developing certain metabolic diseases, the distribution of body fat is more important than overall fatness.[115,116] Even in subjects within the normal BMI range of 20 to 25, there is an inverse relationship between insulin sensitivity and abdominal adipose tissue.[93] Little is known about the factors, including underlying genes, that control where fat is deposited or about the distinguishing features and effects of fat in different locations and its regulation. Recent studies indicate that the fat topography within depots is also important for some metabolic disorders like insulin resistance,[117] and more information is emerging about the biochemical properties that distinguish adipocytes of different origin and about site-related gene expression in fat (e.g.[115,118]). All of this information has given new impetus to research in obesity, focusing on individual fat depots, their regulation and its genetic basis in humans.

Table 5. Change in fatness as correlated effect of selection for growth (modified from [96])

No.	Selection trait. Generations of selection (G)	*Age* (d). Body weight (g)	Correlated resp. in fatness (fat%) at *Age* (d)	CR	*Ages* (d)	Notes on the fat traits	Rf
1	BW60 Goodale's L G43	*42* c29	Sum of fat pads =	0	*140*	Visual; later	97,
	BW60 MacArthur's L G26	c29	15% of BW	+		dissected fat	98,
	BW42 cross G8	c32	(vs. 19% in small	+		depots in	99,
			line)			inbred ex	100
						Goodale's L	
						vs. small line	
2		*42* *84*	*42* *84*		*42*,	Carcass	87,
	BW42 (**NL**) G25-30	30 39	15.1 27.6	+	*63*,	Carcass =	101
	C (**NC**) G11-12	21 27	9.1 17.7		*84*	body – fur	102
	BWG21-42 (**CL**) G15-19	30 43	9.0 15.0	0			
						F	
3		*42* M F	M *42* F I M *56* F		*21–*	Whole body	103
	BW42 (**H$_6$**)	22 19	6.3 8.0 I 10.0 7.5	0	*56*	M & F	
	C (**C$_2$**) G31-36	20 17	6.3 9.5 I 10.0 9.5				
4		*57* M F	M F		*57*	Carcass, =	92
	BWG21-42 (**M16**)	36 30	9.0 7.3	+		body – head,	
	C G9	32 27	6.6 6.1			fur, feet, tail,	
						GI tract	
5		*49*	*49* *98*		*49*,	See [104]	105
	BWG21-42 (**M16**)	c39	ns **15.5**	+	*98*	M & F	104
	C G9	c28	8.3				
	BW42 (**H$_6$**)	27					
	C (C$_2$) G30	22	ns	0			
6		*42* M F				Body – GI	106
	BWG21-42 (**M16**) G30	39 36	4.9 % > C	+	42	tract content	
	BW42 (**W T**) G12-13	35 31	1.5 % > C	+		M & F	
	C	28 26					
7		*42*	*21* *42* *63*		*21*;	See [106]	106
	BWG21-42 (**M16**) G37	c38	**15** **19.3** **24.3**	+	*42*;		107
	BW42 (**H$_6$**)	29	**12** **14.8** 13.6	+,0	*63*	F	
	C (**ICR**)	27	12.4 10.9 13.1				
	C (**C$_2$**)	23	14.4 11.4 11.5				
8	BWG21-42 (**M16**)	*70* 44	18.4	+	*70*	Carcass	108
	C G36	29	11.1				
	--------------------------	-----------------	-----------------------			M & F	
	BW42 (**H$_6$**) G73	32	11.0	0			
	C (**C$_2$**)	23	11.2				
9	BWG28-76	R1 b = 1.08	b = 1.01	+	*112*;	Whole body	109
	(3 replicates, R) 10G	R2 b = 0.93	b = 0.23	0	*150*	M & F	
	C	R3 b = 1.15	b = 1.89	+			

(*Table 5. Continued*)

No.	Selection trait. Generations of selection (G)	*Age* (d). Body weight (g)	Correlated resp. in fatness (fat%) at *Age* (d)	CR	*Ages* (d)	Notes on the fat traits	Rf
10	 BW21 (**T**) BW31 (**F**) BW42 (**S**) 5G C	*42* 28 25 25 24	Abdominal fat(g) *42* 0.30 0.19 0.20 0.17	+ 0 0	*42*	Abdominal fat pad M & F	
11	 BW35 (**H5**) BW70 (**H10**) 15G C (**QC**)	*35 70 147* 31 40 46 27 38 45 20 27 32	*35 70 147* 12.4 **16.8 20.7** **10.5 15.6 18.2** 12.1 12.6 14.7	 + +	*35*; *70*; *147*	Carcass, = body – tail, feet, fur, GI tract, M & F	90, 110
12	 BWG21-42 (**G**) G14 C	*0 42 84* 1.3 24 39 1.3 17 24	*0 42 84* 4.7 11.8 20.3 4.0 8.3 14.7	 +	*0–84*	Carcass, = body – some organs; M	91
13	BWG21-42 (**G**) C G31	*70* 42 23	23.4 9.4	+	*70*	See [91]	111
14	 BWG21-29 (**RGH**) BW56 (**W56H**) C	*56* M F 31 23 c40 38 28 21	*56* M F 13.0 13.4 18.0 19.6 12.8 10.8	 0 +	a) *56* b) *21–105*	Whole body	86
15	 BW56 (H) G14 C G26	*56* G14 | G26 40 | 41 24 | 31	from BW 29g (10% fat) onwards H > C	+	*35–84*; *12–36*	Whole body M & F	112
16	G14 BW21 (**WWL**) (BWG21-42)/21 (**ADGL**) C	*21 42* 10.3 28.2 9.6 33.2 7.7 23.6	*21 42 56* 3.8 6.1 6.6 3.5 6.1 6.9 3.9 5.9 6.4	 0 0	*21*, *42*, *56*	Whole body M	113
17	 G21-42 C	*42* M F 29 24 23 19	*42* M F 4.6 **5.7** 4.7 5.0	 +	*42*	Whole body M & F	114
18	G7/cG60 Lean mass/BW70, **EDH** C	*70* M G7 35 cG60 54 29	 G7 11.6 | cG60 5.0 11.5 | N/A	 0	*70*	M	17, 65
19	 BW42 (**DU-6**) G32 Protein at 42d (**Du-6P**) C (**DU-Ks**)	*42* M 51 46 32	*42 80* **13.6 22.3** 9.0 **14.9** 9.3 11.4	 + 0,+	*0–120*	Whole body – GI tract M	11

Rf: Reference; G: generation; C: control, GI: gastro-intestinal tract, BW*X*, BWG*X-Y*: body weight at *X* days, gain between ages *X* and *Y;* CR: correlated response with + indicates significant increased fatness and 0 no significant change; not all ages are shown; column 2: coding in bold = orig. line names.

In meat-producing livestock, fat is generally an undesired by-product, but not all fat depots can be trimmed off in abattoirs, and so the value of a carcass depends both on the total amount of fat and its distribution between fat depots.[119] Although there is some evidence of a genetic basis for fat distribution from breed differences in different farm animal species,[120–122] the genetic basis of fat partitioning into different depots is not well understood.

Selection experiments in mice provide further evidence that selection on single fat depots is possible (Table 4) and moreover that the distribution of fat in the body can be changed. For example, in the Edinburgh divergent selection experiment on fatness, selection was on gonadal fat pad (GFP) weight as a proportion of body weight for the first 20 generations. The criterion put more selection pressure on this single fat depot than on total fat% and relatively more fat was laid down in the high line and less in the low line. For example, 9.6 and 19.6% of the total fat at 98 d was deposited as GFP in the Lean and Fat lines respectively,[64] data supported by other studies. A study on males in an earlier generation over an age range of 4 to 26 wk showed, however, that the relative weights of GFP and two other fat depots (hindleg and shoulder) had not changed, indicating that the growth of these fat depots is regulated by the same set of genes.[123]

The alteration of the proportion of the GFP in relation to total fat occurred systematically over the first 20 generations (Figure 2. A, B). At the start of this experiment (Table 4-2[17]) it was found that the GFP comprises ca 12.5% of the total fat in the base population, so the generation means for the GFPW are expressed as 8*GFPW/BW in Figure 2. A, B for the first period of the experiment together with the predicted fat% from dry matter[18] from G20 onwards. A fit of a simple exponential model[124] for the first 20 generations and one for subsequent generations shows this prediction of fat% from GFPW works well only in early generations of selection. Subsequently, total fat% is increasingly overestimated in the Fat line and underestimated in the Lean line, presumably because selection on one fat depot, albeit a major one, substantially changed the proportion of total body fat in this depot. Similar observations were made by other groups.[119,125] In conclusion, although there are high positive genetic correlations between various fat depots, there is genetic variation in the distribution between depots and the age pattern of this distribution. This indicates that there must be both a systemic control and a local control of lipogenesis and/or lipolysis at different sites of fat deposition. Consequently, genes responsible for some of the reported fat QTL might act at a systemic level, whereas genes for others mainly affect specific fat depots.

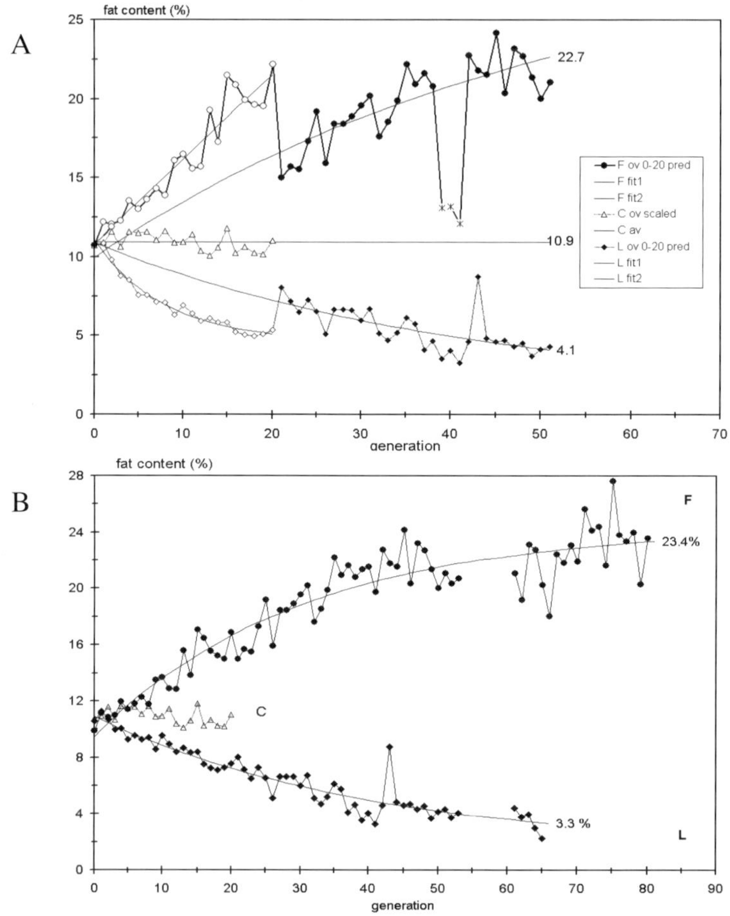

Figure 2. Change of fat content (%) over generations in the F and L line in the Edinburgh experiment.

A: Fat% in period 1 (G0-20) predicted from uncorrected GFPW Fat% = 800*GFPW/BW). Period 2 (after 20G): Fat% predicted from dry matter. Exponential model fitted separately for each period of the experiment (excluding G39-41, where cage effect)[126]

B: Fat% in period 1 predicted from GFPW and corrected for changed proportion of GFP to total body fat using the ratio of exponential curves fitted separately for the two periods. (No observations in G54-60. Mean of G39-41 corrected to expected value from exponential model.)

4. Genetic Basis of Metabolic Rate

As food intake (FI) is related to growth and body composition and is a major cost of animal production, an understanding of its genetic basis is important in livestock breeding programmes as well as having implications for human obesity. Much research has been focused on FI and on feed efficiency of growing animals, which is greatly affected by growth rate, and less on the FI of mature animals, which is closely related to metabolic rate (MR). As the mouse is small, its food intake is high relative to its body weight, and only a small proportion of the food consumed is retained as body tissue even in young animals. The mouse is therefore particularly suitable for studies on the genetic basis and interrelations of FI and MR. Early studies involving measurements of resting and total MR on a series of mouse strains (7 inbred and 4 randombred) at various ambient temperatures (26°C to 36°C) showed that different genotypes had significantly different rates of O_2 consumption,[127] indicative of genetic variation in MR, with further support coming form a similar study involving inbred lines[128] and from selection experiments. For example, selection on high BWG21-42 resulted in a growth difference of 19% and 32% in ca 200-day-old mice, which was accompanied by an 8.5% decrease in MR.[129]

There have been only a few selection experiments aimed directly at a change of the metabolic rate (Table 6), but they show clearly that selection can change MR substantially and therefore that it exhibits genetic variation. In an experiment in Edinburgh (Table 6-1), selection was based on corrected food intake in mature animals (FI 8–10 weeks of age, corrected for mean 8 wk and 10 wk BW), which is highly correlated with MR and is easy and cheap to measure. In another experiment in Lincoln, Nebraska (Table 6-2), selection was directly for heat loss; in a more recently initiated experiment in Białystok, Poland, selection was for MR corrected for BW. These experiments gave very similar patterns of direct and correlated responses. The divergent selection for corrected FI resulted in a substantial high-low divergence of 47% in FI, with only a small divergence in body weight (14%). The High animals also had a 34% higher MR (corrected for body weight), were leaner, more active, had a lower nest building score and had a stronger "hunger drive." There were no differences in gross digestibility between the lines.[130]

In a subsequent test, the elevated resting MR was confirmed, and liver mass was found to be the morphological trait most highly associated with the differences between the lines in resting MR. Length, but not dry mass, of the small intestine and thermal conductance were also greater in the high-intake line.[131,132]

Table 6. Selection for metabolic rate

	Selection trait	Pop. Size (*reps.*)	AoS	Duration	Response	$h^2 \pm SE$	Ref.
1	Adj.FI_F = FI – avBW, adj.FI_M = FI – 1.4avBW	8 P till G23 (*3*) 12 P, G > 23 (*1*)	56–70d	23/38	G21-3: H/L = 1.8 G36-8: H 38% > C > L 39%	0.35 ± 0.05	130, 133
2	Heat loss kcal/kg$^{0.75}$/d over 15 h in males	16 P (*3*)	63–77d	15	H/L = 1.7 H 34% > C > L 20%	0.26 ± 0.01	134
3	MR corrected for BW	15–20 P (*1*)	84–126d	18–19	H/L = 1.18	0.38	135, 136

FI: Food intake; avBW: average BW between 56 and 70d; AoS: age of selection; Duration in generations; reps: replicates; Ref.: references. For further explanation see Tables 4 and 5. Some more h^2 estimates on other rodent species were given by Konarzewski (2004).[136]

Selection directly on heat loss in the experiment in Nebraska produced a difference of 37% between high and low lines (Table 6-2). The high animals had a higher energy intake, were leaner, had a higher core temperature and were much more active.[128,137–139] Estimated differences between divergent lines in feed intake and in heat loss due to locomotor activity were 36% and 11.5%, respectively, showing that variation in activity contributed to variation in FI.[140]

Following selection for MR corrected for BW in the Polish experiment (Table 6-3) there was a response in MR, with high line mice having a similar body weight, higher liver, kidney, small intestine and heart weights, higher food intake, lower VO_{2max}, a higher magnitude of hypothermia elicited by forced swimming, and a smaller brown adipose tissue (BAT) than low line mice.[135] This confirmed and complemented the results of the other two experiments by expanding the range of measured correlated traits.

More detailed genetic analyses have been undertaken on the Nebraska lines. A QTL analysis using F_2 crosses between the selected line and an inbred line showed evidence for significant QTL influencing heat loss, estimated by direct calorimetry, on chromosomes 1, 2, 3 and 7.[141] In both the hypothalamus and BAT, the ribosomal protein *L3* gene was expressed at higher levels in the low line, suggesting the gene has a role in regulation of energy balance; an unknown expressed sequence tag (EST) was also found at higher levels in the hypothalamus of this line of mice.[138]

5. Conclusions

There is substantial genetic variability for body composition, notably fat, as evidenced by single mutant genes, QTL mapping and selection experiments, both those selecting directly on fat or lean traits and indirectly on growth traits. Even though many QTL have been identified for fatness, few have yet led to specific genes. Those identified contribute only a small part of the genetic variance. Overall, little is yet known about the genes involved, but they are likely to be many and of varying effect.

A base value for total body fat in the mature mouse of around 10% can be increased by direct selection to well over 20% or reduced to about 4%. Fat% can also be changed greatly as a correlated response to selection for increased growth, with lines reaching almost 20% fat. Strain differences in the response to high fat diets are known. Introducing either leptin or leptin receptor deficiencies by marker assisted introgression increased fatness further to 30–40% in males and about 40% in females.

The distribution of fat among depots has been changed by selection, indicating that its partition is under partial genetic control. Further selection experiments might be required to fulfill the need for mouse models that have substantial changes only in specified fat depots for use in medicine, as visceral adiposity is an independent risk factor affecting morbidity and mortality in humans. The relation between fatness (total and individual fat depots), age and weight is complicated and needs more systematic studies. New scanning and imaging technologies should help progress in the genetics of body composition, if they can accurately and economically quantify single fat depots on large numbers of live animals. Refining the fat phenotypes using developing technologies to quantify and visualise fat depots and organs *in vivo* and over different stages of development, coupled with high-throughput genotyping, should soon further our understanding of the central and local fat regulation systems and the genes involved.

Food intake can be increased by selection by over 40% and energy intake by up to at least 27%, depending on the diet. These genetic differences cannot be accounted for solely by energy accumulation for growth, particularly energy dense fat, or expenditure on production. Mice with a higher metabolic rate are substantially leaner, making them a good subject for the identification of genes allowing a high intake without increase in fatness. Little information is available yet on the genetics of metabolic rate at the individual gene level.

Selected lines are likely to be of value for QTL-mapping studies and gene identification via fine mapping, gene expression studies and candidate gene approaches. Their value will probably increase further as more is learnt about epistatic interactions and the role of modifying genes to explain why major genes or mutations have different effects on different backgrounds. These selection lines also provide the means to investigate efficiently the metabolic pathways involved in the regulation of energy balance and the deposition and activation of fat to find new therapies and drugs.

The "obesity epidemic" in humans is paralleled by a dramatic increase in the number of scientific and clinical studies on the control of energy homeostasis, and is associated with a rapid and substantive increase in the understanding of molecular pathways, physiological systems and the involved genes underlying the regulation of energy balance. It is now recognised that there are many central and peripheral factors involved in energy homeostasis. The developing molecular and physiological insights into this system offer powerful possibilities for future development of successful therapies.

Acknowledgements

We are grateful to Nik Morton, Chris Kenyon, Simon Horvat, Marek Konarzewski and Gene Eisen for information and helpful comments. LB is grateful to SEERAD for financial support.

References

1. Eisen E. J. 1989. Selection experiments for body composition in mice and rats: a review. *Livest. Prod. Sci.* **23**:17–32.
2. Zhang Y., R. Proenca, M. Maffei, M. Barone, L. Leopold and J. M. Friedman. 1994. Positional cloning of the mouse obese gene and its human homologue. *Nature* **372**:425–432.
3. NIH-National Institutes of Health NHLABI 1998. Clinical guidelines on the identification, evaluation, and treatment of overweight and obesity in adults. *NIH Publication* **98–4083**: USA.
4. Frayn K. N. and S. W. Coppack. 1992. Insulin resistance, adipose-tissue and coronary heart-disease. *Clin. Sci.* **82**:1–8.
5. WHO- World Health Organisation. 2001. Controling the global obesity epidemic. *Website,* http://www. *who. int/nut/obs. htm.*
6. Tataranni P. A., I. T. Harper, S. Snitker, A. Del Parigi, B. Vozarova, J. Bunt, C. Bogardus and E. Ravussin. 2003. Body weight gain in free-living Pima Indians: effect of energy intake vs expenditure. *Int. J. Obes. Rel. Disorders* **27**:1578–1583.

7. Ravussin E., S. Lillioja, W. C. Knowler, L. Christin, D. Freymond, W. G. Abbott, V. Boyce, B. V. Howard and C. Bogardus. 1988. Reduced rate of energy expenditure as a risk factor for body-weight gain. *N. Engl. J. Med.* **318**:467–472.
8. Ravussin E. and J. F. Gautier. 1999. Metabolic predictors of weight gain. *Int. J. Obes. rel. disorders* **23 Suppl 1**:37–41.
9. Roberts S. B., D. S. Krupa and E. Saltzman. 2004. Energy expenditure in obesity. *Am. J. Clin. Nutr.* **79**:181–182.
10. Saltzman E. and S. B. Roberts. 1995. The role of energy expenditure in energy regulation: findings from a decade of research. *Nutr. Rev.* **53**:209–220.
11. Bünger L., U. Renne, G. Dietl and S. Kuhla. 1998. Long-term selection for protein amount over 70 generations in mice. *Genet. Res.* **72**:93–109.
12. Cheek D. B. and A. B. Holt. 1963. Growth and body composition of the mouse. *Am. J. Physiol.* **295**:913.
13. Bailey C. B., W. D. Kitts and A. J. Wood. 1960. Changes in the gross chemical composition of the mouse during growth to the assesment of physiological age. *Can. J. Anim. Sci.* **40**:143–155.
14. Gestrich R. 1972. Untersuchungen zur Gewichtsentwicklung und Körperzusammensetzung der Maus. *PhDThesis, Freie Universiät Berlin.*
15. Renne U., M. Langhammer, E. Wytrwat, G. Dietl and L. Bünger. 2003. Genetic-statistical analysis of growth in selected and unselected mouse lines. *J. Exp. Anim. Sci.* **42**:218–232.
16. Reynolds, D. S., Kunz, T. H. 2001. Standard methods for destructive body composition analysis. In: *Body Composition Analysis of Animals-A Handbook of Non-destructive methods* , ed. J. R. Speakman. pp. 39–55. Cambridge University Press. Cambridge.
17. Sharp G. L., W. G. Hill and A. Robertson. 1984. Effects of selection on growth, body composition and food intake in mice. I. Responses in selected traits. *Genet. Res.* **43**:75–92.
18. Hastings I. M. and W. G. Hill. 1989. A note on the effect of different selection criteria on carcass composition in mice. *Anim. Prod.* **48**:229–233.
19. Speakman, J. R. 2001. *Body Composition Analysis of Animals-A Handbook of Non-destructive Methods*. Cambridge University Press, Cambridge.
20. Rogers P. and G. P. Webb. 1980. Estimation of body fat in normal and obese mice. *Br. J. Nutr.* **43**:83–86.
21. Eisen E. J. and J. M. Leatherwood. 1978. Adipose cellularity and body composition in polygenic obese mice as influenced by preweaning nutrition. *J. Nutr.* **198**:1652–1662.
22. Bünger L. and W. G. Hill. 1997. Effects of leptin administration on long-term selected fat mice. *Genet. Res.* **69**:215–225.
23. Hull P. 1960. Genetic relations between carcass fat and body weight in mice. *J. Agric. Sci.* **55**: 317–321.
24. Eisen E. J. and J. M. Leatherwood. 1976. Effects of early pregnancy in growth, body composition and efficiency in mice. *J. Anim. Sci.* **42**:52–62.
25. McLellan C. R., Jr. and R. R. Frahm. 1973. Direct and correlated responses to two-way selection for hindleg muscle system weight in mice. *J. Anim. Sci.* **36**:442–451.
26. Eisen E. J. 1992. Restricted index selection in mice designed to change body fat without changing body weight: direct responses. *Theor. Appl. Genet.* **83**:973–980.

27. Brecher G. and S. H. Waxler. 1949. Obesity in Albino Mice Due to Single Injections of Goldthioglucose. *Proc. Soc. Exp. Biol. Med.* **70**:498–501.
28. Rutman R. J., F. S. Lewis and W. D. Bloomer. 1966. Bipiperidyl mustard a new obesifying agent in mouse. *Science* **153**:1000-1002.
29. Bourgeois F., A. Alexiu and D. Lemonnier. 1983. Dietary-induced obesity - effect of dietary fats on adipose- tissue cellularity in mice. *Brit. J. Nutr.* **49**:17–26.
30. York B., K. Lei and D. B. West. 1996. Sensitivity to dietary obesity linked to a locus on chromosome 15 in a CAST/Ei x C57BL/6J F2 intercross. *Mamm. Genome* **7**:677–681.
31. Holder R. B. and W. R. Lamberson. 1996. Selection for improved efficiency of lean gain in mice: Populations and Procedures. *AWIC Newsletter* **6**:14–16.
32. Holder R. B., A. S. Moura and W. R. Lamberson. 1999. Direct and correlated responses to selection for efficiency of lean gain in mice. *J. Anim. Sci.* **77**:575–581.
33. Scott, I., Selman, C., Mitchell, P. I., Evans, P. R. 2001. The use of total body electrical conductivity (TOBEC) to determine body composition vertebrates. In: *Body Composition Aanalysis of Animals-A Handbook of Non-destructive Methods*, ed. J. R. Speakman. pp. 127–160. Cambridge University Press. Cambridge.
34. Anonymous. 2004. PIXImus Research Densitometer. *http://www.gehealthcare.com/rad/bonedens/bdcommunity/mediummouse. html.*
35. Sjogren K., N. Hellberg, Y. Bohlooly, L. Savendahl, M. S. Johansson, T. Berglindh, I. Bosaeus and C. Ohlsson. 2001. Body fat content can be predicted in vivo in mice using a modified dual-energy X-ray absorptiometry technique. *J. Nutr.* **131**:2963–2966.
36. Carroll W. F. and T. R. Nagy. 2001. Precision of dual-energy X-ray absorptiometry for determining body composition of rats. *Obes. Res.* **9**:108S.
37. Nagy, T. R. 2001. The use of dual-energy X-ray absorptiometry for the measurement of body composition. In: *Body Composition Analysis of Animals-A Handbook of Non-destructive Methods* , ed. J. R. Speakman. pp. 211–229. Cambridge University Press. Cambridge.
38. Changani K. K., A. Nicholson, A. White, J. K. Latcham, D. G. Reid and J. C. Clapham. 2003. A longitudinal magnetic resonance imaging (MRI) study of differences in abdominal fat distribution between normal mice, and lean overexpressers of mitochondrial uncoupling protein-3 (UCP-3). *Diabetes Obes. Metab.* **5**:99–105.
39. Chatham J. C. and S. J. Blackband. 2001. Nuclear magnetic resonance spectroscopy and imaging in animal research. *ILAR. J* **42**:189–208.
40. Taicher G. Z., F. C. Tinsley, A. Reiderman and M. L. Heiman. 2003. Quantitative magnetic resonance (QMR) method for bone and whole-body-composition analysis. *Anal. Bioanal. Chem.* **377**:990–1002.
41. Tinsley F. C., G. Z. Taicher and M. L. Heiman. 2004. Evaluation of a quantitative magnetic resonance method for mouse whole body composition analysis. *Obes. Res.* **12**:150–160.
42. Brockmann G. A. and M. R. Bevova. 2002. Using mouse models to dissect the genetics of obesity. *Trends Genet.* **18**:367–376.
43. Nobentrauth K., J. K. Naggert, M. A. North and P. M. Nishina. 1996. A candidate gene for the mouse mutation tubby. *Nature* **380**:534–538.
44. Montague C. T., I. S. Farooqi, J. P. Whitehead, M. A. Soos, H. Rau, N. J. Wareham, C. P. Sewter, J. E. Digby, S. N. Mohammed, J. A. Hurst, C. H. Cheetham, A. R. Earley, A. H. Barnett, J. B. Prins and S. Orahilly. 1997. Congenital leptin deficiency is associated with severe early- onset obesity in humans. *Nature* **387**:903–908.

45. Yeo G. S., I. S. Farooqi, S. Aminian, D. J. Halsall, R. G. Stanhope and S. O'Rahilly. 1998. A frameshift mutation in MC4R associated with dominantly inherited human obesity. *Nat. Genet.* **20**:111–112.

46. Holder J. L., Jr., N. F. Butte and A. R. Zinn. 2000. Profound obesity associated with a balanced translocation that disrupts the SIM1 gene. *Hum. Mol. Genet* **9**:101–108.

47. Farooqi I. S. and S. O'Rahilly. 2000. Recent advances in the genetics of severe childhood obesity. *Arch. Dis. Child* **83**:31–34.

48. Yen T. T., A. M. Gill, L. G. Frigeri, G. S. Barsh and G. L. Wolff. 1994. Obesity, diabetes, and neoplasia in yellow a(Vy)/- Mice - ectopic expression of the agouti gene. *FASEB J.* **8**:479–488.

49. Dinulescu D. M., W. Fan, B. A. Boston, K. McCall, M. L. Lamoreux, K. J. Moore, J. Montagno and R. D. Cone. 1998. Mahogany (mg) stimulates feeding and increases basal metabolic rate independent of its suppression of agouti. *Proc. Nat. Acad. Sci. USA* **95**:12707–12712.

50. Huppi K., D. Siwarski, J. R. Pisegna and S. Wank. 1995. Chromosomal localization of the gastric and brain receptors for cholecystokinin (CCKAR and CCKBR) in human and mouse. *Genomics* **25**:727–729.

51. Bi S., K. A. Scott, A. S. Kopin and T. H. Moran. 2004. Differential roles for cholecystokinin a receptors in energy balance in rats and mice. *Endocrin.* **145**:3873–3880.

52. Naggert J. K., L. D. Fricker, O. Varlamov, P. M. Nishina, Y. Rouille, D. F. Steiner, R. J. Carroll, B. J. Paigen and E. H. Leiter. 1995. Hyperproinsulinaemia in obese fat/fat mice associated with a carboxypeptidase-e mutation which reduces enzyme-activity. *Nat. Genet.* **10**: 135–142.

53. Chua S. C., W. K. Chung, X. S. Wupeng, Y. Y. Zhang, S. M. Liu, L. Tartaglia and R. L. Leibel. 1996. Phenotypes of mouse diabetes and rat fatty due to mutations in the ob (leptin) receptor. *Science* **271**:994–996.

54. Eicher E. M. and W. G. Beamer. 1976. Inherited ateliotic dwarfism in mice. *J. Hered.* **67**:87–91.

55. Lin S. C., C. J. R. Lin, I. Gukovsky, A. J. Lusis, P. E. Sawchenko and M. G. Rosenfeld. 1993. Molecular basis of the little mouse phenotype and implications for cell type-specific growth. *Nature* **364**:208–213.

56. Phan J., M. Peterfy and K. Reue. 2004. Lipin expression preceding peroxisome proliferator-activated receptor-gamma is critical for adipogenesis in vivo and in vitro. *J. Biol. Chem.* **279:**29558–29564.

57. He L., A. G. Eldridge, P. K. Jackson, T. M. Gunn and G. S. Barsh. 2003. Accessory proteins for melanocortin signaling: attractin and mahogunin. *Ann. N. Y. Acad. Sci.* **994**:288–298.

58. Miller K. A., T. M. Gunn, M. M. Carrasquillo, M. L. Lamoreux, D. B. Galbraith and G. S. Barsh. 1997. Genetic studies of the mouse mutations mahogany and mahoganoid. *Genetics* **146**:1407–1415.

59. Szabo G., G. Dallmann, G. Müller, L. Patthy, M. Soller and L. Varga. 1998. A deletion in the myostatin gene causes the compact (Cmpt) hypermuscular mutation in mice. *Mamm. Genome* **9**:671–672.

60. Bünger L., G. Ott, L. Varga, W. Schlote, C. Rehfeldt, U. Renne, J. L. Williams and W. G. Hill. 2004. Marker assisted introgression of the *Compact* mutant *myostatin* allele: $Mstn^{Cmpt-dl1Abc}$ into a mouse line with extreme growth - effects on body composition and muscularity. *Genet. Res.* in press.

61. Robinson S. W., D. M. Dinulescu and R. D. Cone. 2000. Genetic models of obesity and energy balance in the mouse. *Annu. Rev. Genet.* **34**:687–745.

62. Chagnon Y. C., T. Rankinen, E. E. Snyder, S. J. Weisnagel, L. Perusse and C. Bouchard. 2003. The human obesity gene map: the 2002 update. *Obes. Res.* **11**:313–367.

63. Snyder E. E., B. Walts, L. Perusse, Y. C. Chagnon, S. J. Weisnagel, T. Rankinen and C. Bouchard. 2004. The human obesity gene map: the 2003 update. *Obes. Res.* **12**:369–439.

64. Bünger L., J. Forsting, K. L. McDonald, S. Horvat, J. Duncan, S. Hochscheid, W. G. Hill and J. R. Speakman. 2002. Long-term divergent selection in mice indicates regulation system for body fatness that is independent of leptin production and reception . *FASEB J. (on-line publication)* 10.1096/fj.02-0111fje.

65. Bünger L. and W. G. Hill. 1999. Role of growth hormone in the genetic change of mice divergently selected for body weight and fatness. *Genet. Res.* **74**:351–360.

66. Bouchard C., A. Tremblay, J. P. Despres, A. Nadeau, P. J. Lupien, G. Theriault, J. Dussault, S. Moorjani, S. Pinault and G. Fournier. 1990. The response to long-term overfeeding in identical twins. *N. Engl. J. Med.* **322**:1477–1482.

67. Horvat S., L. Bünger, V. M. Falconer, P. Mackay, A. Law, G. Bulfield and P. D. Keightley. 2000. Mapping of obesity QTLs in a cross between mouse lines divergently selected on fat content. *Mamm. Genome* **11**:2–7.

68. York B., K. Lei and D. B. West. 1997. Inherited non-autosomal effects on body fat in F2 mice derived from an AKR/J x SWR/J cross. *Mamm. Genome* **8**:726–730.

69. West D. B., J. Goudey-Lefevre, B. York and G. E. Truett. 1994. Dietary obesity linked to genetic loci on chromosomes 9 and 15 in a polygenic mouse model. *J. Clin. Invest.* **94**:1410–1416.

70. York B. and D. B. West. 1995. Genetics of dietary obesity in a cross between Cast/ei and C57bl/6j mice. *FASEB J.* **9**:A 723.

71. Mehrabian M., P. Z. Wen, J. Fisler, R. C. Davis and A. J. Lusis. 1998. Genetic loci controlling body fat, lipoprotein metabolism, and insulin levels in a multifactorial mouse model. *J. Clin. Invest.* **101**:2485–2496.

72. Glazier A., J. H. Nadeau and T. J. Aitman. 2002. Finding genes that underlie complex traits. *Science* **298**:2345–2349.

73. Pomp D. 1997. Genetic dissection of obesity in polygenic animal models. *Behav. Gen.* **27**: 285–306.

74. Corva P. M. and J. F. Medrano. 2001. Quantitative Trait Loci (QTLs) mapping for growth traits in the mouse: A review. *Genet. Sel. Evol.* **33**:105–132.

75. Rocha J. L., E. J. Eisen, L. D. Van Vleck and D. Pomp. 2004. A large-sample QTL study in mice: I. Growth. *Mamm. Genome* **15**:83–99.

76. Rocha J. L., E. J. Eisen, L. D. Van Vleck and D. Pomp. 2004. A large-sample QTL study in mice: II. Body composition. *Mamm. Genome* **15**:100–113.

77. Stylianou I. M., P. D. Keightley, L. Bünger, M. Clinton, G. Bulfield and S. Horvat. 2004. Genetic complexity of an obesity QTL (*Fob3*) revealed by detailed genetic mapping. *Mamm. Genome* **15**:481.

78. Eisen E. J. 1987. Selection for components related to body composition in mice: direct responses. *Theor. Appl. Genet.* **74**:793–801.
79. Eisen E. J. 1992. Restricted selection index in mice designed to change body-fat without changing body-weight - correlated responses. *Theor. Appl. Genet.* **84**:307–312.
80. Weniger H. J., P. Horst, D. Steinhauf, F. Major, M. Wolf and E. S. Tawfik. 1974. Model experiments on selection for endurance and its relation to growth. Part I. Introduction, methods and preliminary investigations on the basic population (in German). *J. Anim. Breed. Genet.* **91**:265–270.
81. Barkemeyer H. 1984. Indirect effects of divergent selection for protein amount on fitness traits in mice (in German). *PhD Thesis,Technische Universität Berlin.*
82. Eisen E. J. 1974. The laboratory mouse as a mammalian model for the genetics of growth. *Proc. First World Congr. Genet. Appl. Livest. Prod. , Madrid* **I**:467–492.
83. Roberts R. C. 1965. Some contributions of the laboratory mouse to animal breeding research. Part I. *Anim. Breed. Abstr.* **33**:339–353.
84. Roberts R. C. 1979. Side effects of selection for growth in laboratory animals. *Livest. Prod. Sci.* **6**:93–104.
85. McCarthy, J. C. 1980. Morphological and physiological effects of selection for growth rate in mice. In: *Selection Experiments in Laboratory and Domestic Animals* , ed. A. Robertson. pp. 100–109. Commonwealth Agricultural Bureaux. Slough,UK.
86. Bakker H. 1974. Effect of selection for relative growth rate and body weight of mice on rate, composition and efficiency of growth. *Mededelingen-Landbouwhogeschool-Wageningen* **74**: 1–94.
87. Fowler R. E. 1958. The growth and carcass composition of strains of mice selected for large and small body size. *J. Agric. Sci* **51**:137–148.
88. Clarke J. N. 1969. Studies on the genetic control of growth in mice. *PhD Thesis, University of Edinburgh.*
89. Falconer D. S. 1973. Replicated selection for body weight in mice. *Genet. Res.* **22**:291–321.
90. Hayes J. F. and J. C. McCarthy. 1976. The effects of selection at different ages for high and low body weight on the pattern of fat deposition in mice. *Genet. Res.* **27**:389–433.
91. Robinson D. W. and G. E. Bradford. 1969. Cellular response to selection for rapid growth in mice. *Growth* **33**:221–229.
92. Timon V. M., E. J. Eisen and J. M. Leatherwood. 1970. Comparison of ad libitum and restricted feeding of mice selected and unselected for postweaning gain. II. Carcass composition and energetic efficiency. *Genetics* **65**:145–155.
93. Ruderman N., D. Chisholm, X. Pi-Sunyer and S. Schneider. 1998. The metabolically obese, normal-weight individual revisited. *Diabetes* **47**:699–713.
94. Vague J. 1947. La differenciation sexuelle, facteur determinant des formes de l'obesite. *Press Med.* **55**:339–340.
95. Vague J. 1956. The degree of masculine differentiation of obesities: a factor determining predisposition to diabetes, atherosclerosis, gout, and uric calculous disease. *Am. J. Clin. Nutr.* **4**:20–34.

96. Bünger L. 1979. Selection on growth in laboratory mice with special consideration of fitness components (in German). *PhD Thesis, Akademie der Landwirtschaftswissenschaften, Berlin, Germany.*
97. Falconer D. S. and J. W. B. King. 1953. A study of selection limits in the mouse. *J. Genet.* **51**: 561–571.
98. Goodale H. D. 1938. A study of the inheritance of body weight in the albino mouse by selection. *J. Hered.* **29**:101–112.
99. MacArthur J. W. 1949. Selection for small and large body size in the house mouse. *Genetics* **34**: 194–209.
100. Ehrich T. H., J. P. Kenney, T. T. Vaughn, L. S. Pletscher and J. M. Cheverud. 2003. Diet, obesity, and hyperglycemia in LG/J and SM/J mice. *Obes. Res.* **11**:1400–1410.
101. Falconer D. S. 1953. Selection for large and small size in mice. *J. Genet.* **51**: 470–501.
102. Falconer D. S. and A. Robertson. 1956. Selection for environmental variability of body size in mice. *Z. indukt. Abstammungs- u. Vererbungslehre* **87**:385–391.
103. Lang B. J. and J. E. Legates. 1969. Rate, composition and efficiency of growth in mice selected for large and small body weight. *Theor. Appl. Genet.* **39**: 306–314.
104. Romsos D. R. and G. A. Leveille. 1972. Effect of intermittent feeding on body fat content and in vitro lipogenesis in mice. *Nutr. Reports Intl.* **5**:131-138.
105. Romsos D. R. and G. A. Leveille. 1974. Body fat and fatty acid synthesis in five lines of mice selected for growth rate. *Proc. Soc. Exp. Biol. Med.* **145**:591–594.
106. Eisen E. J. and T. Bandy. 1977. Correlated responses in growth and body composition of replicated single-trait and index selected lines of mice. *Theor. Appl. Genet.* **49**:133–144.
107. Eisen E. J., H. Bakker and J. Nagai. 1977. Body composition and energetic efficiency in two lines of mice selected for rapid growth rate and their F1 crosses. *Theor. Appl. Genet.* **49**:21–34.
108. Eisen E. J., J. F. Hayes, C. E. Allen, H. Bakker and J. Nagai. 1978. Cellular characteristics of gonadal fat pads. livers and kidneys in two strains of mice selected for rapid growth. *Growth* **42**:7–25.
109. Biondini P. E., T. M. Sutherland and L. H. Haverland. 1968. Body composition of mice selected for rapid growth rate. *J. Anim. Sci.* **27**:5–12.
110. McCarthy J. C. 1974. Insights into genetic variation in the deposition of fat from selection experiments in mice. *Proc. First World Congr. Genet. Appl. Livest. Prod. , Madrid* **I**:529–531.
111. Meyer H. H. and G. E. Bradford. 1974. Estrus, ovulation rate and body composition of selected strains of mice on ad libitum and restricted feed intake. *J. Anim. Sci.* **38**: 271–278.
112. McPhee C. P. and A. R. Neill. 1976. Changes in the body composition of mice selected for high and low eight week weight. *Theor. Appl. Genet.* **47**:21–26.
113. Brown M. A., R. R. Frahm and R. R. Johnson. 1977. Body composition of mice selected for preweaning and postweaning growth. *J. Anim. Sci.* **45**:18–23.
114. Kownacki M., W. Zielinski and T. Jezierski. 1977. Feed efficiency and body composition of selected and unselected mice. *Theor. Appl. Genet.* **50**:79–184.
115. Montague C. T., J. B. Prins, L. Sanders, J. L. Zhang, C. P. Sewter, J. Digby, C. D. Byrne and S. O'Rahilly. 1998. Depot-related gene expression in human subcutaneous and omental adipocytes. *Diabetes* **47**:1384–1391.

116. Niesler C. U., J. B. Prins, S. O'Rahilly, K. Siddle and C. T. Montague. 2001. Adipose depot-specific expression of cIAP2 in human preadipocytes and modulation of expression by serum factors and TNFalpha. *Int. J. Obesity* **25**:1027–1033.
117. Monzon J. R., R. Basile, S. Heneghan, V. Udupi and A. Green. 2002. Lipolysis in adipocytes isolated from deep and superficial subcutaneous adipose tissue. *Obes. Res.* **10**:266–269.
118. Bertile F., F. Criscuolo, H. Oudart, Y. Le Maho and T. Raclot. 2003. Differences in the expression of lipolytic-related genes in rat white adipose tissues. *Biochem. Biophys. Res. Commun.* **307**:540–546.
119. Allen P. and J. C. McCarthy. 1980. The effects of selection for high and low body weight on the proportion and distribution of fat in mice. *Anim. Prod.* **31**:1–11.
120. Shahin K. A., O. Y. Abdallah and A. R. Shmeis. 1990. Genetic influences on growth and partition of fat between depots and its distribution in fowl carcasses. *Reprod. Nutr. Dev.* **30**: 673–681.
121. Kolstad K. 2001. Fat deposition and distribution measured by computer tomography in three genetic groups of pigs. *Livest. Prod. Sci.* **67**:281–292.
122. Ermias E., A. Yami and J. E. O. Rege. 2002. Fat deposition in tropical sheep as adaptive attribute to periodic feed fluctuation. *J. Anim. Breed. Genet.* **119**:235–246.
123. Hastings I. M., J. Yang and W. G. Hill. 1991. Analysis of lines of mice selected on fat content. 4. Correlated responses in growth and reproduction. *Genet. Res.* **58**:253–259.
124. Bünger L. and G. Herrendörfer. 1994. Analysis of a long-term selection experiment with an exponential model. *J. Anim. Breed. Genet.* **111**:1–13.
125. Eisen E. J. and M. T. Coffey. 1990. Correlated responses in body composition based on selection for different indicator traits in mice. *J. Anim. Sci.* **68**:3557–3562.
126. Hastings I. M. and W. G. Hill. 1993. The effect of cage type on murine body composition. *Mouse Genome* **91**:329–330.
127. Pennycuik P. R. 1967. A comparison of the effects of a variety of factors on the metabolic rate of the mouse. *Aust. J. Exp. Biol. Med. Sci.* **45**:331–346.
128. Moody D. E., D. Pomp and M. K. Nielsen. 1997. Variability in metabolic rate, feed intake and fatness among selection and inbred lines of mice. *Genet. Res.* **70**:225–235.
129. Kownacki M. and J. Keller. 1978. The basal metabolic rate in selected and unselected mice. *Genet. Pol.* **19**:338–344.
130. Bünger L., M. G. Macleod, C. A. Wallace and W. G. Hill. 1998. Direct and correlated effects of selection for food intake corrected for body weight in the adult mouse. *Proc. 6th World Congr. Appl. Livest. Prod.* **26**:97–100.
131. Selman C., T. K. Korhonen, L. Bünger, W. G. Hill and J. R. Speakman. 2001. Thermoregulatory responses of two mouse mus musculus strains selectively bred for high and low food intake. *J. Comp. Physiol. B.* , **171**:661–668.
132. Selman C., S. Lumsden, L. Bünger, W. G. Hill and J. R. Speakman. 2001. Resting metabolic rate and morphology in mice (Mus musculus) selected for high and low food intake. *J. Exp. Biol.* **204**:777–784.
133. Hastings I. M., S. M. Moruppa, L. Bünger and W. G. Hill. 1997. Effects of selection on food intake in the adult mouse. *J. Anim. Breed. Genet.* **114**:419–433.

134. Nielsen M. K., L. D. Jones, B. A. Freking and J. A. DeShazer. 1997. Divergent selection for heat loss in mice .1. selection applied and direct response through fifteen generations. *J. Anim. Sci.* **75**:1461–1468.
135. Ksiazek A., M. Konarzewski and I. B. Lapo. 2004. Anatomic and energetic correlates of divergent selection for BMR in laboratory mice. *Physiol. Biochem. Zool.* in press.
136. Konarzewski M., A. Ksiazek and I. B. Lapo. 2004. Artificial selection on metabolic rates and related traits in rodents. *Integrative and Comparative Biology.* submitted.
137. Nielsen M. K., B. A. Freking, L. D. Jones, S. M. Nelson, T. L. Vorderstrasse and B. A. Hussey. 1997. Divergent selection for heat loss in mice .2. correlated responses in feed intake, body mass, body composition, and number born through fifteen generations. *J. Anim. Sci.* **75**: 1469–1476.
138. Allan M. F., M. K. Nielsen and D. Pomp. 2000. Gene expression in hypothalamus and brown adipose tissue of mice divergently selected for heat loss. *Physiol. Genomics* **3**:149–156.
139. Wesolowski S. R., M. F. Allan, M. K. Nielsen and D. Pomp. 2003. Evaluation of hypothalamic gene expression in mice divergently selected for heat loss. *Physiol. Genomics* **13**:129–137.
140. Mousel M. R., W. W. Stroup and M. K. Nielsen. 2001. Locomotor activity, core body temperature, and circadian rhythms in mice selected for high or low heat loss. *J. Anim. Sci.* **79**: 861–868.
141. Moody D. E., D. Pomp, M. K. Nielsen and L. D. Van Vleck. 1999. Identification of quantitative trait loci influencing traits related to energy balance in selection and inbred lines of mice. *Genetics* **152**:699–711.

CHAPTER 8

GENETICS OF REPRODUCTION

M. K. Nielsen

Department of Animal Science, University of Nebraska,
Lincoln, NE, USA
mnielsen1@unl.edu

1. Introduction

Numerous studies have been conducted to understand genetic variation in aspects of female reproduction in the mouse; studies directed at uncovering genetic variation in components of male reproduction have been less numerous. Because female reproductive performance is so critical to economic performance of all livestock species, understanding genetic control of many female reproductive measures is a high priority. Studies in the mouse have provided considerable basis for further research and development in the target livestock species. Besides its use as a laboratory species in research leading to future livestock applications, the mouse has also served as a laboratory model for many wild species, as well as for humans.

Many scientists have worked with outbred populations, especially with lines selected for increased reproduction. Many of the studies and results that will be cited in this chapter come from selection experiments. Underlying genetic variances and covariances that explain direct and correlated selection responses (these are usually expressed in some standardized statistics like realized heritabilities and genetic correlations) have been estimated. In addition, covariances between various relatives have also been used to derive genetic variances and covariances.

Most reproductive characteristics that have been studied fall in the classification of quantitative. And both genetic and environmental variations contribute to the phenotypic variation observed in a population. Underlying additive and, to a limited extent, epistatic genetic variation are illuminated in our estimates of heritabilities and genetic correlations. Evidence of additive variation is usually the main topic of many studies. Dominance and possible epistatic genetic variation are uncovered in studies of line crosses and evaluation of inbreeding. We will look at all of these in this chapter.

2. Direct Selection for Larger Litter Size

Falconer[1] reported that after 20 generations of within full-sister selection, response reached a plateau at 1.6 pups, and realized heritability was 0.15 up to the point of plateau. Joakimsen and Baker[2] practiced mass selection and reported no plateau of response through 15 generations. The response to that point was about 4 pups and the realized heritability was 0.18; litter size was standardized to 8 pups at birth. Mass selection was also practiced and reported by Eisen;[3] litters were standardized to 8 pups, and after 12 generations, the response was 4 pups and the realized heritability was 0.19.

Bakker and coworkers[4] reported no plateau in response at Generation 29. Response to that point was about 6 pups, and the realized heritability was 0.11; no standardization of litter size for rearing was practiced. The Wageningen experiment has one very interesting and unique aspect. The original selections were made by choosing the top 24 females of approximately one thousand, and then mass selection began. Thus, a large population screening program, analogous to hyper-prolific swine screening and multiple-birth cattle screening, was used to increase intensity for the first generation. Luxford and Beilharz[5] also reported selection for increased first litter size without standardization of litters. Their report only covered 7 generations, and the rate of response was 0.13 per generation with a realized heritability of 0.10.

Bradford[6,7] has described the results from three lines, with litter size standardized to 10 pups in all lines. Mass selection in a population derived from crossing 8 inbred lines showed no plateau at generation 11 after a gain of 3 pups. Realized heritability was 0.22. Mass selection in another line derived from 4 inbred lines was quite good after 15 generations (realized heritability of 0.16). After Generation 16, rate of response tapered off and reached a plateau after 30 generations; total response was 4.3 pups. In another line derived from the 4-line base, Bradford practiced within full-sister selection. There was no plateau at Generation 8 when the line was terminated, and realized heritability, adjusted to a mass selection basis, was 0.25. However, due to working with only the within family variability, response was only 0.8 pup.

Vangen[8,9] gave results from a selection study comparing different levels (4 pups = H4, 8 pups = H8, 12 pups = H12, or no standardization = H1) of standardization of litter size at birth and subsequent effectiveness of selection; the control (K) was standardized at 8 pups. Phenotypic variability and selection differentials were similar in all selection lines. After 10 generations, response in H4 was large (response of about 4 pups and realized heritability of 0.37), similar to other studies in H8 and H12 (response of about 2 pups and realized heritability of 0.20), and non-existent in H1 where no standardization was practiced. However, at the end of 16 generations, realized heritabilities were 0.12 in H4, < 0.0 in H8, 0.16 in H12, and 0.08 in H1. In the long term, there was no clear benefit from standardizing litters to gain more selection response.

Vangen[10] reported on continued selection in the high (H) line developed at Ås and the high litter size B line imported from Wageningen. The H and B were crossed to form a new population (X) for selection. Line H was at Generation 21 and line B at Generation 33 when the study period of 20 more generations began. Litters were standardized to 8 pups in all lines. Little further response was attained in lines H and B; they reached their maximum within 5 generations of extra selection. The X line showed more response and had an average advantage of 2.4 pups over the average of L and B during the first 10 generations of selection, but only 1.1 pups average advantage for the subsequent 10 generations. Realized heritability was only positive for the X selection (0.02); the decrease in performance of the B and L lines gave negative regressions of response on selection applied.

In our laboratory, we[11,12] have also done direct selection for increasing litter size (criterion LS). We standardized litters to 10 pups at birth and practiced selection in three replications with contemporary control lines (criterion LC). Response after 13 generations averaged 1.7 pups with a realized heritability of 0.09. After 21 generations, the response was 3.5 pups, and realized heritability through all generations of selection was 0.10. Selection ceased at Generation 21, and response maintained during the following 6 generations of relaxed selection was higher at 4.1 pups. The response as a deviation from the control line is shown in Figure 1.

Realized heritability in selection for increased litter size has generally been in the range of 0.10 to 0.20. Some have suggested that a negative maternal effect (mice from larger litters are smaller in size and this suppresses size of litters they produce) had clouded expression of genetic variation, explaining some of the difference between heritability estimates in experiments that have or have not standardized litter size for rearing. The work by Vangen[9] disputes the importance of size of rearing group in expression of genetic variation. Whether in standardized litters or not, females selected for breeding are reared under similar competition for nutrients; thus, they would have similar maternal environment within an experiment.

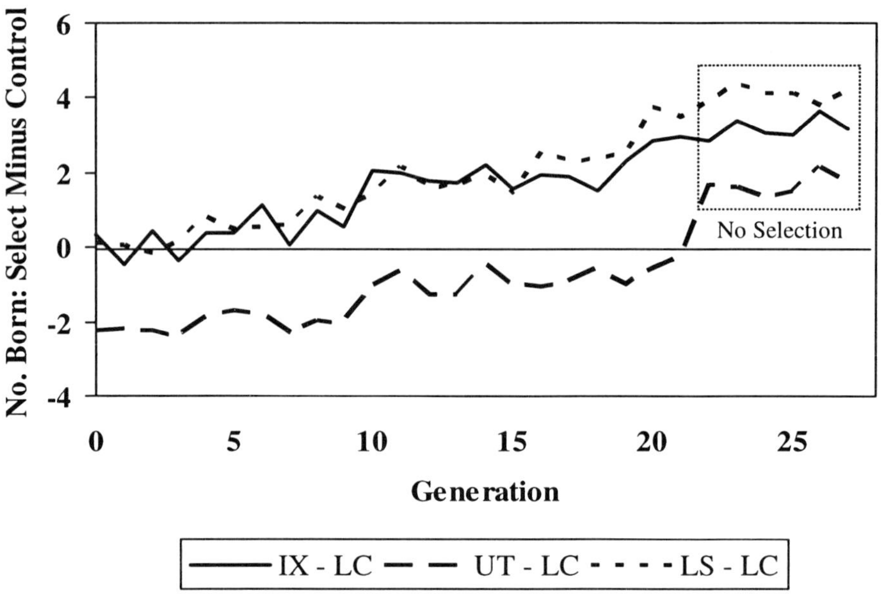

Figure 1. Number born (deviated from control) for the three selection criteria studied at Nebraska[12] over 21 generations of selection followed by 6 generations of relaxed selection. IX = index of ovulation rate and ova success, UT = uterine capacity (no. born to unilaterally ovariectomized females), and LC = unselected control. During the no selection phase, UT line females were left intact.

3. Direct Selection for Smaller Litter Size

Falconer,[1] in a line contemporary to the large litter size selection line, selected for smaller number born within full-sister families. After 20 generations, response reached a decrease of 1.6 pups; realized heritability was 0.40 on a mass-selection basis. Bradford[7] initiated selection for small litter size (CN). Mean performance dropped by 2.3 pups in number born after 17 generations, but then increased so that the final response observed was about 1.5 pups at the plateau. Joakimsen and Baker[2] selected for small litter size in conjunction with their line for large litters. Response reached a plateau for them after 17 generations as well (decrease of about 3 pups), and the realized heritability was 0.22. Continued selection in the low line (L) at Ås[10] produced little additional response, the lowest mean occurring in Generation 28. Thereafter, its performance increased through Generation 41.

4. Selection for Components of Litter Size

Selection for ovulation rate has received more attention than other components of litter size. Land and Falconer[14] practiced high and low selection for ovulation rate. Primiparous females were measured and existing litters were retrospectively selected, reared, and mated. After 12 generations, realized heritability for divergence was 0.31, with response of 5 ova in the high line and a loss of 2 ova in the low line. There was no corresponding response in litter size, which is in disagreement with the following experiments.

Bradford[15] produced two parities from females, and based on the ovulation rates of the first-parity daughters, second-parity litters were selected to become the next generation parents. Realized heritability with upward selection after 11 generations was 0.10, and response was 2.6 ova with no corresponding change in litter size. However, at Generation 15, a 5-ova response had been achieved with a concomitant response of 2 pups in litter size.

Bradford[15] also practiced selection for increasing "embryonic survival." He actually practiced selection on the criterion (number of fetuses)^2/(number of corpora lutea) to avoid putting negative selection on ovulation rate. Like the selection for ovulation rate described above, females produced two litters. Females and males in the second litters were selected based on data from the measurement in the first-time females. After 11 generations of selection, (number of fetuses)/(number of corpora lutea) increased from 0.8 to 0.9, and there was little response in ovulation rate. Consequently, the response in litter size was 2 pups.

Along with selection for increased litter size in our laboratory, we have also done replicated selection[11,12,16] for a linear index (criterion IX) of I = 1.21 * ovulation rate + 9.05 * ova success (proportion of ova shed resulting in fully formed pups) and for "uterine capacity" (criterion UT). Selection in UT was for number born to unilaterally (right excised) ovariectomized females, and the index in IX was the one that would maximize response in litter size using base population parameters. Response in litter size of intact animals with UT selection was 0.8 pups after 13 generations and 1.7 pups during the 6 generations of relaxed selection following the full 21 generations of selection (Figure 1). Realized heritability for uterine capacity as defined here was 0.08.

Selection under criterion IX yielded a response in litter size of 1.9 after 13 generations and 2.9 pups after 21 generations. The response increased to 3.2 pups during the period of relaxed selection. Realized heritability, broadly defined here as the regression of response in litter size on the cumulative selection differential (standard deviation of the index intended to be same as litter size) was 0.12 at Generation 13 but dropped to 0.10 over the 21 generations. Through 13 generations, IX selection produced more response than selection on litter size (LS) as was predicted from the index derived from the base-generation estimates of parameters. However, through the next 8 generations, LS selection was better, resulting in more total response by Generation 21 than the IX selection. We never changed the index, and with changing parameters, the effectiveness of the index became poorer than the natural index of number born.

5. Selection for Testes Mass

High and low selection for testes mass measured at 11 wk to possibly change ovulation rate and hence litter size[17] and at 5 wk as part of a larger project to study selection to change maturing rate[18] has been done in replicated lines at Edinburgh. Direct selection for testes mass at 11 wk had a high realized heritability (0.52) as did selection at 5 wk (0.44). Response after 5 generations in the experiment selecting for testes size at 11 wk was 1.6 ova, but there was no corresponding change in litter size. The difference between the high and low in the experiment selecting at 5 wk was 2.5 in first litter size with a similar difference in ovulation rate. It appears that either selection at a less mature age (5 versus 11 wk) or selection for more generations, thus producing more correlated response in ovulation rate, is needed to show ultimate response in number born.

Adding further to the confusion are the correlated responses in testes mass observed after selection for litter size. Joakimsen and Baker[2] reported significant positive correlated responses in males after sexual maturity (greater testes mass in the high litter selection and the opposite in the low selection). Eisen and Johnson[18] reported a positive genetic correlation (0.60) between testes weight and litter size, and this correlation was only reduced to 0.42 after accounting for variation in body weight. In comparison to our LS and LC selection[19] (difference of about 4 pups in litter size at birth), we found no difference in testes mass at 12 wk of age (after mating).

6. Selection for Multiple-Parity Litter Size

Selection for increased (L+) and decreased (L–) total number born in the first 3 parities (85 d of production) has been practiced and reported through 13 generations in León.[20,21] Litters were standardized to 8 pups, and those from the first parity became the replacements if their dam was selected. After 8 generations, the response was 3.8 pups in L+ and –3.4 pups in L–; realized heritabilities were 0.28 for L+ and 0.21 for L–. Because the characteristic is the sum of three parities, we would expect, barring negative covariances, the heritability to be higher than for a single parity. Response in first-parity litter size was 1.9 in L+ and –2.7 pups in L–. During Generations 9 through 13, selection was less effective, especially in L–, and realized heritabilities through 13 generations were reduced to 0.18 in L+ and 0.06 in L–. Responses for the selection criteria were 5.1 and –2.1 pups in L+ and L–, respectively.

7. Response in Lifetime Reproduction from First-Parity Selection

In the high litter size and control lines after 25 generations of selection in the Wageningen work, Wallinga and Bakker[22] compared 308-d litter production of females under interval (out before pupping and in after weaning) and continuous exposure to males. The high line was superior under interval exposure to males, but the control was better under continuous exposure to males. Performance in the high line dropped with each parity with continuous exposure. The Wageningen and Ås high litter size lines have been studied in a report by Luxford and others;[23] 10-wk-old females were exposed continuously to males for 240 d. Mean litter size in the two selected lines was over 3 pups higher than the control, but the control had a shorter interval between successive litters and produced 0.6 more litters. Total number born in this management was not higher in the two selected lines.

8. Limits Observed in Selection for Litter Size

Many experiments have hit plateaus in response near 30 generations of selection for larger litter size.[24,25] Reasons given for cessation of response with selection for larger litter size have been: (a) little additive genetic variation left to apply selection, (b) increased embryonic/fetal mortality due to a negative genetic correlation with ovulation rate that has increased with selection, and (c) magnified negative maternal effect expressed in large litters limiting body size and hence ovulation rate and litter size. Losses in fertility or in variance have not been factors. Pregnancy rates have dropped slightly, less than might be expected with the accumulated inbreeding, but they are offset by the selection for reproduction. Most experiments report rates of near 90%; lines at Lincoln showed pregnancy rates of 93 to 95% through 21 generations. In our three selection lines,[12] variation in number born has increased significantly over generations of selection.

In the long-term selection at Wageningen, including selection for smaller litters in the line that had already reached a plateau following high selection, Buis[25] concluded that additive variation still remained because new selection in the opposite direction yielded significant response or there were segregating recessives affecting some component (embryonic mortality?). The latter, genetic variation present due to recessive alleles for lowering litter size at low but not extinguishable frequencies, seems to be a plausible explanation. This downward response in a line at a plateau following high selection was also seen in earlier work at Davis.[26]

9. Elaboration on Responses in the Nebraska Selection

In addition to the responses in number born through our 21 generations of selection on alternative criteria (IX, UT, LS and LC) that were described above, we have also studied other characteristics during the relaxed selection phase. Differences created by selection have remained consistent during 14 generations of relaxed selection. In Generations 22 and 23, we recorded left- or right-side ovulation rate and uterine capacity in females from all lines that were unilaterally ovariectomized at 4 wk and mated at 9 wk and then measured at 17 d of gestation.[27] Ovulation rate increased 3.0, 2.8 and 0.9 ova in LS, IX and UT relative to LC. Uterine capacity, estimated from an exponential equation, increased 2.2, 0.9 and 1.6 fetuses on the left side and 2.5, 1.9 and 1.3 on the right side for LS, IX and UT compared to LC. Uterine capacity was greatest on the right side compared to the left, and the largest response to selection usually occurred on the right side. Ova success in this unilateral model (defined as number of fetuses at 17 d divided by number of corpora lutea) was clearly higher on the right side as compared to the left side (0.76 versus 0.70 in LC control), and the UT selection exceeded all other criteria in ova success on both sides (0.75 versus 0.67 to 0.71 on left, and 0.81 versus 0.76 to 0.77 on right). We had seen a similar effect for intact animals in a Generation 13 evaluation.[11]

Two experiments were done using Generation 27 animals from all lines, 6 generations after selection ceased. Pre-implantation (d 3.5) embryonic development was investigated in a cooperative project at Stillwater.[28] Selection (IX, LS and UT) increased the average stage of embryonic development on the left side of the uterus and tended to do the same on the right side. There was an increase in the proportion of expanded blastocysts and a concomitant decrease in the proportion of two-cell embryos. The variation in stage of embryonic development was significantly less on the right side in the IX, LS and UT animals compared to the LC; this variation was most pronounced in the UT. This result is consistent with those from an extensive study of follicular development by Spearow[29] at Davis comparing a high litter size line with a control. Selection for litter size increased ovarian weight and rate of follicular growth, thus reducing time to follicular maturity. Increasing rate of follicular development helps explain, at a physiological level, the increase in ovulation rate, which explains much of the responses observed in litter size.

The second experiment,[30] done with Generation 27 animals, measured uterine mass and uterine blood volume at either 3 or 6 d of gestation in females mated at 10 wk of age. Uterine mass and uterine blood volume, whether expressed relative to body size or not, were significantly larger in the three selection groups than in the control. Females derived from IX and LS selection also had greater uterine mass and blood volume than those from UT selection.

Another study[31] was done (Generation 34) to measure pup birth weight and its variation within litter. There were no differences between selection criteria in mean pup weight, even though there were 3.9 more pups per litter in LS compared to LC. The regression of mean pup weight on litter size was only –0.027 g; thus, an increase of 4 pups (37%) from the level of LC would only result in a decrease in mean weight of 0.108 g (–7%). The variation within litters was less in UT compared to IX and LS; however, UT was not different from LC. Thus, differences in stage of development seen in the earlier experiment may affect number born but not variation of those pups reaching term.

10. Modeling Litter Size

The "ovulation rate-potentially viable embryo-uterine capacity model," described and simulated for pigs by Bennett and Leymaster,[32] merited further study in mice. We were in a unique position to do this study, given that we had estimated genetic and phenotypic parameters for left and right-side uterine capacity in addition to ovulation rate[33] and that we had evaluated differences in embryonic survival to 6 d in our selection lines.[34] Uterine capacity is heritable, and the genetic correlation between sides of the uterus is ~0.9.

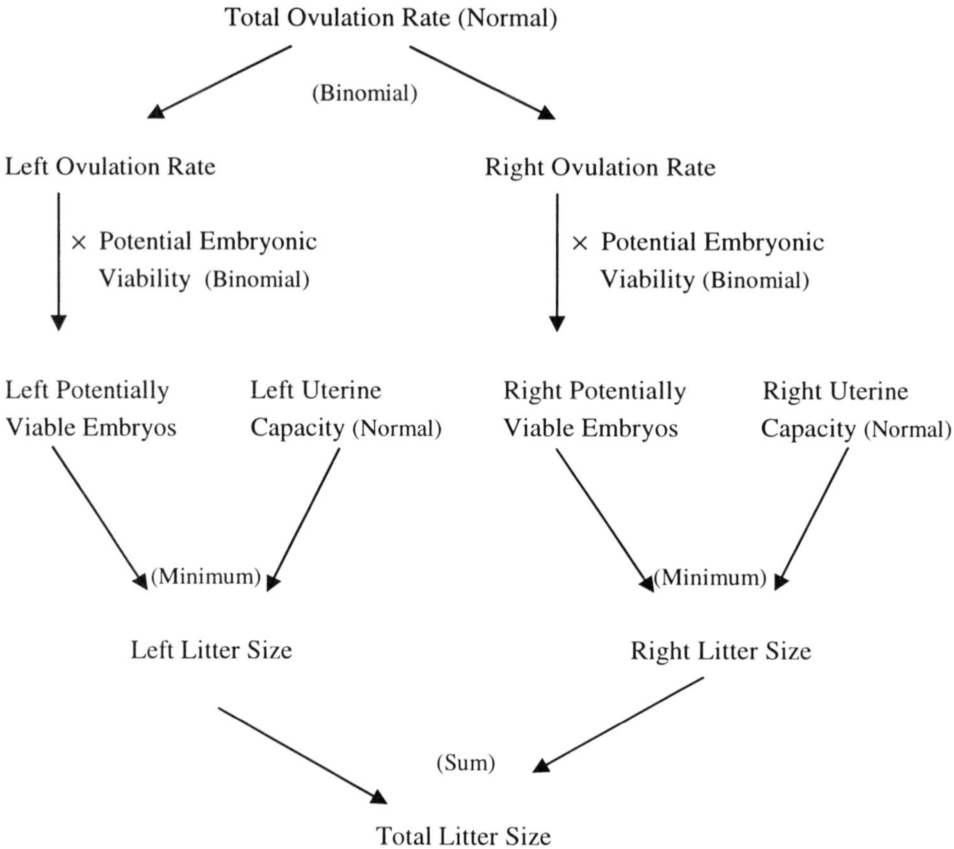

Figure 2. Simulation model for litter size in mice.[35] Ovulation rate is simulated by a normal distribution and is separated into sides by the binomial probability (P = 0.52 for right side). Of these ova, 0.92 is the probability they are potentially viable embryos. Uterine capacity on each side is simulated by normal distributions with different means and standard deviations. The minimum of potentially viable embryos and uterine capacity defines the resulting litter size of each side of the reproductive tract.

We simulated a phenotypic model[35] at Lincoln for litter size that was verified by showing it would mimic phenomena observed in our LS, IX, UT and LC selections. The model is shown in Figure 2. Ovulation rate must be split between left and right sides, and uterine capacity for both sides must also be modeled to account for the duplex uterus in mice versus pigs. Using our observed means and variances for ovulation rate and embryonic survival to implantation, plus parameters for uterine capacity means and variances derived through iteration, we have produced litter size records (left and right sides) that have the means, variances and correlations as observed in our selected groups (Generation 13 data). Genetic parameters for this model that support our empirical selection results are: (a) heritabilities of 0.25 for ovulation rate and 0.065 for left- and right-side uterine capacities,(b) genetic correlation of 0.0 between ovulation rate and the two sides' uterine capacities, and (c) 0.92 between left and right uterine capacity.

Responses to selection for various selection criteria, based on this model and its underlying parameters, have also been modeled and reported by Ribeiro.[36] Response in litter size would be about 24% greater using an index of ovulation rate and total uterine capacity as compared to direct selection on only litter size, but this would be difficult to implement. Selection on an index of ovulation rate and ova success would be about 13% more effective than direct selection on litter size.

11. Litter Size and Body Weight Correlation

Eisen[3] reported a positive genetic (+0.5) correlation between litter size and body weight and demonstrated this in his independent litter size and body weight selection lines. Selection for litter size increased body weight, and selection for body weight increased litter size. Fuente and others[36] have also demonstrated this positive genetic relationship between size of the first three litters and body weight, using their selection lines. Others[38–41] have established the positive relationship between ovulation rate and body weight, and this provides the basis for the relationship between litter size and body weight. Additionally, the group at Raleigh[18] has shown that testis size, like ovulation rate, is related positively to body size.

12. Heterosis in Line Crosses

Working at Wageningen, van der Nieuwenhuizen *et al.*[42] crossed two lines with especially large litter sizes. They observed limited heterosis at birth and at weaning due to offspring genotype. However, heterosis attributed to dam's genotype was great at birth and even greater at weaning for litter size. At Raleigh, Hörstgen-Schwark *et al.*[43] reported similar (~5 to 6%) heterosis for litter size at birth due to both pup and dam genotypes in a diallel study using crosses of five lines. They also found that by weaning, heterosis for litter size had increased to over 7%.[44]

Newman *et al.*[45,46] have reported results from pure-line, first-cross, backcross, and three-way crosses of mice. Litter size at birth was greater in backcross and three-way cross litters as compared to pure-line and first-cross litters. Maternal heterosis (cross-line dam vs. pure-line dam) was an important effect; however, individual heterosis (cross-line vs pure-line pups) had essentially no effect. This same phenomenon was observed in continuity of re-breeding, or longevity of reproductive life, age at first litter (reduced in cross-line dams), and reproductive rate, which was defined as the ratio of number born alive to the number of days elapsed between birth of litters. At weaning, litter size was affected by maternal heterosis and individual heterosis, the latter through greater survival of cross-line pups.

McGloughlin,[47] working at Dublin, produced females with 0, 25, 50, 75 and 100% F_1 heterozygosity for the cross of two lines, and these females were subsequently crossed to an unrelated third line of sire. Heterosis was 12% for dam effects on litter size at birth, and there was a linear increase in heterosis with percent F_1 heterozygosity. Thus, a dominance model adequately explained the phenomenon of heterosis in dams.

13. Inbreeding Depression

Falconer[1] practiced intentional inbreeding (repeated generations of full-sib matings) in 20 family lines. All 20 lines survived to an average inbreeding coefficient of 0.44, but only three lines survived to average inbreeding of 0.76. There was a loss of ~0.5 pups per 10% inbreeding in the dam, and about two-fifths of the loss was attributed to inbreeding in the dam and three-fifths to inbreeding depression in potential offspring. Losses were mainly in pre-implantation with no loss in ovulation rate.

Inbreeding, accumulated after many generations of selection, has been reported from projects at Edinburgh and Davis. Falconer[48] practiced inbreeding and selection in family lines derived from a line selected for large litter size for 42 generations that had reached a plateau in response. After several generations of inbreeding, there was a loss of 1.5 to 2 pups even though selection was still practiced. Eklund and Bradford[26] gave results from a similar situation. In their work, the line selected for larger litter size was at Generation 35, and after several generations of inbreeding with selection, litter size dropped ~0.5 pup.

14. Conception Rate and Fertility

Although conception rates are usually high with outbred populations of mice, changes in fertility may result with extended intense selection. Bradford[49] reported that infertility of matings and length of estrous cycles were 5%, 5.6 d; 25%, 6.3 d; 13%, 6.4 d; 6%, 7.5 d; and 4%, 8.8 d in their embryo survival line, post weaning gain line, litter size line, control line and low litter size line, respectively. By contrast at Raleigh,[43] little difference was found in percentage infertile matings among the lines selected for litter size, 6-wk weight or an antagonistic combination of litter size and 6-wk weight.

15. Conclusions

Much of the insight into genetic variation in reproductive characteristics in mice comes from selection experiments and spin-off research using lines that have a history of selection. Litter size and its components, including ovulation rate, embryonic survival and uterine capacity, have been the focus of most studies. This prominence of research concentrating on litter size is mainly due to the major role of number born per female in economically important species that are raised as livestock to serve man. The major components of variation in litter size are all heritable; hence, litter size is heritable and has responded to selection both for increased as well as for decreased litter size. Response in litter size due to single-parity selection shows lifetime response in numbers of progeny per breeding female. Similar findings were found for selection on multiple-parity performance. Predicting phenotypic expression for litter size by modeling genetic variation in its components has been successful in helping to describe the complex genetic architecture of litter size.

Selection studies for litter size have revealed many aspects of genetic variation. Limits to selection response, changes in a wide array of physiological characteristics, and changes in testis size in males have all been observed and reported. Reductions in conception rate can occur with advancing generations of selection. Evidence of dominance genetic variation in litter size and measures of reproductive fitness (longevity, conception) comes from desirable heterosis observed in line crossing experiments and inbreeding depression resulting from intentional inbreeding.

References

1. Falconer, D.S. 1960. The genetics of litter size in mice. *J. Cell. Comp. Phys.* **56** (1):153–167.
2. Joakimsen, O. and R.L. Baker. 1977. Selection for litter size in mice. *Acta Agric. Scand.* **27**:301–318.
3. Eisen, E.J. 1978. Single-trait and antagonistic index selection for litter size and body weight in mice. *Genetics* **88**:781–811.
4. Bakker, H., J.H. Wallinga and R.D. Politiek. 1978. Reproduction and body weight of mice after long-term selection for larger litter size. *J. Anim. Sci.* **46**:1572–1580.
5. Luxford, B.G. and R.B. Beilharz. 1990. Selection response for litter size at birth and litter weight at weaning in the first parity of mice. *Theor. Appl. Genet.* **80**:625–630.
6. Bradford, G.E. 1968. Selection for litter size in mice in the presence and absence of gonadotropin treatment. *Genetics* **58**:283–295.
7. Bradford, G.E. 1979. Genetic variation in prenatal survival and litter size. *J. Anim. Sci.* **49**:66–74.
8. Vangen, O. 1986. Genetic control of reproduction in pigs from parturition to puberty *Proc. 3rd World Cong. Genet. Appl. Livest. Prod.* **11**:168–179.
9. Vangen, O. 1990. Realized heritabilities for first parity litter size in mice and effects of maternal environments and response in lifetime performance. *Proc. 4th World Cong. Genet. Appl. Livest. Prod.* **13**:341–344.
10. Vangen, O. 1993. Results from 40 generations of divergent selection for litter size in mice. *Livest. Prod. Sci.* **37**:197–211.
11. Gion, J.M., A.C. Clutter and M.K. Nielsen. 1990. Alternative methods of selection for litter size in mice: II. Response to thirteen generations of selection. *J. Anim. Sci.* **68**:3543–3556.
12. Kirby, Yvonne Kochera and M.K. Nielsen. 1993. Alternative methods of selection for litter size in mice: III. Response to twenty-one generations of selection. *J. Anim.* Sci. **71**:571–578.
13. Land, R.B. and D.S. Falconer. 1969. Genetic studies of ovulation rate in the mouse. *Genet. Res.* **13**:25–46.
14. Bradford, G.E. 1969. Genetic control of ovulation rate and embryo survival in mice. I. Response to selection. *Genetics* **61**:905–921.
15. Clutter, A.C., M.K. Nielsen and R.K. Johnson. 1990. Alternative methods of selection for litter size in mice: I. Characterization of base population and development of methods. *J. Anim. Sci.* **68**:3536–3542.
16. Islam, A.B.M., W.G. Hill and R.B. Land. 1976. Ovulation rate of lines of mice selected for testis weight. *Genet. Res.* **27**:23–32.
17. Hill, W.G., P.J. Marks, J.C. Jenkins and R.B. Land. 1990. Selection on testis size as an indicator of maturity in growing animals. II. Correlated responses in reproductive rate. *Genet. Sel. Evol.* **22**:247–255.
18. Eisen, E.J. and B.H. Johnson. 1981. Correlated responses in male reproductive traits in mice selected for litter size and body weight. *Genetics* **99**:513–524.
19. Ribeiro, E.L. de A., R.J. Kittok and M.K. Nielsen. 1994. Serum cholesterol concentration of mice selected for litter size and its relationship to litter size and testis mass. *J. Anim. Sci.* **72**:2943–2947.

20. Fuente, L.F. and F. San Primitivo. 1985. Selection for large and small litter size of the first three litters in mice. *Genet. Sel. Evol.* **17**:251–264.
21. Bayon, Y., L.F. Fuente and F. San Primitivo. 1988. Selection for increased and decreased total number of young born in the first three parities of mice. *Genet. Sel. Evol.* **20**:259–266.
22. Wallinga, J.H. and H. Bakker. 1978. Effect of long-term selection for litter size in mice on lifetime reproduction rate. *J. Anim. Sci.* **46**:1563–1571.
23. Luxford, B.G., R.C Buis, R.G. Beilharz, R.G. 1986. Lifetime reproductive performance of lines of mice after long-term selection for first-parity litter size at birth. *J. Anim. Breed. Genet.* **103**:249–254.
24. Eisen, E.J. 1980. Conclusions from long-term selection experiments with mice. *Z. Tierzűcht. Zűchtungsbiol.* **97**:305–319.
25. Buis, R.C. 1988. Investigation of a selection limit for litter size in mice. *Livest. Prod. Sci.* **20**:161–172.
26. Eklund, J. and G.E. Bradford. 1977. Genetic analysis of a strain of mice plateaued for litter size. *Genetics* **85**:529–542.
27. Clutter, A.C., Y.L. Kochera-Kirby and M.K. Nielsen. 1994. Uterine capacity and ovulation rate in mice selected 21 generations on alternative criteria to increase liter size. *J. Anim. Sci.* **72**:577–583.
28. Al-Shorepy, S. A., A.C. Clutter, R.M. Blair and M.K. Nielsen. 1992. Effects of three methods of selection for litter size in mice on preimplantation embryonic development. *Biol. Reprod.* **46**:958–963.
29. Spearow, J.L. 1986. Changes in the kinetics of follicular growth in response to selection for large litter size in mice. *Biol. Reprod.* **35**:1175–1186.
30. Nielsen, M.K., R.J. Kittok and T.L.K. Kirby. 1995. Uterine mass and uterine blood volume in mice selected 21 generations for alternative criteria to increase litter size. *J. Anim. Sci.* **73**:2243–2248.
31. van Engelen, M.A.J., M.K. Nielsen and E.L. de A. Ribeiro. 1995. Differences in pup birth weight, pup variability within litters, and dam weight of mice selected for alternative criteria to increase litter size. *J. Anim. Sci.* **73**:1948–1953.
32. Bennett, G.L. and K.A. Leymaster. 1989. Genetic implications of a simulation model of litter size in swine based on ovulation rate, potential embryonic viability and uterine capacity: I. Genetic theory. *J. Anim. Sci.* **68**:980–986.
33. Nielsen, M.K., Y.L.K. Kirby and A.C. Clutter. 1996. Estimates of heritabilities and genetic and environmental correlations for left- and right-side uterine capacity and ovulation rate in mice. *J. Anim. Sci.* **74**:529–534.
34. Ribeiro, E.L. de A., M.A.J. van Engelen and M.K. Nielsen. 1996. Embryonal survival to 6 days in mice selected on different criteria for litter size. *J. Anim. Sci.* **74**:610–615.
35. Ribeiro, E.L. de A., M.K. Nielsen, G.L. Bennett and K.A. Leymaster. 1997. A simulation model including ovulation rate, potential embryonic viability, and uterine capacity to explain litter size in mice. I. Model development and implementation. *J. Anim. Sci.* **75**:641–651.
36. Ribeiro, E.L. de A., M.K. Nielsen, K.A. Leymaster and G.L. Bennett. 1997. A simulation model including ovulation rate, potential embryonic viability, and uterine capacity to explain litter size in mice. II. Responses to alternative criteria of selection. *J. Anim. Sci.* **75**:652–656.
37. Fuente, L.F., F. San Primitivo and Y. Bayon. 1986. Genetic correlation between litter size and body weight in mice. *J. Anim. Breed. Genet.* **103**:249–254.

38. Land, R.B. 1970. Genetic and phenotypic relationships between ovulation rate and body weight in the mouse. *Genet. Res.* **15**:171–182.
39. Durrant, B.S., E.J. Eisen and L.C. Ulberg. 1980. Ovulation rate, embryo survival and ovarian sensitivity to gonadotrophins in mice selected for litter size and body weight. *J. Reprod. Fertil.* **59**:329–339.
40. Brien, F.D., G.L. Sharp, W.G. Hill and A. Robertson. 1984. Effects of selection on growth, body composition and food intake in mice. II. Correlated responses in reproduction. *Genet. Res.* **44**:73–85.
41. Bayon, Y., L.F. Fuente and F. San Primitivo. 1986. Effects of selecting for litter size and body weight on the components of litter size in mice. *Livest. Prod. Sci.* **14**:195–203.
42. van der Nieuwenhuizen, J., H. Bakker and R.C. Buis. 1982. Genetic differences in reproduction and growth rate between two lines of mice selected for litter size. *Z. Tierzűcht. Zűchtungsbiol.* **99**:292–302.
43. Hörstgen-Schwark, G., E.J. Eisen, A.M. Saxton and T.R. Bandy. 1984. Reproductive performance in a diallel cross among lines of mice selected for litter size and body weight. *J. Anim. Sci.* **58**:846–862.
44. Eisen, E.J., G. Hörstgen-Schwark, T.R. Bandy and A.M. Saxton. 1984. Postpartum performance in a diallel cross among lines of mice selected for litter size and body weight. *J. Anim. Sci.* **58**:863–877.
45. Newman, S., D.L. Harris and D.P. Doolittle. 1985. Lifetime parental productivity in twenty-seven crosses of mice. I. Birth traits. *J. Anim. Sci.* **61**:358–366.
46. Newman, S., D.L. Harris and D.P. Doolittle. 1985. Lifetime parental productivity in twenty-seven crosses of mice. II. Weaning traits reflecting reproduction and lactation. *J. Anim. Sci.* **61**:367–375.
47. McGloughlin, P. 1980. The relationship between heterozygosity and heterosis in reproductive traits in mice. *Anim. Prod.* **30**:69–77.
48. Falconer, D.S. 1971. Improvement of litter size in a strain of mice at a selection limit. *Genet. Res.* **17**:215–235.
49. Bradford, G.E., M.S. Barkley and J.L. Spearow. 1980. Physiological effects of selection for aspects of efficiency of reproduction. In: *Selection Experiments in Laboratory and Domestic Animals.* ed. A. Robertson. pp.161–175. Commonwealth Ag. Bur., Slough.

CHAPTER 9

GENETICS AND BEHAVIOR

R. J. Hitzemann

Department of Behavioral Neuroscience
Oregon Health & Science University, Portland, OR, USA &
Veterans Affairs Medical Center, Portland, OR, USA
hitzeman@ohsu.edu

"phrases oft repeated finally ossify into conviction and
utterly dull the organs of intuitive perception"

Goethe (1796)

1. Introduction

This chapter focuses on the genetics of complex behavioral traits and emphasizes the role quantitative trait loci (QTL) analysis has played in our understanding of the genetic architecture. The behavioral phenotypes discussed will be those that are believed to have relevance for understanding human behavior, and here the emphasis will be on phenotypes modeling various aspects of psychiatric disorders, including alcohol and substance abuse. Unfortunately, many readers will come to this chapter with a healthy dose of skepticism regarding the integration of quantitative genetics and mouse behavior. Their reasons for this skepticism are easily understood. The number of laboratories actually using the tools of quantitative genetics to investigate mouse behavior is relatively small, and thus, there is the unfamiliarity with the theory, concepts and nuances of behavioral paradigms. Further, accepting the idea that most behaviors are quantitative traits, no different from body weight or blood pressure, has been difficult. Advances made through transgenic technologies have long reinforced a Mendelian view of one gene - one behavior.

Until relatively recently, clinical and preclinical research was focused on finding *the* gene for human diseases such as schizophrenia or depression. It was not until the early 1990s that a more quantitative approach was taken, largely led by mouse geneticists. Here the work of Plomin, McClearn and colleagues was seminal as they demonstrated that the BXD recombinant inbred panel could be used to map QTL for behavioral phenotypes.[1,2] The first QTL study mapping a behavioral trait in a mouse F_2 intercross appeared in 1995;[3] it is of interest to note that the behavior analyzed was open-field behavior, which was originally developed by Hall[4,5] in the 1930s as a test of rodent emotionality that could be used for selective breeding. In 2003, Flint[6] reported more than 125 different behavioral QTL that had been mapped in mice with LOD scores greater than 3. Many of these QTL have been independently confirmed multiple times, e.g., see Belknap and Atkins,[7] and recently we have begun to see some of these QTL turned into quantitative trait genes (QTG).[8,9] The question of whether or not what we learn from mouse behavior will be useful in understanding human behavior will be dealt with elsewhere in this chapter, but here we simply note that there are some promising signs.[10]

Skepticism about the integration of quantitative genetics and mouse behavior also stems from the view in some circles that the genetic effects will be overwhelmed by environmental factors. While the issue of the environment has long been discussed (and a concern of the behavioral geneticist), it was a paper published in 1999 by Crabbe, Whalsten and Dudek,[11] which unfortunately polarized the conversation. The context of this paper is important. Increasing numbers of targeted mutations and the development of large mutagenesis centers strongly argued for a standardized behavioral testing paradigm. The question that was largely unanswered is whether or not standardized testing would produce equivalent results across laboratories. To examine this issue, Crabbe *et al.*[11] decided to examine eight different phenotypes in eight different mouse genotypes in three different laboratories. Every effort was made to equate laboratory environments and breeding history. Despite these attempts for consistency, significant effects of site were found for nearly all variables.

The authors concluded, "It is not clear whether standardization of behavioral assays would markedly improve future replicability of results across laboratories. Standardization will be difficult to achieve because most behaviorists seem to have differing opinions about the 'best' way to assay a behavioral domain." The article stimulated considerable discussion about behavioral measures, including a fair number of misinterpretations.[12] For some the entire enterprise of mouse behavioral genetics came into question. Unfortunately, a point missed by many was the observation that large genetic differences were replicated across laboratories; e.g., open field activity was consistently low in the A/J strain but high in the C57BL/6J strain.[13] The subtle genetic differences were difficult to replicate. On this and related points, the reader is referred to the detailed follow-up article.[13]

Fortunately, there are numerous behavioral examples of where genetic effects appear to trump the gene × environment interaction. A favorite example is illustrated in Figure 1. The phenotype is ethanol-induced locomotor activation, which is measured in a two-step procedure. On day 1 animals are administered saline and on day 2 ethanol (1.5 g/kg); the difference is the ethanol score. The figure shows data that were collected for 22 strains of the BXD recombinant inbred (RI) series in 1983 and in 1999.[14,15] The laboratories were in Portland, OR and Stony Brook, NY; the testing apparatus were completely different, and the animals differed by more than 30 generations. The correlation between the strain means is remarkable at 0.68 ($P < 0.001$). There are other examples of such "long range " correlation but first an introduction to WebQTL.

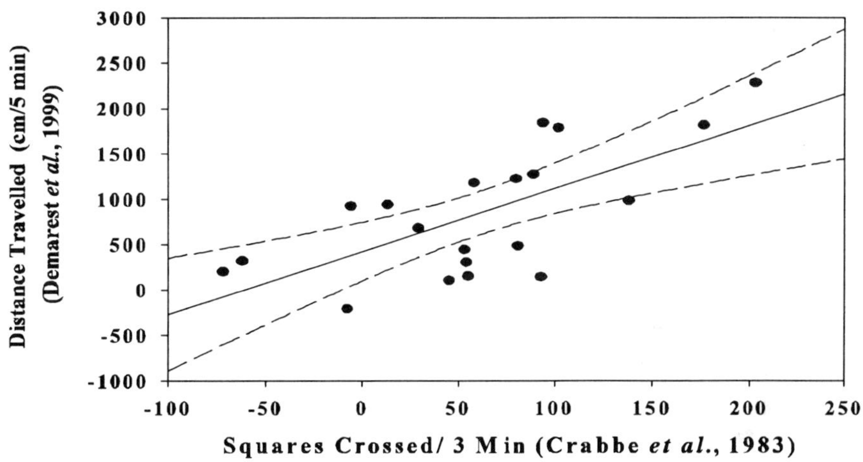

Figure 1. Comparison of two independent measures of the ethanol-induced activation response in the BXD RI panel. Data were taken from Crabbe *et al.*[14] and Demarest *et al.*[15] Note the measurements were made 16 years apart and used entirely different apparatus.

2. WebQTL and Behavioral Phenotypes

WebQTL is an assembly of databases and tools that exploits sophisticated gene mapping and related statistical methods to rapidly perform genome-wide analyses of gene expression and behavioral data.[16] The databases focus on the original BXD RI panel,[17] although data for some of the newer RI strains are available. Important for the immediate discussion is that the database contains over 650 published phenotypes, with the largest single segment focusing on drug related behavioral phenotypes. Although much of the behavioral data was obtained at a single institution,[6] there are some multiple site datasets. The question that needs to be addressed is whether or not the correlation illustrated in Figure 1 is an outlier, or, in fact, do we see trends for behavioral measures to be repeatable across sites. One advantage here is that the number of genotypes (~ 20) is considerably larger than the number tested by Crabbe *et al.*,[11] and thus, there is a marked increase in statistical power.

Two phenotypes will be considered: open-field/basal locomotor activity and ethanol preference. As some measures of basal activity are common to several phenotypes, this is a frequent entry into the database. The database was queried to see what phenotypes were correlated with basal activity after an injection of saline (record ID 9880575-11-BXD); phenotypic details are found elsewhere.[15] The query yielded 82 significant correlations ($P < 0.05$) including significant correlations obtained across seven different laboratories and four different institutions. There are entries for ethanol preference (2-bottle choice) from two different institutions;[18,19] a comparison of records #7978106-7-BXD and #7625571-2-BXD reveals a correlation of $r = 0.82$ ($N = 18$ strains), $P < 0.000007$. The interested reader can continue this exercise for other phenotypes at the WebQTL site (www.WebQTL.org). The point to be made is that given sufficient genetic diversity and at least a moderate number of genotypes, data appear to generally be highly repeatable (across laboratories). In addition, this cumulative BXD RI dataset illustrates the importance of using a common reference population.

The question arises as to whether or not behavioral phenotypes are more or less variable than other quantitative traits. One answer to this question can be found if one is willing to go back and carefully examine the original data that generated the BXD RI phenotypic strain means. From these data it is possible to estimate the heritability (h^2). For phenotypes studied in our laboratory, we have found, not surprisingly, a wide range of heritability. For some phenotypes the heritability is fairly low (< 0.3); examples of phenotypes in this category would be prepulse inhibition (PPI) of the acoustic startle response (ASR)[20] and the ethanol activation response.[15]At the opposite extreme, we find that the heritabilty for open-field/basal activity is > 0.5, and the heritability for haloperidol-induced catalepsy is > 0.7.[21] This range compares favorably with other "neural" phenotypes that have been studied in the laboratory, including the density of dopamine D_2 receptors[22] and the number of tyrosine hydroxylase and choline

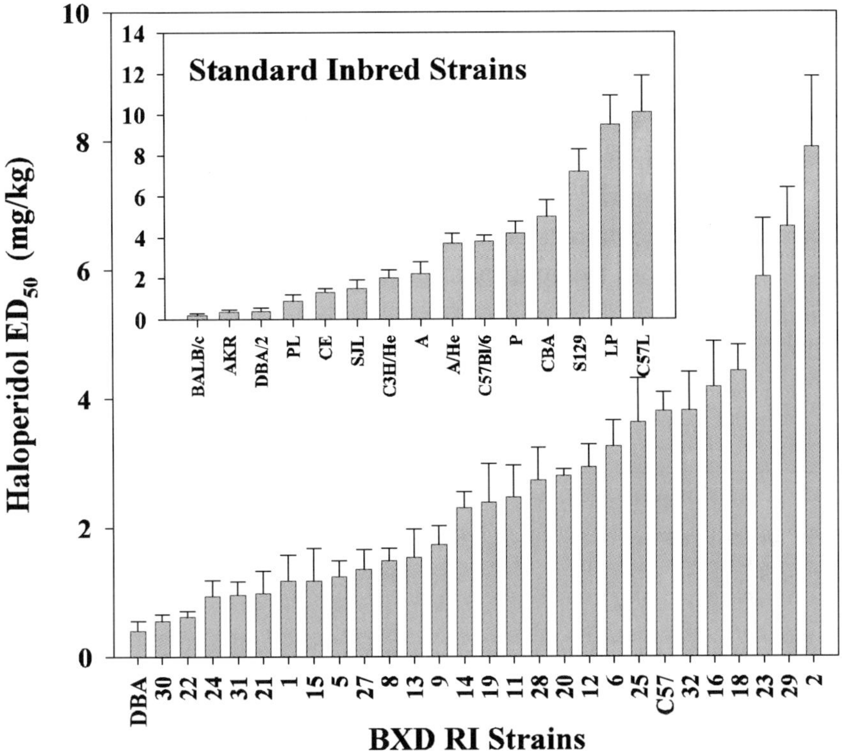

Figure 2. Comparison of the ED_{50} values for haloperidol-induced catalespy among a panel of standard inbred strains and a panel of the BXD recombinant inbred strains. Data are adapted from Kanes *et al.*[21] and Hitzmann *et al.*[33]

acetyl transferase positive neurons.[22,23] What is somewhat remarkable is that regardless of the estimated narrow sense heritability, the effect size of the QTL associated with these phenotypes is fairly similar, moderate, and generally h^2_{QTL} < 0.10.[24–26] For ethanol preference, Belknap and Atkins[7] were able to combine data from several studies to estimate the effect size of 4 different QTL. Effect sizes ranged from $h^2_{QTL} = 0.04$ to $h^2_{QTL} = 0.07$. There are, however, rare examples of behavioral QTL with large effect sizes, e.g., saccharin preference.[27]

The details of how to detect QTL are described elsewhere in this volume. Here we note that for many "behaviorists," mapping QTL in RI panels was the common portal to quantitative genetics. The WebQTL site provides state of the art analysis tools for RI data. However, the main statistical problem has not changed from the first use of such data. That is, the sample sizes are too small to detect anything but relatively large effects (accounting for ~ 25% of the genetic variance) with relatively little precision. Recently, new and larger RI panels have become available,[28,29] which will improve detection and precision; however, what is required, especially for low heritability phenotypes, are very large RI panels.[30] Such large RI panels would also be useful for fine mapping and for detecting first order and higher order interactions (epistasis).

3. Inbred Strains and Behavior

As with the investigation of other phenotypes, there is a long history of behavioral geneticists using panels of standard inbred mouse strains to establish heritability, the range of genetic variation and genetic correlation among phenotypes. One example of such data is shown in Figure 2. Here the phenotype is haloperidol-induced catalepsy, which is the murine equivalent of the extrapyramidal symptoms (EPS) induced by the "typical" antipsychotic drugs. EPS has many of the same symptoms encountered with Parkinson's disease. Variation among mouse strains in this phenotype was first noted by Reis and colleagues.[31] Kanes *et al.*[32] expanded the number of strains examined to eight and observed that while the inbred strains showed a 30-fold variation in sensitivity to haloperidol (a D_2 dopamine receptor antagonist), there was essentially no variation in the catalepsy response to SCH 23390 (a D_1 dopamine receptor antagonist), suggesting that the variation in response to haloperidol was not due to a general deficit in extrapyramidal function.

It was also observed that the differences in haloperidol response were not due to differences in brain drug uptake or metabolism. Since many behavioral phenotypes are measured in response to psychoactive drugs, the pharmacokinetic explanation always needs to be carefully considered. In Figure 2, the number of inbred strains has been expanded to fifteen;[33] drug sensitivity ranged from 0.2 mg/kg (BALB/cJ) to ~10 mg/kg for the LP/J and C57L/J strains. Figure 2 also illustrates the range of drug sensitivity in the BXD RI panel; no RI strain was more sensitive than the DBA/2J, but several strains were significantly less sensitive than the C57BL/6J strain. This phenomenon, sometimes called transgressive segregation, must in the RI panel be associated with genetic effects. For example, the DBA/2J strain may contribute some non-responsive alleles. Thus, in a mapping study using a C57BL/6J × DBA/2J intercross, it would not be unexpected to detect a QTL in which the DBA/2J allele is associated with non-response, and, indeed, such a QTL has been detected.[34]

Although the range of genetic variation among the standard inbred strains in Figure 2 is somewhat exceptional, it is generally possible for most behaviors to find several strains at the phenotypic extremes which could be used to generate an F_2 or backcross population that would be suitable for QTL mapping, i.e., the extreme strains differ by several standard deviations. Through the use of multiple crosses, one may detect additional QTL, examine genetic background effects, and develop a haplotype structure for the QTL (more on this concept in the following sections). However, for some phenotypes, multiple breeding scheme options are not available. Consider the ethanol preference phenotype; it is only the C57BL/6J strain or very closely related substrains that will voluntarily consume large quantities of ethanol.[35] C57BL/6J mice have been reported to consume up to 20 g/kg/day when the alcohol is presented in the two-bottle choice paradigm (water vs 10% v/v ethanol). In contrast, other inbred strains, including the DBA/2J drink < 5g/kg/day and rarely reach a measurable blood alcohol concentration. Using a panel of 10 inbred strains, fear-potentiated startle was detected only in the DBA/2J strain despite using a wide variety of conditioning paradigms.[36]

Such isolated, unusual and/or unexplained phenotypes have probably led all behavioral geneticists to express at some time the following sentiment, which is also part of the title of a very informative book,[37] "What's Wrong with My Mouse?" For the behaviorist it is important to remember that inbred strains of mice were developed for a variety of reasons, but in essentially all cases it was important that they thrive in the colony environment. Thus, from the breeding perspective, the strains which have survived are largely those which are docile, have a high fecundity, and exhibit good maternal instincts. Given that this selection process was largely random and unsupervised, the behavioral repertoires that were fixed differ markedly. However, these differences are relatively small when compared to the differences between the standard laboratory strains and the wild strains from which they were derived.[38] Finally, breeding for hundreds of generations in the laboratory environment has led to the random fixation of spontaneous mutations which have little effect on success in the colony environment, but which can markedly affect behavioral responses. These mutations include the loss of high frequency hearing, deafness, blindness and abnormalities in brain development, e.g., the partial development of the corpus callosum.[39–45]

Despite what is known about the diversity of response and the special limitations among standard inbred strains, the majority of behavioral QTL mapping studies have been limited to a very few strains. Prominent are the C57BL/6J, DBA/2J, BALB/cJ, A/J and closely related substrains, e.g., the BALB/cByJ. There are several reasons for this narrow perspective. *One*, these strains were used to develop several panels of RI strains, which have been widely used for behavioral research. In addition to the BXD RI panel, these include the AXB-BXA and BXC RI panels (behavioral data from these panels can also be found at the WebQTL site). When an RI panel is used as a preliminary QTL screen, it is not surprising that a cross from the parental strains would be used for more detailed analyses. While this two-step procedure,[46] widely used in the 1990s, is now less popular, the ability to build upon previous experiments continues the popularity of these strains. *Two*, these four strains differ markedly on a number of phenotypes that are of interest to the behavioral scientist. These phenotypes include open-field activity, fear conditioning, sensory gating, learning tasks and a wide variety of psychoactive drug responses.[47] The latter include responses to drugs of abuse (alcohol, cocaine, morphine/heroin, phencyclidine/ketamine) as well as responses to therapeutic agents (e.g., the typical and atypical antipsychotic drugs, sedative/hypnotic agents and anesthetics). *Three*, as noted above, some phenotypes are uniquely associated with one of these strains. *Four*, there are a number of additional genetic resources available for use with these strains. These include sequence data for the C57BL/6J, DBA/2J and A/J strains, an advanced intercross line and chromosome substitution strains formed from the C57BL/6J and A/J strains,[48,49] the recent production of 50 additional BXD RI strains,[28] extensive bacterial artificial chromosome (BAC) libraries for the C57BL/6J and DBA/2J strains and the availability of extensively tested microsatellite maps.[50] Overall, it seems unlikely that the dependence on these four strains for behavioral genetic analyses will change any time in the near future.

4. Genes and Phenotypic Complexity

Behavioral geneticists tend to think that their phenotypes are the most complex, the most difficult to understand and the most difficult to genetically dissect. Others are likely to disagree and argue that phenotypes such as growth, metabolism or cancer are equally, if not more, complex. Nonetheless, behavioral complexity is an important issue. We have already discussed in some detail that the gene × environment interaction can, under certain circumstances, be quite marked.[11–13] Here the environmental intervention was a different laboratory. However, what about gene × environmental effects that occur during the course of an experiment? We return to the ethanol-induced activation experiment described previously; this phenotype is a measure of acute ethanol sensitivity, which has been shown to be a risk factor for the development of alcoholism.[51,52] Saline (10ml/kg — i.p.) data are collected on day 1, ethanol (1.5 g/kg — i.p.) data are collected on day 2. The ethanol response is day 2 minus day 1. Activity data (distance traveled) are collected in 5 min bins for 20 min.

Figure 3 illustrates the results that are obtained on day 1 and for the ethanol score. The data clearly illustrate the marked time × genotype interaction. On the saline control day, the C57BL/6J strain was clearly the most active and showed the greatest habituation of response. Also, low activity appeared to be dominant. The ethanol response was more complex. At the 5 min interval, the DBA/2J strain and the F_1 hybrid were the most active, the F_2 intercross was intermediate between the DBA/2J and C57BL/6J strains, and the latter was essentially non-responsive.

As time progressed, the F_1 hybrid and the F_2 intercross rapidly habituated and assumed the C57BL/6J phenotype. The results from QTL mapping of the F_2 intercross are instructive.[15,24,26] For basal activity, two different QTL (LOD >5) were detected on chromosome 1; both QTL were essentially time-independent. The more proximal QTL centered at approximately 80 Mb was additive, while the more distal QTL centered at approximately 175 Mb showed the expected dominance of the DBA/2J (low activity) allele. For the ethanol response, the 0–5 min interval was associated with a QTL on Chr 1 (LOD > 7 and different from those for basal activity); this QTL was not present in the 5–20 min intervals. The QTL was completely additive, and it was the C57BL/6J allele that was associated with increased response.

During the 5–20 min period two QTL were detected on Chr 2; both of these QTL were additive, with the DBA/2J alleles being associated with increased ethanol response. Both of these QTL now have cumulative LOD scores from multiple experiments of >15.[15,24] Focusing on time as the environmental variable, it can seen to have relatively little effect on the genetic architecture of basal activity, but a marked effect on the ethanol response. The time × QTL interaction shows that the ethanol response can be parsed into what may well be two distinct behaviors – the initial "explosive" response to the rapid uptake of ethanol (the peak brain concentration is reached by ~ 2 min) and habituation to this response. Importantly, if one measured the ethanol response for only the first 5 min, the "genetic" conclusions reached would be far different from focusing on the 5–20 min interval. Time is just one of many variables that can affect the outcome of an experiment; the example given here of using time as the dissector has revealed the underlying complexity of the behavior and the associated genetics.

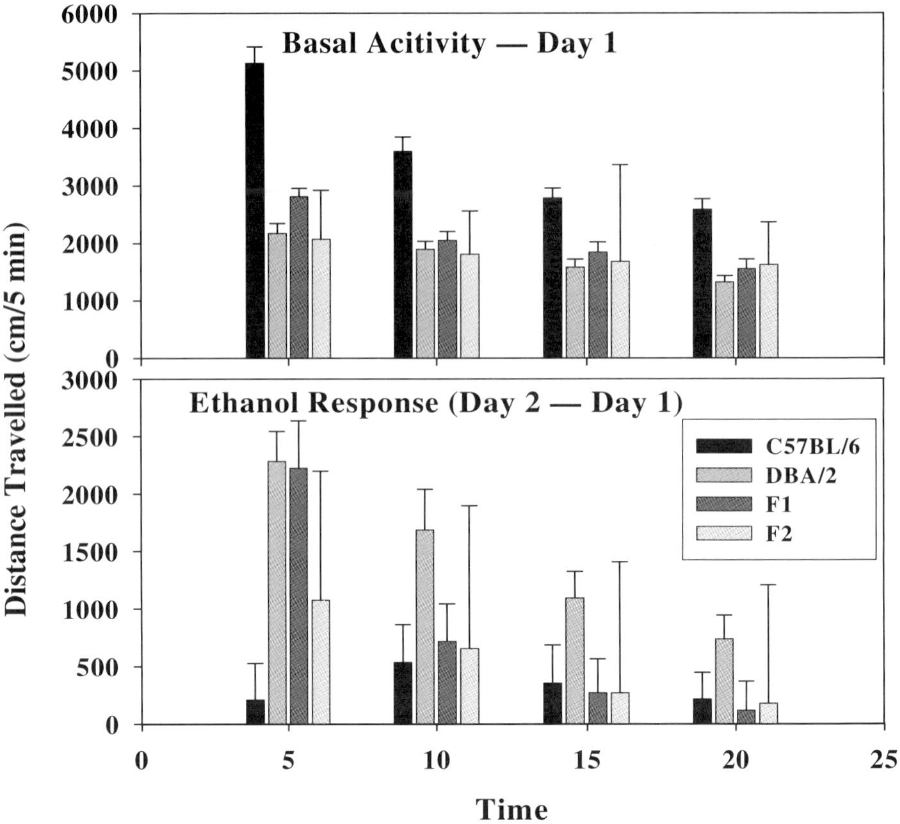

Figure 3. Time course of basal activity and ethanol response for the C57BL/6 and DBA/2 inbred strains and the associated F_1 and F_2 hybrids. Experimental details in text. N = 24–30 except for F_2 (N = 900).Data reported as mean ± SD. Note the expected large increase of the variance in the F_2 group. Data adapted from Demarest *et al.* [15]

5. QTL Analysis and Behavior — The Problem

In this and the following section, the argument is made that our ability to move QTL analysis to the next level will require an integration of QTL and gene expression data. This integration is not unique to behavioral phenotypes, but behavioral geneticists have contributed significantly to the process. This section specifically looks at the factors which have impeded progress in moving from QTL to QTG.

QTL analysis for any phenotype may be viewed as a series of sequential steps that begins with QTL detection, followed by isolation of the QTG (or genes) and finishes with a determination of which polymorphisms within these gene(s) are the relevant quantitative trait nucleotides (QTNs).[53] As noted in the Introduction, QTL detection for behavioral phenotypes in mice, has been enormously successful.[6] As shown elsewhere in this volume, the same could be said for other phenotypes. There are several reasons for this success. QTL detection benefitted greatly from the development of dense genetic maps with easily genotyped microsatellite markers.[54,55] Equally important was the adoption of stringent statistically-based thresholds for calling a QTL "present."[56,57] Methods of analyzing QTL data have continued to evolve; here it is of some interest to note the improvements from Mapmaker/QTL[61] to more recent tools like R/qtl.[59]

The second step in QTL analysis, isolating the QTGs, has not been particularly successful and has been the subject of some debate.[60,61] Although the number of QTGs that have been detected (for all phenotypes) is increasing,[62] the backlog of uncharacterized QTL is quite large and may well number in the thousands. Some of the reasons for the backlog seem clear. Contrary to QTL detection, which was able to rapidly incorporate technical advances, the process of isolating a QTG has remained decidedly conventional and time-consuming. The process necessarily must begin with reducing the QTL interval to a size that is practical for the subsequent analyses, e.g., positional cloning or expression based analyses. Darvasi[63] summarized the advantages and disadvantages of four different strategies for QTL fine mapping: selective phenotyping, recombinant progeny testing, the recombinant inbred segregation test (RIST) and the production of interval specific congenic strains (ISCS). The last procedure, which was generally viewed as having the greatest promise and wide-spread application, borrowed from Drosophila geneticists the technique of genetic chromosome dissection.[64–67] With the advent of dense microsatellite maps, Darvasi[68] formalized the design based on the systematic marker-directed construction of a series of congenic strains, each with a recombinant point in a specific interval out of a series of small tandem intervals, which cover the chromosomal region of interest. The QTL would first be captured in a congenic strain, followed by the production of overlapping recombinant (interval specific) congenic strains. To our knowledge, the application of this technique to behavioral traits has been limited.[8,69–71]

Problems with the ISCS approach are considerable. The QTL may not be initially captured in a congenic strain due to a variety of factors, including the break up of closely linked QTL and/or the loss of unlinked epistatic loci, e.g., the interaction between QTL on different chromosomes. The recombination rate in the congenic interval may be low for some strain combinations, delaying the production of the interval specific strains. Finally, producing the congenic and recombinant congenic strains can be expected to take several years and thus is vulnerable to experimental attrition.

Assuming the QTL interval has been reduced to 1 or 2 cM, the next step in the process involves sorting through 10, 20 or more candidate genes within the interval and attempting to determine which is the QTG. This step is daunting because more than one gene may be involved, and the function of some genes within the interval may be unknown. Until quite recently, this step emphasized the detection of polymorphisms within coding sequence;[61,62] for a polymorphism that produces an amino acid substitution, one can often infer and then test for a functional consequence. For some QTL the discovery phase of detecting coding sequence polymorphisms can now be replaced by "in silico" analysis.[72] The application of this approach suggests that the rate of sequence polymorphisms within known QTL intervals may be substantial[71,72] and may well be as prevalent as differences in gene expression (~10%).[73–77] The advantage of the gene expression based approach to QTL dissection is that it is not limited to the few strains for which sequence data are available, and since the expression data are genome-wide, they can be applied to all known QTL. A gene-expression-based approach is inherently high-throughput and has the potential to substantially increase the rate of converting QTL to QTG. However, it is important not to lose sight of the idea that a significant proportion of QTL may be associated with coding sequence polymorphisms.

The application of the gene expression based approach to behavioral phenotypes is not without potential problems. The target organ for behavioral studies is the brain, a complex structure composed of thousands of nuclei. Potentially, the QTL could be generated from differences in gene expression in only a few neurons. Where does one begin to look for differences in gene expression? For some behaviors, the associated neural circuitry is well established. Examples here would include fear-potentiated startle, prepulse inhibition and haloperidol-induced catalepsy.[33,78,79] However, for most behaviors, understanding the neural circuitry remains a "work in progress."

How does one proceed? One approach has been simply to measure, with great precision, gene expression in the whole brain.[77] Given the large dynamic range of most microarray platforms, it is possible to measure accurately genes, which are known to be expressed in only one nuclei. The interested reader should go to the WebQTL site and look at one or more of the whole-brain expression databases. One will find, for example, that it is possible to reliably measure gene expression for neuropeptides, which are known to be expressed only in hypothalamic nuclei. The problem with the whole-brain approach is the "what if" problem — what if gene expression is up in one region and down in another, or what if gene expression is increased in a critical region, but the change is masked by the noise of expression in the remainder of the brain. "What if" can lead to experimental paralysis, which is not helpful but it is wise to keep in mind the limitations of measuring whole-brain gene expression.

Another approach is to try to determine the likely circuit(s) associated with the genetic differences in the behavioral response. A strategy we and others have used, which is particularly well suited to drug-induced responses, is to map the immediate early gene (IEG) response. For example, we have shown that the differences in the acute ethanol response between the C57BL/6J and DBA/2J strains (Figure 3) is associated with a differential activation of the extended amygdala (as measured by the expression of c-Fos).[80]

Other strategies for mapping an unknown circuitry include neurophysiological and lesion (electrolytic and chemical) based strategies. The importance of understanding the circuitry associated with a particular phenotype cannot be overstated. Not only is it important for detecting differences in gene expression, but, as will be discussed in the next section, its greatest importance may be in the "end game," proving that a candidate QTG is indeed the QTG.

6. QTL Analysis and Behavior — Some Solutions

As noted in the previous section, QTL analysis for behavioral phenotypes (and other phenotypes as well) has stalled at the level of reducing the QTL interval to a size (1–2 cM) which is suitable for detailed analysis. In this section, various options for reducing the QTL interval are discussed. The emphasis here will be on developing a high-throughput strategy which provides information about the haplotype structure of the QTL.

Darvasi and Soller[81] proposed that an advanced intercross line (AIL) could be used to reduce the QTL interval. The AIL was generally conceptualized as originating from a cross between two inbred strains, followed by random and sequential intercrossing. With each generation, the genetic map is stretched such that at the F_4 generation the 95% confidence interval (CI) for a QTL would be reduced two-fold and at F_{10}, 5-fold. For a moderately sized QTL (d = 0.25), 12,000 F_2 animals would be required to reduce the QTL 95% confidence interval to 2 cM [from CI = $1500/Nd^2$, where N = the sample size and d = the standardized gene effect; the constant, 1500, was determined empirically from computer simulations and is not related to genome size]. For an AIL at F_{10}, the same resolution can be obtained with 2,400 animals. To our knowledge, there has been only one widely used population of AIL mice (formed from a B6 × A intercross),[82–84] and these lines have never been used to map a behavioral trait. However, the concept was established and was subsequently modified to fine map behavioral QTL.

Talbot *et al.*,[85] building from the AIL concept, provided evidence that advanced generation heterogeneous stock (HS) animals could be used to reduce the QTL interval. [Two populations of HS animals (mice) have been independently maintained for > 50 generations — HS_{Ibg} and HS_{Npt}. Both HS populations were formed by crossing 8 inbred mouse strains.] Talbot *et al.*[85] were the first to document that HS animals (here HS_{Ibg} at G_{58}) could be used to map a moderately sized QTL for open-field activity (effect size ~6%) to a resolution of ~1cM. This QTL for open-field response on chromosome 1 and another open-field QTL mapped with somewhat less precision on chromosome 12 were clearly not the same QTL that had been previously detected on these chromosomes in a C57BL/6J × BALB/cJ intercross.[3] For both HS QTL, the markers associated with significant effects did not distinguish between the C57BL/6J and BALB/cJ strains. Although detecting new QTL in the HS animals was of interest, obviously the greater value would be to fine map known QTL. With this idea in mind, the dataset of Talbot *et al.*[85] was revisited using a multipoint analysis that assigns the probability that an allele descends from each of the HS progenitors,[86] i.e., provides a haplotype signature for the QTL. Ten cM regions around each of four QTL (chromosomes 1, 10, 12 and 15) previously identified in the C57BL/6J × BALB/cJ intercross were analyzed.

Three of the QTL previously undetected in the HS animals were mapped (chromosomes 1, 10 and 15) to the expected loci; the chromosome 12 QTL appeared to split in the HS animals, suggesting two closely linked QTL. Importantly, the analysis predicted that for the chromosome 1 QTL, the BALB/cJ, DBA/2J and A/J alleles were associated with decreased open-field activity, consistent with the published F_2 intercross data.[3,15,87] Mott *et al.*[86] concluded that the reason the QTL were not previously detected was the failure of the single marker association approach to discriminate between opposing phenotypic effects when they occur on the same marker allele. The multipoint analysis strategy was effective in detecting and fine mapping the QTL because information on the progenitor strains' haplotype structure was known.

Paralleling these developments in fine mapping, we began constructing a QTL strategy, which we termed Multiple Cross Mapping (MCM). MCM was built from what seemed at the time (1999–2000) to be several independent observations. The first was the common observation that the microsatellite markers, which had greatly facilitated QTL detection, were not randomly distributed. There were clearly extended regions, sometimes up to 10 or more centimorgans, in which pairs of strains in the MIT catalog were not polymorphic. The usual explanation for this phenomenon was that the density of the microsatellite map was not high, and that if the density was increased, polymorphisms would be found. The alternative explanation was that the strains because of their largely common heritage, were actually identical over these extended regions.[88]

The second observation was built from the collection of QTL data for the same phenotype across multiple diallele crosses. Here open-field activity was especially important, and data were available for multiple intercrosses, especially multiple intercrosses involving the C57BL/6J strain.[3,15,87] For some of the crosses, the QTL were similar, and for others the QTL were different. The usual explanation for the differences was the effect of "genetic background." However, the alternative explanation was that these data were actually telling us something important about the structure of the QTL. We synthesized elements of the alternative arguments and suggested that multiple cross information could be useful for improving mapping precision; QTL would not be expected where strains showed no polymorphisms (in the microsatellites).[88] It is of interest retrospectively that in this publication the term "haplotype" was never used.

The problem with the multiple strain argument was that, to our knowledge, there were no data for a balanced series of intercrosses. Why was this important? Again, consider the case for open-field activity. On Chr 1, QTL had been detected in C57BL/6J vs BALB/cJ, DBA/2J and A/J strains. Therefore, would a QTL be detected in DBA/2J × BALB/cJ, DBA/2J × A/J or A/J × BALB/cJ intercrosses? If not, one could conclude for these strains that the C57BL/6J allele was unique and capable of generating a QTL, independent of the genetic background. On the other hand, if one of the non-C57BL/6J intercrosses generated a QTL in the same region, the interpretation of the data would become much more complex. Given that we had considerable data for QTL in C57BL/6J × DBA/2J and BALB/cJ × LP/J intercrosses,[88] the decision was made to complete a balanced data set, i.e., interrogate C57BL/6J × BALB/cJ, C57BL/6J × LP/J, DBA/2J × LP/J and DBA/2J × BALB/cJ intercrosses for behavioral QTL. The behaviors examined included open-field activity, catalepsy, ethanol activation, the acoustic startle response and prepulse inhibition of the startle response. The first report to indicate that the approach may well work as intended focused on the Chr 1 QTL for ethanol response described previously;

using the MCM algorithm, i.e., sorting the microsatellite markers based on whether or not a particular cross generated a QTL, it was possible to reduce the QTL interval to approximately 3 cM, which was confirmed by fine mapping in HS_{Npt} mice.[89] However, perhaps more important than the improvement in QTL mapping, it was recognized that MCM could also be used as a haplotype-based approach for interrogating gene expression and sequence data. The argument here was as follows. Our ability to reduce the QTL interval will generally reach some practical limitation, leaving an interval with numerous potential candidates. Knowing that the QTG must differ in expression and/or sequence between the two strains which were first used to detect the QTL will provide some reduction in the number of candidates. However, for some intervals, numerous candidates will remain — a point confirmed experimentally.[90] By expanding our characterization of the QTL to include additional "haplotype" information, the number of candidate genes should be substantially decreased. Further, there should be an inverse relationship between the amount of haplotype information acquired and the need to reduce the QTL interval.

Supporting the MCM approach, detailed SNP analyses of laboratory mice have detected an embedded and relatively simple haplotype structure derived from the common lineage of the inbred strains.[91,92] These authors emphasized that such data would be of particular value for QTL analysis via strategy we have termed Multiple Strain Mapping (MSM),[77] which, in concept, was first suggested by Grupe *et al.*[93] MSM is based on the principle that if one had detailed SNP and/or microsatellite maps for a sufficiently large number of inbred mouse strains, it would be possible to map QTL with great precision, simply by determining the phenotypic strain means. The number of strains that would be required has been an issue of some controversy,[94,95] but the basic principle is sound and has the distinct advantage that strain means can be determined with great precision (unlike the phenotypic value of a single animal in a segregating population). Progress in effectively testing the MSM strategy has been slowed by lack of needed databases. However, this may soon change. A detailed SNP map for several hundred mouse strains is now under construction (R. Mott, personal communication). In addition, the needed phenotypic datasets are being developed [e.g., the Mouse Phenome Project (www.jax.org)].

The earliest proof of principle (in rodent models) for the integration of QTL and gene expression data was two studies which identified genes involved in insulin resistance[96,97] and airway hyper-responsiveness.[98] The first application of the approach to neural phenotypes is found in Sandberg *et al.*[99] These authors reported marked differences in brain gene expression between two inbred mouse strains, C57BL/6J and 129S1/SxImJ, in whole brain and discrete brain regions (cortex, midbrain, hippocampus and cerebellum). These data led the authors to the salient observation that some of these differences appeared to coincide with the known location of "behavioral" QTL; a particular note was made of the fact that *Kcnj9* (which encodes for Kir3.3, a G-protein coupled inwardly rectifying potassium channel) had a markedly lower expression in the C57BL/6J strain and was located in the behavioral QTL-rich region on distal Chr 1.[6]

Subsequent publications from this and other laboratories[100–106] continued the argument and provided additional data for combining analyses of transcript levels (using expression arrays) with information from QTL mapping to nominate candidate genes. Three recent studies[74,75,107] confirmed the underlying basis for this argument, namely marked strain-dependent differences in gene expression. For example, Schadt *et al.*[75] have found that ~25% of hepatic genes are differentially expressed between the C57BL/6J and DBA/2J strains; perhaps more important, the authors found ~1000 genes that appear to exhibit significant cis-regulation. Differential brain gene expression between the C57BL/6J and DBA/2J strains appears to be smaller than the magnitude of the differences seen in liver, but the magnitude is still considerable, ~ 10%.[90]

The multiple cross algorithm was first used to interrogate gene expression data for the isolation of a QTG for open-field activity.[77] QTL were generated only in the C57BL/6J crosses, but not in the residual crosses. Sorting microsatellite and SNP markers suggested that the QTL was centered at ~ 175Mb +/– 2 Mb; this location was confirmed by fine mapping in HS animals. Using the algorithm which simplifies to the C57BL/6J allele that is not found in the other three strains, it was concluded that *Kcnj9* was the only gene within the interval of interest, which had the appropriate gene expression profile (the C57BL/6J strain has low expression compared to the other strains — Affymetrix U74Av2 array). The pattern of differential expression was detected in whole brain, the dorsomedial striatum and extended amygdala. *Kcnj9* also showed significant cis-regulation (LOD >15, WebQTL analysis tools and databases), which is expected for an expression QTG to generate a QTL at the same site. *In silico* analysis of the 5' regulatory region (Celera and Ensembl databases), revealed a polymorphism between the C57BL/6J and DBA/2J strains in a regulatory motif that was predicted to disrupt the binding of three transcription factors (*Ikaros 1, MZF1* and *C/EBPbeta*) in the C57BL/6J strain. Sequence analysis of the BALB/cJ and LP/J strains revealed identity, as expected, in the motif region with the DBA/2J strain. Overall, the multiple cross strategy led directly to a candidate QTG and a candidate QTN. These conclusions must be tempered by noting that expression data were available for only 50% of the known and predicted genes within the QTL interval. Thus, the possibility must be considered that other genes may show a similar pattern of differential expression. The experiments described above were robust because all of the C57BL/6J crosses generated a QTL; balance was maintained in that no QTL was generated among the DBA/2J, BALB/cJ and LP/J strains.

The success of these experiments blinded us somewhat to the obvious problem with the MCM strategy, namely how does one interpret "QTL not detected." While the meaning of no QTL detected is clear from a statistical perspective, it is less clear from a biological perspective. For example, trans-acting features (epistasis) may suppress a QTL such that we would conclude the strains are identical at the QTG when in fact they are different. We began to critically reconsider the elements of the MCM strategy, especially given that the vast majority of the QTL we have detected are "singles," i.e., these are detected in only one cross. The value of knowing the QTL haplotype for interrogating the expression and sequence data was established; MCM as a strategy for determining the QTL haplotype appeared most suitable for allele effects unique to a single strain. However, for many QTL, it appeared that the MCM strategy would yield data that had the potential for (QTL haplotype) assignment errors.

Fortunately, a solution to this problem was on hand: advanced intercross mapping using HS. As noted above, the multipoint analysis of the HS data[86] calculates the relative contribution of each strain to the QTL (i.e., the QTL haplotype), which in turn can be used to interrogate gene expression (and sequence) databases. Importantly, the contribution of each strain is calculated simultaneously and not in separate experiments as with the MCM approach. In addition, QTL formerly detected in different intercrosses may now be detected in the same mapping population. HS mapping substantially reduces the QTL intervals and dissects closely linked QTL without disrupting the interactions between these QTL. Finally, HS mapping provides a "high throughput" solution to the problem on how to best determine the QTL haplotype. The question arises as to which HS population is most suitable. We favor an HS we have termed HS_4, as it was formed by crossing the C57BL/6J, DBA/2J, BALB/cJ and LP/J strains. Mapping in the HS_4 allows direct comparison with the MCM data we have already collected.

In addition, it is often possible to precisely know the genotype, given microsatellites polymorphic among all four strains. Figure 4 illustrates this point. The QTL for open-field activity on distal Chr 1 was mapped in the HS_4 population (now at G_{20}). All 10 genotypes can be dissected, and, as expected, it is only the C57BL/6 homozygous animals which show high open-field activity. However, for some applications, the 8-way cross such as the HS_{Npt} may be more suitable, i.e., the known QTL was generated by a strain in the 8- and not the 4-way cross. Our HS_{Npt} cross was formed from the same four strains as the HS_4 plus the AKR/J, CBA/J, C3H/HeJ and A/J strains. Both HS_4 and HS_{Npt} mice are maintained in colonies of 50 families and are available to investigators; a second colony of the HS_{Npt} mice has been established at Oxford, UK.

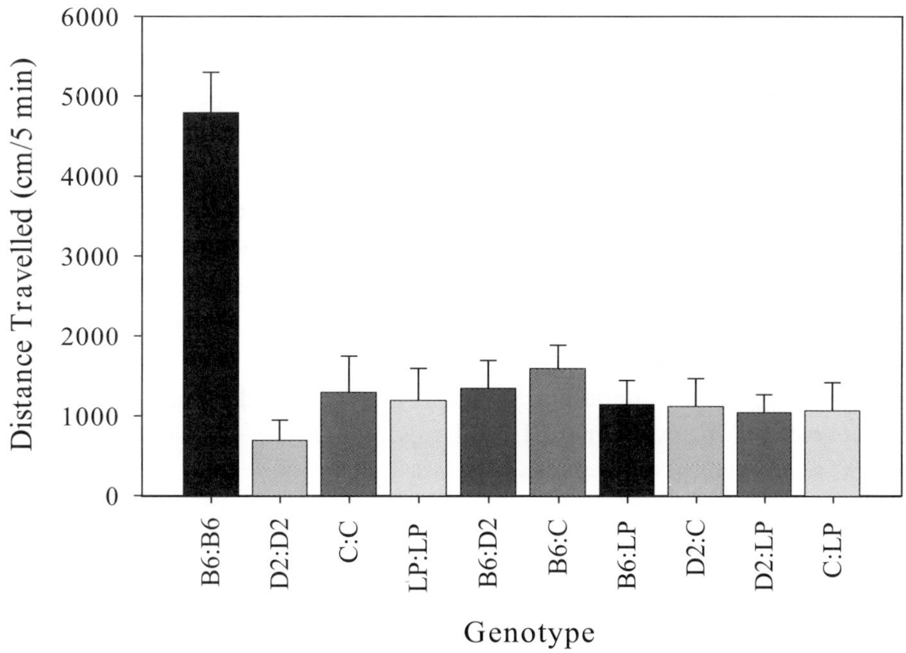

Figure 4. The relationship between genotype and phenotype in HS_4 mice at a QTL for open-field activity on distal chromosome 1 (~ 175Mb). The four strains used to form the HS_4 are C57BL/6J (B6), DBA/2J (D2), BALB/c (C) and LP/J (LP). Data are unpublished observations (Hitzemann and Malmanger)

The "end game" for QTL analysis has been the subject of some controversy.[60] What are the criteria that we are willing to accept as proof for a QTG? If our goal is to identify genetic targets for human research, how best to proceed? In some cases the QTG will be the same in mice and humans.[10] However, we assume that in many cases QTL research will help us to identify the relevant gene network; the source of genetic variation within that network may not be homologous. We have argued previously that there is no one solution to the "end game"[103] and that it is best to consider all possible approaches. In some cases it will be possible to rescue a phenotype through the construction of a BAC transgenic. Flint and Mott [105] have argued that the genetic complimentation test is an effective proof. Pharmacological approaches may well prove useful and here there is the potential of both preclinical and clinical testing. With behavioral phenotypes, the burden of proof assumes some additional complexity.

As noted previously, some QTL may be associated with differences in gene expression that only occur in very discrete brain regions. Increasing gene expression via the transgenic approach or decreasing gene expression in knockout mice, generally will not provide the needed specificity. Conditional knockin or knockout mice can provide the needed specificity if one is fortunate to find a tissue specific promoter. Viral vectors have been used to deliver genes to discrete brain regions in order to modify behavior.[108,109]Assuming that the underlying circuitry is known, this approach is certainly the simplest and least time-consuming. A final point regarding behavioral phenotypes. In this chapter, we have assumed that the differences in gene expression are ones that we can measure in the adult. However, there is no reason not to believe that a transient change in gene expression during development is the source of the QTL. Genes that are cis-regulated in the embryo may be trans-regulated in the adult and excluded from our consideration. With this point in mind, we can expect to see the construction of developmental databases, which may well revolutionize our concepts of behavioral QTL analysis.

7. Conclusions

The "phrase oft repeated" is that the study of behavioral genetics will always be confounded by the complexity of the phenotypes and the overwhelming influence of environmental factors. Fortunately, many behavioral geneticists, and especially those working with mice, have not "ossified," but instead have been working diligently to solve the relevant problems. Importantly, their work has clearly demonstrated that the natural variation in most behaviors is associated with the effects of multiple genes, each with generally small to modest effects. Detecting such gene effects in clinical populations is inherently difficult; some have said impossible (another "phrase oft repeated"). One solution to this problem is to examine within genetically defined mouse populations phenotypes which model human risk factors. Some examples of this approach are given in this chapter. To date, we have no examples of where a behavioral QTL, first detected in the mouse, led to the discovery of the relevant gene or gene network, which was then used either to uncover the etiology and/or improve the therapy for human disease. However, it is important to remember that this discovery process is actually less than 15 years old. Furthermore, there is already solid evidence that the incorporation of tools from genomics and functional genomics already have and will continue to greatly speed this process of discovery.

Acknowledgements

The author thanks the many students, fellows and collaborators that have contributed in some way to the research described in this chapter. Special thanks to Barbara Hitzemann, John Belknap and Shannon McWeeney. This work was supported in part by grants from the US Public Health Service (AA 11034, AA 13484 and MH 51372) and support from the Department of Veterans Affairs Medical Research Program.

References

1. Plomin, R., G.E. McClearn, G. Gora-Masalk and J.M. Neiderhiser. 1991. Use of recombinant inbred strains to detect quantitative trait loci associated with behavior. *Behav. Genet.* **21**:9–116.
2. Plomin, R. and G.E. McClearn. 1993. Quantitative trait loci (QTL) analyses and alcohol-elated behaviors. *Behav. Genet.* **23**:197–211.
3. Flint, J., R. Corley, J.C. DeFries, D.W. Fulker, J.A. Gray, S. Miller and A.C. Collins. 1995. A. simple genetic basis for a complex psychological trait in laboratory mice. *Science* **269**:1432–1435.
4. Hall, C.S. 1934. Emotional behavior in the rat. I. Defecation and urination as measures of individual differences in emotionality. *J. Comp. Psychol.* **18**: 385–403.
5. Hall, C.S. 1938. The inheritance of emotionality. *Sigma Xi Quarterly* **26**:17–27.
6. Flint, J. 2003. Analysis of quantitative trait loci that influence animal behavior. *J. Neurobiol.* **54**:46–77.
7. Belknap, J.K. and A.L. Atkins. 2001. The replicability of QTLs for murine alcohol preference drinking behavior across eight independent studies. *Mamm. Genome.* **12**:893–899.
8. Fehr, C., R.L. Shirley, J.K. Belknap, J.C. Crabbe and K.J. Buck. 2002. Congenic mapping of alcohol and pentobarbital withdrawal liability loci to a <1 centimorgan interval of murine chromosome 4: Identification of Mpdz as a canidate gene. *J. Neurosci.* **22**:3730–3738.

9. Shirely, R.L., N.A. Walter, M.T. Reilly, C. Fehr and K.J. Buck. 2004. Mpdz is a quantitative trait gene for drug withdrawal seizures. *Nat. Neurosci.* **7**:699–700.
10. Mogil, J.S., S.G. Wilson, E.J. Chesler, A.L. Rankin, K.V. Nemmani, W.R. Lariviere, M.K. Groce, M.R. Wallace, L. Kaplan, R. Staud, T.J. Ness, T.L. Glover, M. Stankova, A. Mayorov, V.J. Hruby, J.E. Grisel and R.B. Fillingim. 2003. The melanocortin-1 receptor gene mediates female-specific mechanisms of analgesia in mice and humans. *Proc. Natl. Acad. Sci. USA.* **100**:4867–4872.
11. Crabbe, J.C., D. Wahlsten and B.C. Dudek. 1999. Genetics of mouse behavior: Interactions with the laboratory environment. *Science* **284**:1670–1672.
12. Immen, W. 1999. Laid-back Edmonton mice have scientists puzzled. *Globe Mail* **June 4**: A1–A6.
13. Wahlsten, D., P. Metten, T.J. Phillips, S.L. Boehm, S. Burkhart-Kasch, J. Dorow, S. Doerksen, C. Downing, J. Fogarty, K. Rodd-Henriks, R. Hen, C.S. McKinnon, C.M. Merrill, C. Nolte, M. Schalomon, J.P. Schlumbohm, J.R. Sibert, C.D. Wenger, B.C. Dudek and J.C. Crabbe. 2003. Different data from different labs: lessons from studies of gene-environment interaction. *J. Neurobiol.* **54**:283–311.
14. Crabbe, J.C., A. Kosobud, E.R. Young and J.S. Janowsky. 1983. Polygenic and single-gene determination of responses to ethanol in BXD/Ty recombinant inbred mouse strains. *Neurobehav. Toxicol. Teratol.* **5**:181–187.
15. Demarest, K., J. McCaughran Jr., E. Mahjubi, L. Cipp, and R. Hitzemann. 1999. Identification of an acute ethanol response quantative trait locus on mouse chromosome 2. *J. Neurosci.* **19**:549–561.
16. Chesler, E.J., L. Lu, J. Wang, R.W. Williams and K.F. Manly. 2004. WebQTL: rapid exploratory analysis of gene expression and genetic networks for brain and behavior. *Nat. Neurosci.* **7**:485–486.
17. Taylor, B.A. 1978. Recombinant inbred strains: use in gene mapping. In: *Origins of Inbred Mice.* ed. H.C. Morse. pp. 423–438. Academic Press, New York.
18. Tarantino, L.M., G.E. McClearn, L.A. Rodriguez and R. Plomin. 1998. Confirmation of quantitative trait loci for alcohol preference in mice. *Alcohol. Clin. Exp. Res.* **22**:1099–1105.
19. Phillips, T.J., J.K. Belknap, K.J. Buck and C.L. Cunningham. 1998. Genes on mouse chromosomes 2 and 9 determine variation in ethanol consumption. *Mamm. Genome.* **9**:936–941.
20. McCaughran, J. Jr., J. Bell and R. Hitzemann. 1999. On the relationships of high frequency hearing loss and cochlear pathology to the acoustic startle response (ASR) and prepulse inhibition of the ASR in the BXD recombinant inbred series. *Behav. Genet.* **29**:21–30.
21. Kanes, S., K. Dains, L. Cipp, J. Gatley, B. Hitzemann, E. Rassmussen, S. Sanderson, M. Silverman and R. Hitzemann. 1996. Mapping the genes for haloperidol-induced catalepsy. *J. Pharmacol. Exp. Ther.* **277**:1016–1025.
22. Dains, K., B. Hitzemann and R. Hitzemann. 1996. Genetics, neuroleptic response and the organization of cholinergic neurons in the mouse striatum. *J. Pharmacol. Exp. Ther.* **279**: 1430–1438.
23. Hitzemann, R., B. Hitzemann, S. Rivera, J. Gately, P. Thanos, L.L. Shou, R.W. Williams. 2003. Dopamine D2 receptor binding, Drd2 expression and the number of dopamine neurons in the BXD recombinant inbred series: genetic relationships to alcohol and other drug associated phenotypes. *Alcohol Clin. Exp. Res.* **27**:1–11.

24. Demarest, K., J. Koyner, J. McCaughran Jr., L. Cipp and R. Hitzemann. 2001. Further characterization and high-resolution mapping of quantitative trait loci for ethanol-induced locomotor activity. *Behav. Genet.* **31**:79–91.
25. Rasmussen, E., L. Cipp, and R. Hitzemann. 1999. Identification of quantitative trait loci for haloperidol-induced catalepsy on mouse chromosome 14. *J. Pharmacol. Exp. Ther.* **290**:1337–1346.
26. Koyner, J., K. Demarest, J. McCaughran Jr., L. Cipp and R. Hitzemann. 2000. Identification and time dependence of quantitative trait loci for basal locomotor activity in the BXD recombinant inbred series and a B6D2 F2 intercross. *Behav. Genet.* **30**:159–170.
27. Bachmanov, A.A., X. Li, D.R. Reed, J.D. Ohmen, S. Li, Z. Chen, M.G. Tordoff, P.J. de Jong, C. Wu, D.B. West, A. Chatterjee, D.A. Ross and G. K Beauchamp. 2001. Positional cloning of the mouse saccharin preference (Sac) locus. *Chem. Senses.* **26**:925–933.
28. Peirce, J.L., L. Lu, J. Gu, L.M. Silver and R.W. Williams. 2004. A new set of BXD recombinant inbred lines from advanced intercross populations in mice. *BMC Genetics.* **5**: 7.
29. Williams, R.W., B. Bennett, L. Lu, J. Gu, J.C. DeFries, P. Carosone-Link, B. Rikke, J.K. Belknap, and T.E. Johnson. 2004. Genetic structure of the LXS panel of recombinant inbred mouse strains: A powerful resource for complex trait analysis. *Mamm. Genome* **15**:637–647.
30. Threadgill, D.W., K.W. Hunter and R.W. Williams. 2002. Genetic dissection of complex and quantitative traits; from fantasy to reality via a community effort. *Mamm. Genome.* **13**:175–178.
31. Fink, J.S., A. Swerdloff, and D.J. Reis. 1982. Genetic control of dopamine receptors in the mouse caudate nucleus: relationship on cataleptic response to neuroleptic drugs. *Neurosci. Lett.* **32**:301–306.
32. Kanes, S., B.A. Hitzemann and R.J. Hitzemann. 1993. On the relationship between D2 receptor density and neuroleptic-induced catalepsy among eight inbred mouse strains. *J. Pharmacol. Exp. Ther.* **267**:538-547 .
33. Hitzemann, R., Y. Qian, S. Kanes, K. Dains, and B. Hitzemann. 1995. Genetics and the organization of the basal ganglia. *Int. Rev. Neurobiol.* **438**:43–94.
34. Patel, N.V. and R.J. Hitzemann. 1999. Detection and mapping of quantitative trait loci for haloperidol-induced catalepsy in a C57BL/6J x DBA/2J F2 intercross. *Behav. Genet.* **29**: 303–310.
35. Belknap, J.K., J.C. Crabbe, and E.R. Young. 1993. Voluntary consumption of ethanol in 15 inbred mouse strains. *Psychopharmacology (Berl).* **112**:503–510.
36. McCaughran, J.A. Jr., J. Bell and R.J. Hitzemann. 2000. Fear potentiated startle response in mice: genetic analysis of the C57BL/6J and DBA/2J intercross. *Pharmacol. Biochem. Behav.* **65**:301–312.
37. Crawley, J.N. 2000. *What's Wrong With My Mouse?: Behavioral Phenotyping Of Transgenic and Knockout Mice*. Wiley-Liss, New York.
38. Wahlsten, D., P. Metten and J.C. Crabbe. 2003. A rating scale for wildness and ease of handling laboratory mice: results for 21 inbred strains tested in two laboratories. *Genes Brain Behav.* **2**:71–79.
39. Willott, J.F., S. Carlson and H. Chen. 1994. Prepulse inhibition of the startle response in mice: relationship to hearing loss and auditory system plasticity. *Behav. Neurosci.* **108**:703–713.

40. Willott, J.F. and L.C. Erway. 1998. Genetics of age-related hearing loss in mice. IV. Cochlear pathology and hearing loss in 25 BXD recombinant inbred mouse strains. *Hear. Res.* **119**:27–36.
41. Wahlsten, D. and H.S. Ozaki. 1994. Defects if the fetal forebrain in acallosal mice. In: *Callosal Agenesi.* eds. Lassonde M., and M. Jeeves. pp 125–133. New York, Plenum Press.
42. Henry, K.R. and M.D. McGinn. 1992. The mouse as a model for human audition. A review of the literature. *Audiology* **31**:181–189.
43. Festing, M.F.W. 1999. Inbred strains of mice. *Mouse Genome Informatics*. The Jackson Laboratory, Bar Harbor, Maine (www.informatics.jax.org).
44. Yoshimure, T., M. Nishio, M. Goto and S. Ebihara. 1994. Differences in circadian photosensitivity between retinally degenerate CBA/J mice (rd/rd) and normal CBA/N mice (+/+). *J. Biol. Rhythms.* **9**:51–60.
45. Keeler, C. 1966. Retinal degeneration in the mouse is rodless retina. *J. Hered.* **57**:47–50.
46. Belknap, J.K., S.R. Mitchell, L.A. Toole, M.L. Helms and J.C. Crabbe. 1996. Type I and type II error rates for quantitative trait loci (QTL) mapping studies using recombination inbred mouse strains. *Behav. Genet.* **26**:149–160.
47. Crawley, J.N., J.K. Belknap, A. Collins, J.C. Crabbe, W. Frankel, N. Henderson, R.J. Hitzemann, S.C. Maxson, L.L. Miner, A.J. Silva, J.M. Whener, A. Wynshaw-Boris and R. Paylor. 1997. Behavioral phenotypes of inbred mouse strains: implications and recommendations for molecular studies. *Psychopharmacology (Berl).* **132**:107–124.
48. Iraqi, F. 2000. Fine mapping of quantitative trait loci using advanced intercross lines of mice and positional cloning of the corresponding genes. *Exp. Lung Res.* **26**:641–649.
49. Nadeau, J.H., J.B. Singer, A. Matin and E.S. Lander. 2000. Analyzing complex genetic traits with chromosome substitution strains. *Nat. Genet.* **24**:221–225.
50. Dietrich, W.F., J. Miller, R. Steen et al. 1996 A comprehensive genetic map of the mouse genome. *Nature.* **380**:149–152.
51. Schuckit, M.A. 1994. A clinical model of genetic influences in alcohol dependence. *J. Stud. Alcohol.* **55**:5–17.
52. Ehringer, M.A., J. Thompson, O. Conroy, D. Goldman, T.L. Smith, M.A. Schuckit, J.M. Sikela. 2002. Human alcoholism studies of genes identified through mouse quantitative trait locus analysis. *Addict. Biol.* **7**:365–371.
53. Mackay, T.F. 2001. The genetic architecture of quantitative traits. *Ann. Rev. Genet.* **35**:303–339.
54. Dietrich, W., H. Katz, S.E. Lincoln, H. Shin, J. Friedman, N. Dracopoli, and E.S. Lander. 1992. A genetic map of the mouse suitable for typing intraspecific crosses. *Genetics* **131**:423–447.
55. Dietrich, W.F., J.C. Miller, R.G. Steen et al. 1994. A genetic map of the mouse with 4,006 simple sequence polymorphisms. *Nat. Genet.* **7**:220–245.
56. Churchill, G.A. and R.W. Doerge. 1994. Empirical thresholds for quantitative trait mapping. *Genetics.* **138**:963–971.
57. Lander, E. and L. Kruglyak. 1995. Genetic dissection of complex traits: guidelines for interpreting and reporting linkage results. *Nat. Genet.* **11**:241–247.
58. Lander, E.S., P. Green, J. Abrahamson, A. Barlow, M.J. Daly, S.E. Lincoln and L. Newberg. 1987. MAPMAKER: an interactive computer package for constructing primary genetic linkage maps of experimental and natural populations. *Genomics.* **1**:174–181.

59. Broman, K.W., H. Wu, S. Sen and G.A. Churchill. 2003. R/qtl: QTL mapping in experimental crosses. *Bioinformatics.* **19**(7):889–890.
60. Nadeau, J.H., W.N. Frankel. 1999. The roads from phenotypic variation to gene discovery: mutagenesis versus QTLs. *Nat. Genet.* **25**:381–384.
61. Glazier, A.M., J.H. Nadeau and T.J. Aitman. 2002. Finding genes that underlie complex traits. *Science.* **298**:2345–2349.
62. Korstanje, R. and B. Paigen. 2002. From QTL to gene: the harvest begins. *Nat. Genet.* **31**:235–236.
63. Darvasi, A. 1998. Experimental strategies for the genetic dissection of complex traits in animal models. *Nat. Genet.* **18**:19–24.
64. Breese, E.L. and K. Mather. 1957. The organization of polygenic activity within a chromosome in drosophila 1. Hair characters. *Heredity.* **11**:373–395.
65. Davies, R.W. 1971. The genetic relationship of two quantitative characters in Drosophila *melanogaster* II. Location of the effects. *Genetics.* **69**:363–375.
66. Shrimpton, A.E. and A. Robertson. 1988. The isolation of polygenic factors controlling bristle score in *Drosophila melanogaster.* I. Allocation of third chromosome sternopleural bristle effects to chromosome sections. *Genetics* **188**:437–443.
67. Shrimpton, A.E. and A. Robertson. 1988. The isolation of polygenic factors controlling bristle score in *Drosophila melanogaster.* II. Distribution of third chromosome bristle effects within chromosome sections. *Genetics* **118**:445–459.
68. Darvasi, A. 1996. Interval-specific congenic strains (ISCS): an experimental design for mapping a QTL into a one centimorgan interval. *Mamm. Genome.* **8**:163–167.
69. Kernek, K.E., J. A. Trofatter, A.R. Mayeda and J.R. Hofstetter. 2004. A locus for circadian period of locomotor activity on mouse proximal chromosome 3. *Chronobiol. Int.* **21**:343–352.
70. Bennett, B., M. Beeson, L. Gordon, P. Carosone-Link and T.E. Johnson. 2002. Genetic dissection of quantitative trait loci specifying sedative/hypnotic sensitivity to ethanol: mapping with interval-specific congenic recombinant lines. *Alcohol Clin. Exp. Res.* **26**:1615–1624.
71. Ferraro, T.N., G.T. Golden, G.G. Smith, J.F. Martin, F.W. Lohoff, T.A. Gieringer, D. Zamboni, C.L. Schwebel, D.M. Press, S.O. Kratzer, H. Zhao, W.H. Berrettini and R.J. Buono. 2004. Fine mapping of a seizure susceptibility locus on mouse Chromosome 1: nomination of *Kcnj10* as a causative gene. *Mamm. Genome.* **15**:239–251.
72. Marshall, K.E., E.L. Godden, F. Yang, S. Burgers, K.J. Buck and J.M. Sikela. 2002. In silico discovery of gene-coding variants in murine quantitative trait loci using strain-specific genome sequence databases. *Genome Biol.* **3**: RESEARCH 0078.
73. Darvasi, A. 2003. Genomics: Gene expression meets genetics. *Nature* **20**:269–70.
74. Brem, R.B., G. Yvert, R. Clinton and L. Krugylak. 2002. Genetic dissection of transcriptional regulation in budding yeast. *Science.* **296**:752–755.
75. Schadt, E.E., S.A. Monks, T.A. Drake, A.J. Lusis, N. Che, V. Colinayo, T.G. Ruff, S.B. Milligan, J.R. Lamb, G. Cavet, P.S. Linsley, M. Mao, R.B. Stoughton and S.H. Friend. 2003. Genetics of gene expression surveyed in maize, mouse and man. *Nature.* **422**: 297–302.
76. Tabakoff, B., S.V. Bhave and P.L. Hoffman. 2003. Selective breeding, quantitative trait locus analysis and gene arrays identify candidate genes for complex drug-related behaviors. *J. Neurosci.* **23**:4491–4498.
77. Hitzemann, R., B. Malmanger, C. Reed, M. Lawler, B. Hitzemann, S. Coulombe, K. Buck, B. Rademacher, N. Walter, Y. Polyakov, J. Sikela, B. Gensler, S. Burgers, R.W. Williams, K.

Manly, J. Flint and C. Talbot. 2003. A Strategy for the Integration of QTL, gene expression and sequence analyses. *Mamm. Genome.* **14**:733–747.

78. Swerdlow, N.R., M.A. Geyer and D.L. Braff. 2001. Neural circuit regulation of prepulse inhibition of startle in the rat: current knowledge and future challenges. *Psychopharmacology (Berl)* **156**:194–215.
79. Davis, M., D. Rainnie, M. Cassell. 1994. Neurotransmission in the rat amygdala related to fear and anxiety. *Trends Neurosci.* **17**:208–214.
80. Hitzemann, B. and R. Hitzemann. 1997. Genetics, ethanol and the Fos response: a comparison of the C57BL/6J and DBA/2J inbred mouse strains. *Alcohol Clin. Exp. Res.* **21**:1497–1507.
81. Darvasi, A., M. Soller. 1995. Advanced intercross lines, an experimental population for fine genetic mapping. *Genetics* **141**:1199–1207.
82. Iraqi, F.A., J.M. Behnke, D.M. Menge, A.M. Lowe, A.J. Teale, J.P. Gibson, L.R. Baker and D.R. Wakelin. 2003. Chromosomal regions controlling resistance to gastro-intestinal nematode infections in mice. *Mamm. Genome.* **14**:184–191.
83. Iraqi, F. 2000. Fine mapping of quantitative trait loci using advanced intercross lines of mice and positional cloning of the corresponding genes. *Exp. Lung Res.* **26**:641–649.
84. Wang, M., W.J. Lemon, G. Liu, Y. Wang, F.A. Iraqi, A.M. Malkinson and M. You. 2003. Fine mapping and identification of candidate pulmonary adenoma susceptibility 1 genes using advanced intercross lines. *Cancer Res.* **63**:3317–3324.
85. Talbot, C.J., A. Nicod, S.S. Cherny, D.W. Fulker, A.C. Collins and J. Flint. 1999. High resolution mapping of quantitative trait loci in outbred mice. *Nat. Genet.* **21**:305–308.
86. Mott, R., C.J. Talbot, M.G. Turri, A.C. Collins and J. Flint. 2000. A method for fine mapping quantitative trait loci in outbred animal stocks. *Proc. Natl. Acad. Sci. USA* **97**:12649–12654.
87. Gershenfeld, H.K., P.E. Neumann, C. Mathis, J.N. Crawley, X.H. Li and S.M. Paul. 1997. Mapping quantitative trait loci for open-field behavior in mice. *Behav. Genetics.* **27**:201–210.
88. Hitzemann, R., K. Demarest, J. Koyner, L. Cipp, N. Patel, E. Rasmussen and J. McCaughran Jr. 2000. Effects of genetic cross on the detection of quantitative trait loci and a novel approach to mapping QTLs. *Pharmacol. Biochem. Behav.* **67**:767–772.
89. Hitzemann, R., B. Malmanger, S. Cooper, S. Coulombe, C. Reed et al. 2002. Multiple cross mapping (MCM) markedly improves the localization of a QTL for ethanol-induced activation. *Genes Brain and Behavior.* **1**:214–222.
90. Hitzemann, R., C. Reed, B. Malmanger, M. Lawler, B. Hitzemann, B. Cunningham, S. McWeeney, J. Belknap, C. Harrington, K. Buck, T. Phillips and J. Crabbe. 2004. On the integration of alcohol-related quantitative trait loci and gene expression analyses. *Alcohol Clin. Exp. Res.,* **28**:1437–1448.
91. Wade, C.M., E.J. Kulbokas, A.W. Kirby, M.C. Zody, J.C. Mullikin, E.S. Lander, K. Lindblad-Toh and M.J. Daly. 2002. The mosaic structure of variation in the laboratory mouse genome. *Nature.* **420**:574–578.
92. Wiltshire, T., M.T. Pletcher, S. Batalov, S.W. Barnes, L.M. Tarantino, M.P. Cooke, H. Wu, K. Smylie, A. Santrosyan, N.G. Copeland, N.A. Jenkins, F. Kalush, R.J. Mural, R.J. Glynne, S.A. Kay, M.D. Adams and C.F. Fletcher. 2003. Genome-wide single-nucleotide polymorphism analysis defines haplotype patterns in mouse. *Proc. Natl. Acad. Sci. USA* **100**:3380–3385.
93. Grupe, A., S. Germer, J. Usuka *et al.* 2001. In silico mapping of complex disease-related traits in mice. *Science.* **292**:1915–1918.
94. Darvasi, A. 2001. In silico mapping of mouse quantitative trait loci. *Science* **294**:2423.

95. Chesler, E.J., E.J. Rodriguez-Zas, J.S. Mogil *et al.* 2001. In silico mapping of mouse quantitative trait loci. *Science* **294**:2423.

96. Collison, M., A.M. Glazier, D. Graham, J.J. Morton, J.H.Dominiczak, T.J. Aitman, J.M. Connell, G.W. Gould and A.F.Dominiczak. 2000. *Cd36* and molecular mechanisms of insulin resistance in the stroke-prone spontaneously hypertensive rat. *Diabetes.* **49**:2222–2226.

97. Aitman, T.J., A.M. Glazier, C.A. Wallace, L.D. Cooper *et al.* 1999. Identification of Cd36 (Fat) as an insulin-resitance gene causing defective fatty acid and glucose metabolism in hypertensive rats. *Nat. Genet.* **21**:76–83.

98. Karp, C.L., A. Grupe, E. Schadt, S.L. Ewart *et al.* 2000. Identification of complement factor 5 as a susceptibility locus for experimental allergic asthma. *Nat. Immunol.* **1**:221–226.

99. Sandberg, R., R. Yasuda, D.G. Pankratz, T.A. Carter, J.A. Del Rio, L. Wodicka, M. Mayford, D.J. Lockhart and C. Barlow. 2000. Regional and strain-specific gene expression mapping in the adult mouse brain. *Proc. Natl. Acad. Sci. USA* **97**:11038–11043.

100. Carter, T.A., J.A. Del Rio, J.A. Greenhall, M.L. Latronica et al. 2001. Chipping away at comples behavior: Transcriptome/phenotype correlations in the mouse brain. *Physiol. Behav.* **73**:849–857

101. Lockhart, D.J. and C. Barlow 2001. Expressing what's on your mind: DNA arrays and the brain. *Nat. Rev. Neurosci.* **2**:63–68.

102. Geschwind, D.H. 2000. Mice, microarrays, and the genetic diversity of the brain. *Proc. Nat. Acad. Sci. USA.* **97**:10676–10678.

103. Belknap, J.K., R. Hitzemann, J.C. Crabbe, T.J. Phillips, K.J. Buck and R.W. Williams. 2001. QTL analysis and genome wide mutagenesis in mice: complementary genetic approaches to the dissection of complex traits. *Behav. Genet.* **31**:5–15.

104. Phillips,T., J. Belknap, R. Hitzemann, K. Buck *et al.* 2002. Harnessing the mouse to unravel the genetics of human disease. *Genes, Brain and Behavior.* **1**:14–26.

105. Flint, J. and R. Mott. 2001. Finding the molecular basis of quantitative traits: Successes and pitfalls. *Nat. Rev. Genet.* **2**:437–445.

106. Wayne, M.L. and L.M. McIntyre. 2002. Combining mapping and arraying: An approach to candidate gene identification. *Proc. Nat. Acad. Sci. USA.* **99**:14903–14906.

107. Yvert, G., R.B. Brem, J. Whittle, J.M. Akey, E. Foss, E.N. Smith, R. Mackelprang, and L. Kruglyak. 2003. Trans-acting regulatory variation in Saccharomyces cerevisiae and the role of transcription factors. *Nat. Genet.* **35**:57–64.

108. Thanos, P.K., N.B. Taintor, S.N. Rivera, H. Umegaki, H. Ikari, G. Roth, D.K. Ingram, R. Hitzemann, J.S. Fowler, S.J. Gately, G.J. Wang, and N.D. Volkow. 2004. Drd2 gene transfer into the nucleus accumbens of the alcohol preferring and nonpreferring rats attenuates alcohol drinking. *Alcohol Clin. Exp. Res.* **28**:720–728.

109. Thanos, P.K., N.D. Volkow, P. Freimuth, H. Umegaki, H. Ikari, G. Roth, D.K. Ingram, and R. Hitzemann. 2001. Overexpression of dopamine D2 receptors reduces alcohol self-administration. *J. Neurochem.* **78**:1094–1103.

CHAPTER 10

GENETICS OF DISEASE RESISTANCE

[1,3]Susan L. Ewart and [2,4]Ravisankar A. Ramadas

[1]*Department of Large Animal Clinical Sciences,* [2]*Comparative Medicine and Integrative Biology Program, College of Veterinary Medicine, Michigan State University, East Lansing, Michigan, USA*

[3]*ewart@cvm.msu.edu*

[4]*ramadasr@cvm.msu.edu*

1. Introduction

Disease resistance and susceptibility are two sides of a very important coin that determine individual outcomes to a variety of diseases. Attention has recently turned to consideration of the specific factors that contribute to disease resistance. This new look at disease resistance has been made possible by advances in biomedical knowledge, which have begun to reveal cellular and disease-specific mechanisms for a number of conditions. While many factors likely interact to determine disease outcome, one of the more recently recognized features contributing to disease resistance is the genetic constitution of the host.

The degree of influence that the host genetic makeup has on disease outcome varies with both the disease and the individual. For example, disease outcome is entirely dependent on the host genotype at the disease gene in simply inherited Mendelian traits with high penetrance, such as Duchenne muscular dystrophy, cystic fibrosis, and Huntington's disease. Quite simply, individuals carrying the disease-causing mutation in one gene will acquire the disease (susceptible), and those without it are not affected (resistant). This is the definitive illustration of the genetics of disease resistance.

In contrast to Mendelian diseases, complex genetic diseases are those that result from the interaction between multiple genes and various environmental factors. The genetic underpinnings of complex genetic diseases are demonstrated by their tendency to run in families, but without a simple mode of inheritance. Examples of complex genetic diseases include asthma, stroke, hypertension, diabetes and a number of cancers. Because multiple genetic and environmental factors contribute to disease outcomes, the study of these traits is complicated, and our understanding of disease mechanisms is consequently limited. Unlike Mendelian traits, mutations in single genes do not cause complex genetic diseases; thus, we are not searching for causative mutations, but instead for predisposing polymorphisms. Indeed, the genes underlying complex traits are termed susceptibility loci. Disease predisposing alleles at susceptibility loci can and do occur in many unaffected individuals, and may even be common in a population. The contribution of a single susceptibility locus may be quite small. Not only does the presence of susceptibility alleles appear to underlie disease outcomes in relationship to environmental exposures, but the interactions between these genes and the environment work in complex ways to determine final outcomes.

Susceptibility to disease is also determined by the genetic background in which the disease predisposing genotype is located. This has been experimentally confirmed by designing appropriate congenic strains of mice having different genes or chromosomal segments, suspected to predispose to susceptibility, on a non-disease predisposing background. For example, congenic mice containing *Sle1/Sle2/Sle3* loci on a C57BL/6 background have severe lupus nephritis. However, these same genes on the NZW strain background result in significantly suppressed disease, suggesting that the genetic background affects gene expression.[1] Thus, the inheritance of disease susceptibility is likely to be determined by the cumulative content of susceptibility loci that an individual inherits.

Aside from Mendelian or complex diseases with recognized genetic underpinnings, there has been growing interest in the concept that genetic makeup plays a role in susceptibility to diseases previously thought to be strictly of environmental origins, such as infectious diseases. This interest stems from the recognition that genetic susceptibility underlies the differential response to pathogens such that microbial exposure leads to disease in some individuals and not others. Thus, individual variation in the genetic control of the immune response to pathogen exposure has been the focus of recent studies of disease resistance. Furthermore, while familial patterns of disease resistance or susceptibility may not be readily apparent for most infectious conditions, it is likely that they exist.

In this chapter we review the contributions of mouse models to the study of disease resistance. The information presented here is an overview of this research-intense area. We have selected examples of diseases with genetic bases for resistance with the aim of discussing a diverse selection of diseases.

2. Utility of Mouse Models

The mouse is the major model for the study of human disease and normative biology. The study of inbred mouse strains has yielded immense advances in many research areas, including the fields of immunology, physiology, and genetics. Although many limitations are inherent to the study of humans or non-rodent animals, these limitations can often be minimized, if not avoided entirely, by the use of murine models. The use of laboratory mice affords investigators an extensive level of control over the genetic makeup of the study population. Genetic variation is minimal within inbred strains and the inheritance of phenotypes, genes, and genetic sequences can be tracked in hybrid crosses. Additionally, more sophisticated manipulation of genetic material of embryos and genomes can be performed in mice using congenic and consomic breeding strategies and transgenic and targeted gene deletion techniques. This precise level of control over the genetic makeup of the study population is invaluable and previously unparalleled in the mouse.

A note of explanation regarding genetic nomenclature may be helpful at this point. The symbols for gene names are italicized. Mouse gene symbols begin with an uppercase letter followed by all lowercase letters (e.g,. *Gene*), while gene symbols for humans and other species are written in uppercase letters (e.g., *GENE*).

The haploid content of mammalian DNA is constant across species at about three billion basepairs. The genomic organization is largely conserved across species as well, with typical segments of 10–20 million basepairs conserved virtually intact between mice, humans, and other species.[2] This homology between species allows direct genomic comparisons based on mouse data. Furthermore, the extent of knowledge of the murine genome closely parallels that known for humans and far surpasses that of other mammals.

Laboratory mice are routinely housed in strictly controlled environmental conditions that further minimize phenotypic variation between individuals within an inbred strain. Factors such as ambient temperature, humidity, light:dark cycle, air quality, and diet can be controlled throughout an animal's life. Additional sources of variation in the expression of various diseases including diversity in life style (diet, smoking status), occupational exposures, and community setting (rural *vs* urban) can be essentially eliminated in a mouse model.

One of the most compelling reasons to study mouse models of disease is the ability to test hypotheses by performing functional studies in mice. Clearly, ethical considerations strictly limit the amount and type of testing that can be done in humans. While the results obtained from the study of murine models do not necessarily predict specific genes or mechanisms involved in diseases of other species, murine models may be useful to indicate common pathways that contribute to these phenotypes. Mice have long been used to study mechanisms of immunologic and physiologic processes, which has resulted in the development of many reagents, antibodies, and recombinant proteins specific for the mouse. Using these murine-specific tools, investigators are able to manipulate levels of target proteins and test for their influence on the phenotype of interest. In summary, inbred murine models have proven to be valuable in biomedical research and have gained respect and acceptance as important tools for studying human disease.

3. The Genetics of Resistance to Pathogens

The immunologic response to challenge by various microbes or other foreign material is determined by the presenting threat and the host response. Responses include mechanical removal of the agent, which occurs by a variety of means: intact skin as an effective barrier, filtration and entrapment in nasal passages, deposition in conducting airways with removal by the mucociliary apparatus, and subjection to digestive enzymes. If a foreign particle penetrates these barriers, then the foreign particle will come in contact with the innate immune system composed of circulating and tissue specific monocytes, alveolar macrophages, and dendritic cells as antigen presenting cells, and other white blood cells. Cells of the innate immune system rely heavily on their pattern recognition receptors to identify a few highly conserved structures shared by many microbes, such as lipopolysaccharide (LPS), bacterial DNA, double stranded RNA, mannans, and glucans. In this way, limited genetic diversity is required to produce receptors that recognize the wide array of pathogens, and a response can be generated immediately without the delay due to clonal expansion. The innate immune system is critical in the decision regarding whether activation of an immune response is appropriate, as in the case of microbial pathogens, or inappropriate, as in the case of innocuous environmental antigens or self antigens.

Genetic alteration of the innate immune system components can lead to inactivation of innate immunity. For example, C3H/HeJ and C57BL/10ScCr mice carry loss-of-function mutations in the toll-like receptor 4 (*Tlr4*) gene,[3] a pattern recognition receptor for LPS, rendering these mice highly susceptible to gram-negative bacterial infections. Conversely, because mice with *Tlr4* defects do not respond to endotoxin, they are resistant to endotoxic shock. Another innate immune response gene that affects disease resistance is the solute carrier family 11 (proton-coupled divalent metal ion transporters), member 1 gene (*Slc11a1*), previously termed *Nramp*. The *Slc11a1* gene affects macrophage activation and determines antimicrobial activity through its expression on endosomes and lysosomes within macrophages.[4] Disease resistance to pathogens that survive in macrophages, such as *Salmonella, Leishmania* and *Mycobacteria* species, is influenced by genetic variation in *Slc11a1*.[4]

Along with providing immediate host defense, the innate immune system primes the adaptive immune system when a pathogen is present. The adaptive immune response is organized around B and T cells; antigens are recognized by either immunoglobulins secreted by B-cells or by antigen-specific receptors on T-cells. Antigen-presenting cells present the processed foreign pathogenic peptide to the T-cell receptors and natural killer (NK) cells with the help of specialized host-cell glycoproteins called major histocompatibility (MHC) molecules. These MHC molecules were first discovered in mice,[5,6] and they are the foundation of self/non-self discrimination. There are two major types of MHC molecules that are related in structure and function, MHC-I and MHC-II. Every receptor in the effector cells can recognize only a specific combination of MHC and antigenic peptide. This process, called MHC restriction, adds to the specificity of the immune response by enabling host organisms to respond to different types of antigens. The interaction between T cells and the MHC is critical in determining the host response to pathogens.

3.1. *The Major Histocompatibility Complex Affects Resistance to Microbes*

The majority of knowledge to date regarding individual susceptibility or resistance to pathogens is based on variation in the MHC molecules. The MHC molecules are encoded by a set of linked genes located on chromosome 17 in mice. The MHC genes in humans are located on chromosome 6 and are referred to as human leukocyte antigen (HLA). This multigene family includes loci that are among the most polymorphic known in any organism.[7] The remarkable polygenic genetic variability in the MHC complex greatly extends the range of peptides that can be presented to T cells by each individual and each population at risk from an infectious pathogen.[8-10] Studies in coisogenic mouse strains (mice that differ by a single defined locus as a result of targeted mutation) have shown that MHC polymorphisms significantly shape the immune defense by allowing selection of the most efficient precursors from a diverse T cell receptor (TCR) repertoire.[11] This intrathymic positive selection results in the diversification of the immune repertoire and an increase of the defensive fitness of the species through protective immunity.[12]

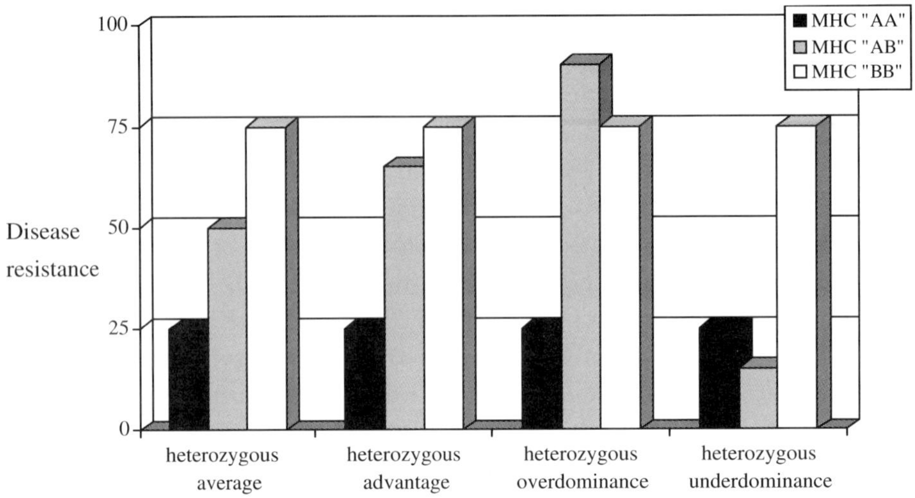

Figure 1. The extent of disease resistance may be influenced by heterozygosity at MCH loci. The response to pathogens of individuals heterozygous at MHC loci as compared to homozygous states may be midway between the homozygotes (heterozygous average), above the homozygote average (heterozygous advantage), greater than the most resistant (heterozygous overdominance) or less than the least resistant homozygote (heterozygous underdominance).

There are several theories regarding the relationship between the MHC genotype and disease resistance. One is that individual MHC alleles may be more or less effective in presenting antigens from particular infections, so that the presence of a given MHC allele predisposes a pathogen-exposed individual to a more or less favorable outcome. This outcome has been repeatedly documented in human populations by various methods of genetic epidemiology. For example, specific alleles at HLA genes have been linked to resistance to AIDS (HLA-B14, HLA-C8), hepatitis B (HLA-DR2), and hepatitis C (HLA-DRB1^{P}1101, HLA-DQB1^{P}0301),[13,14] while genetic susceptibility to these diseases has been associated with different HLA alleles: HLA-A29 and HLA-B22 for AIDS; HLA-DR7 for hepatitis B; and HLA-DRB1^{P}0701 and HLA-DRB4^{P}0101 for hepatitis C.[13,14]

MHC heterozygosity is also postulated to be critical in maintaining the ability to recognize diverse antigens presented to T cells by generating a diverse T cell repertoire. Since each MHC allele provides an individual with the ability to present a particular set of antigens, individuals who are heterozygous (e.g., genotype "AB") at a particular MHC locus may mount a more vigorous immune response to a given infection, resulting in a better outcome than individuals who are homozygous for either of the corresponding alleles ("AA" or "BB"). If the pathogen resistance of heterozygotes is more than the average of the resistance of the homozygotes, it is referred to as heterozygote advantage (Figure 1). The phenomenon of heterozygous advantage is demonstrated in the studies of Penn *et al.* in which they infected different MHC haplotype-containing congenic strains of mice with avirulent *Salmonella* strains and found that MHC heterozygosity conferred a selective advantage against infection compared to the average of homozygotes, but was never significantly more resistant than the most resistant parental homozygote.[15]

In other circumstances MHC heterozygotes have higher fitness than both the homozygotes, a phenomenon referred to as heterozygote overdominance[15] or superiority (Figure 1). Allele-specific overdominance has been suggested by several animal studies. In experimental coinfections of *Salmonella enterica* (serovar *typhirmurium* C5TS) and Theiler's murine encephalomyelitis virus in MHC-congenic mice, heterozygotes were superior in terms of disease resistance to both dominant and recessive homozygotes, suggesting that heterozygote superiority may be the result in the case of simultaneous or sequential infection by more than one pathogen.[16]

Alternatively, and less common than MHC heterozygosity-driven heightened immune response, individuals heterozygous for MHC alleles may be less effective at antigen presentation with resultant worse disease than either of the homozygous states, as the result of allele-specific underdominance (Figure 1).[17]

The mode of inheritance of the resistance phenotype can follow traditional inheritance patterns of dominance, when the outcome of the heterozygote is more like the disease-resistant homozygote, or resistance, in which the heterozygote outcome is more similar to that of the disease-prone homozygote (Table 1).

Variation at the MHC locus is also maintained by selection factors like maternal-fetal interaction and negative-assortative mating. Negative-assortative mating occurs when individuals mate with others that differ in MHC alleles more often than the rate expected under random mating conditions. The support for negative-assortative mating based on MHC comes mainly from experimental work with mice.[23] It has been suggested that the negative-assortative mating observed in populations of house mice results in increased heterozygosity, which has been implicated both in resistance to diseases and in the avoidance of inbreeding depression.[24]

Table 1. MHC heterozygosity-associated resistance in congenic mice

Pathogen	MHC heterozygote	Resistance	Reference
Mouse hepatitis virus	*b/f*	Dominant	18
Cytomegalovirus	*b/d, b/k, d/k*	Dominant	18
Theiler's virus	*b/q, f/k*	Dominant	16
Murine AIDS retrovirus	*a/b, b/d, b/k*	Recessive	19
Murine leukemia virus	*a/b*	Recessive	20
Murine leukemia virus	*a/b, b/pa*	Dominant	21
Salmonella enterica (Typhimurium)	*d/k, d/q, f/k* males	Dominant	16
Salmonella enterica (Typhimurium)	*d/k, d/q, f/k* females	Recessive	16
Streptococcus pyogenes	*b/d, d/k*	Dominant	22
Streptococcus pyogenes	*b/k*	Recessive	22

Table adapted from. [16]

A heritable basis for disease resistance to microbial pathogens has been observed in domestic animal species, as well as mice. Studies in mice have greatly facilitated our understanding of parallel processes in other species, especially with regard to immune mechanisms of pathogen resistance and candidate genes for heritable traits. Of the domestic species, understanding of the genetics of disease resistance is most advanced in poultry. Breed and strain-specific resistance to a variety of poultry pathogens including *Salmonella* serotypes, Mareks disease, Newcastle disease, and lymphoid leukosis virus, has long been established, and a number of specific susceptibility genes have been identified.[25,26] This knowledge is commonly used in poultry production and management decisions. Other notable examples of heritable pathogen resistance in domestic animals include the inheritance of receptors for *E. coli* fimbrae (K88ab and K88ac) that influence diarrheal disease in pigs[27] and heritability of somatic cell counts and clinical mastitis in dairy cattle.[28]

3.2. *Host Resistance to Parasitic Infection*

Both MHC and non-MHC genes have been implicated in resistance against parasites. The pathology of gastrointestinal nematodes has been extensively examined in mice, and several common features with human inflammatory bowel disease, celiac disease, and other gastrointestinal diseases have been found. Different strains of mice exhibit varying degrees of resistance to parasitic infections, as indicated by worm burden or time to clearance.

The majority of genes that have been found to influence susceptibility to parasitic diseases have been MHC genes (Table 2). For example, quantitative trait loci for worm burdens in *Heligmosomoides polygyrus* have been identified on murine chromosomes 1, 2, 13 and 17, using CBA (H-2^k) and SWR (H-2^q)

mice.[29] In this study a significant locus for worm burden was found on chromosome 17 in the region of the MHC locus; however, non-MHC regions were identified as well. Other non-MHC parasite resistance genes include a gene on murine chromosome 4 linked to the *Fv1* locus and the gene for interferon-γ (*IFNG*) on chromosome 3 in the ovine genome (Table 2).

Table 2. Resistance to gastrointestinal nematodes

Nematode	Host	Susceptibility locus	Alleles[a] R	Alleles[a] S	Reference
Heligmosomoide polygyrus	Mice	*H-2* near *I-Ea*	q, f, s	k	30
Trichuris muris	Mice	*H-2, I-A*	q, b	k, d	31
Trichuris muris	Mice	*H-2, D* end	q	b, d	31
Trichinella spiralis	Mice	*H-2, I-Ab*	s, q, f, b	k	32
Trichinella spiralis	Mice	near *Fv1*			33
Haemonchus contortus	Sheep	*IFNG*	A	B	34
Teladorsagia circumcincta	Sheep	MHC-*DRB1*	G2	I	35
Teladorsagia circumcincta	Sheep	*IFNG*	126	130	36

[a]R indicates resistance alleles and S indicates susceptibility alleles.
Table adapted from. [7]

T helper 2 (Th2) cytokines play a balancing act with Th1 cytokine responses, and both Th1 and Th2 cells have been found to mediate resistance to parasites at the cytokine level. Indeed, the initial insights into the immune regulatory role of the balance between Th1 and Th2 cells was first identified in a murine model of infection with the protozoan parasite *Leishmania major*. Genetic predisposition to *L. major* disease susceptibility, seen in BALB/c mice, is associated with sustained Th2 cell responses and inadequate Th1-derived interleukin (IL)-12 and IFN-γ to clear the infection. Alternatively, mouse strains that can generate a strong Th1 response following an initial Th2 response, such as the C3H/He and C57BL/6 strains, are able to clear the parasite.[37,38]

In contrast to *L. major*, clearance of nematode infections relies more on a sustained Th2 response. Among the Th2 cytokines, IL-5 mediates the parasitic-induced eosinophilic response, IL-4 upregulates immunoglobulin (Ig)E levels, and IL-13 is critical in worm expulsion.[29] In susceptible mouse strains, overexpression of the Th1-type cytokines IFN-γ and IL-12 causes delayed worm expulsion. The role of these cytokines has been explained in knockout mouse models by infection with *Trichinella spiralis*, *Nippostrongylus brasiliensis* and *Trichuris muris* (reviewed by Mahida[39]).

The information on genetics of disease resistance obtained in mouse models of parasitic infections can be extended to observations made in domestic animal species (Table 3). The precise mechanisms of parasitic disease resistance in livestock species have not been fully elucidated, but their occurrence deserves highlighting. Selection of breeds or individual animals based on genetic resistance to parasites may be widely applicable and has economic merit.[40]

Table 3. Breed-specific resistance to parasite infestations in sheep and cattle

Parasite	Relative resistance of sheep and cattle breeds	Reference
Hemonchus contortus	St. Croix > Barbados blackbelly > Finn-Dorset	41
Nematodirus spathiger	Lacaune > Romanov	42
Amblyomma variegatum	N'Dama *(Bos taurus)* > Zebu *(Bos indicus)*	43
Boophilus decoloratus	Nguni > Brahman > Santa Gertrudis > Simmentaler	44
Babesia spp.	*Bos indicus* > *(Bos indicus* x *Bos taurus)* cross	45
Hyalomma spp.	Nguni > Bonsmara or Hereford	46
Psoroptes ovis	Holstein-Friesian > Belgian White and Blue	47

Thus, the immunologic basis of susceptibility or resistance to various parasitic diseases is well documented. Furthermore, the evidence for a genetic predisposition to these various immune responses is clear. What remains to be identified are the specific genes that determine the immune responses and ultimately the disease outcomes.

3.3. *The Genetics of Prion Disease Susceptibility*

Prion diseases, also known as transmissible spongiform encephalopathies (TSEs), are a group of fatal neurodegenerative disorders characterized by long incubation periods. They include Creutzfeldt-Jakob disease (CJD), variant CJD, Gerstmann-Straussler-Schienker disease, fatal familial insomnia, and kuru in humans, bovine spongiform encephalopathy (BSE) in cattle, scrapie in sheep, and chronic wasting disease (CWD) in deer and elk. These diseases are believed to be caused by prions, whose infectivity is thought to be due to a conformational alteration of a normal host protein. This host protein is a glycophosphatidylinositol anchored cellular protein, termed PrP. The host gene encoding the PrP protein is *PRNP*. The *PRNP* gene contains 3 exons and resides on murine chromosome 2, 75 cM from the centromere, and on human chromosome 20pter–p12. Only *PRNP* exon 3, which is preceded by a 10 kb intron, is coding.[48] Two isoforms of PrP have been identified: the normal isoform PrPc, which has an alpha helical configuration and the disease-associated isoform PrPsc, which is characterized by a high β-sheet content. The central lesion of TSEs occurs when the normal isoform of the host PrPc protein is modified into the abnormal PrPsc isoform. The mechanism for this protein modification is unknown, but appears to involve direct interaction of abnormal PrPsc with the normal PrPc isoform to enact the pathologic conformational change. Subsequent accumulation of the abnormal PrPsc protein in the brain leads to the pathology of spongiform encephalopathy.

Prion diseases are generally classified as inherited, sporadic, or acquired in nature. Inherited forms result from germline mutations in the *PRNP* gene. In these familial forms of TSEs, the *PRNP* gene mutation allows spontaneous conversion of PrPc into PrPsc at a rate that results in disease.[49] More than 20 mutations have been identified in families suffering from inherited prion diseases.[48] Alternatively, somatic mutations may arise in the *PRNP* gene resulting in sporadic disease. About 85% of the human prion diseases occur as sporadic CJD, and about 15% are associated with autosomal dominant pathogenic mutations in *PRNP*.[50] Other forms of TSEs are not linked to host gene mutations, but rather result from rare sporadic events that can generate *de novo* PrPsc due to spontaneous conversion of PrPc to PrPsc. TSE diseases can also be acquired subsequent to exposure to diets containing the infectious form of the prion protein.[49,51] This appears to have been the mechanism of transmission causing the outbreaks of kuru in Papua New Guinea in the 1950s, BSE in the U.K. during the 1980s–90s, and variant CJD in the U.K. in the mid-1990s. Lastly, horizontal transmission appears to be the main method of disease propagation of scrapie in sheep and CWD in deer and elk.

3.3.1. *Mice in Prion Disease Research*

Murine models have greatly facilitated our understanding of prions as unique infective proteins. Mice have been extensively used to investigate the prion disease process, beginning with a series of knockout and transgenic mice designed to study prion protein with regard to its physiological role in TSE diseases (reviewed by Weissmann and Flechsig[48]). *Prnp* knockout mouse models for scrapie have shown that mice lacking the normal PrP isoform ($Prnp^{o/o}$) are resistant to the pathology and clinical signs of scrapie.[52] Furthermore, these $Prnp^{o/o}$ mice showed no gross phenotype and exhibited normal development and behavior.[53] This result demonstrated that the normal prion protein isoform, PrPc, is required for the conversion to PrPsc mediated by the abnormal isoform, but it is apparently not indispensable in normal physiology. Moreover, *in vitro* experiments showed that just the presence of PrPc together with PrPsc in a cell-free environment could result in the transformation of the normal PrPc to the abnormal PrPsc isoform.[54] Similar experiments in yeast showed that the yeast prion-like protein $Sup35p^{psi+}$ alone can cause a prion-like conversion of the normal $Sup35p^{psi-}$ in a cell-free system.[55]

These studies provide ample evidence for the nucleic acid-free, protein-only infectious mechanism proposed by Prusiner,[56,57] which is a widely accepted view of the disease process. However, while these results expand our knowledge of the molecular workings of prions, they do not fully explain the TSE disease process. For example, chronic deposition of PrPsc did not damage brains of *Prnp* knockout mice;[58] therefore, accumulation of PrPsc alone does not cause disease.

Incubation times in experimentally infected mice, characterizing the rate of progression of the disease, are widely used indices for susceptibility to prion diseases in experimental mouse models. Incubation times can be affected by both host factors and properties of the infecting prion strain. Strain variation in prion proteins is characterized by differences in biochemical properties, length of incubation period between infection and the development of clinical disease, pathological changes in the brains of infected animals and the ease of transmission to other species. Thus, strain variation in the prion protein contributes to the diversity of TSEs. A host factor that affects the course of disease is genetic variation in the *Prnp* gene. Two of the most common alleles present in the murine *Prnp* gene arise from dimorphisms in codons 108 (leucine/phenylalanine) and 189 (threonine/valine).[59] The haplotypes *Prnp*^{a} (leucine-108, threonine-189) and *Prnp*^{b} (phenylalanine-108, valine-189) are associated with short and long incubation times, respectively.[60] Furthermore, the effect of the *Prnp* genotype on incubation time varies depending on the strain of prion,[57,61] the effect being related to the rate of progression of the disease rather than to differing susceptibilities of infection.[62]

Incubation times are also influenced by the genetic background on which the *Prnp* gene is expressed. Studies in inbred lines showed that large differences in incubation times exist between different inbred strains of mice, even if they have the same amino acid sequence of the prion protein. This suggested that apart from the *Prnp* locus, other genes contribute to the disease process. Several quantitative trait loci (QTL) affecting prion incubation time in mice have been characterized (Table 4). Mapping these quantitative trait loci was aided by the dramatic difference in incubation time among different inbred strains of mice, and the consistency of incubation times within inbred strains following intracerebral inoculation with PrP molecules.[59,63,64] The specific TSE-related genes within these candidate regions await identification.

Thus, mouse models have largely been useful in elucidating the pathological process of TSEs over time, transmissibility across species, and the isolation of different strains. They provide valuable information to understand the disease process and the possible intervention strategies to select against the disease for future breeding purposes in livestock.

Table 4. Quantitative trait loci affecting prion disease

Mouse strain[a]					
Resistant	Susceptible	Trait	Inoculum	QTL location[c]	Ref.
CAST/Ei (172)	SJL/J (105)	Incubation time	Scrapie strain	9, 11	65
CAST/Ei, (188)	NZW/OlaHSd (108)	Incubation time	Scrapie strain	2, *6*, *7*, 11, 12	66
C57BL (163–173)[b]	RIII (160–165)[b]	Susceptibility	Scrapie strain	5, *4*, *6*, 7, *8*, *17*	67
CAST/Ei, (188)	NZW/OlaHSd (108)	Incubation time	BSE strain	2, 11	68
C57BL (541)	RIII (441)	Incubation time	BSE homogen.	2, 4, 8, 15	69

[a]Numbers in parentheses are days to trait manifestation.
[b]This study reported susceptibility by gender: C57BL males 173 d, females 163 d; RIII males 165 d, females 160 d.
[c]Numbers indicate murine chromosomes containing quantitative trait loci (QTL); italicized chromosomes contain suggestive evidence of linkage (lod > 1.9), remaining chromosomes had significant evidence for linkage (lod > 3.2).

3.3.2. *Scrapie in Sheep*

Scrapie is the TSE affecting sheep. Spread of disease is caused by horizontal transmission of infective particles among animals, with exposure of newborn lambs and herdmates to infected placentas being a presumptive major source of transmission. Scrapie is endemic in most parts of the world, but has been eradicated in flocks in Australia and New Zealand by means of strict importation policies and long quarantines. Scrapie is a disease of economic importance due to its clinical course in affected animals and the fact that flock depopulation is one of the means of control.

Table 5. *PRNP* common genotypes and associated risk of scrapie

PRNP genotype[a]	Risk group	Interpretation
ARR/ARR	R1	Genetically resistance to scrapie, low risk in 1st generation progeny
ARR/ARQ ARR/ARH ARR/AHQ	R2	Sheep and 1st generation progeny are at low risk, not as low as R1
AHQ/AHQ ARH/ARH ARQ/ARH AHQ/ARH ARQ/AHQ ARQ/ARQ	R3	Genetically limited resistance to scrapie, progeny may be susceptible depending on genotype of other parent
ARR/VRQ	R4	Genetically susceptible to scrapie
ARQ/VRQ ARH/VRQ AHQ/VRQ VRQ/VRQ	R5	Highly susceptible to scrapie

[a]Genotypes at codons 136, 154 and 171. Amino acid symbols: A, alanine; H, histidine; Q, glutamine; R, arginine; V, valine.

Scrapie is unique in the fact that susceptibility to this disease in sheep is under such strong genetic influence. Indeed, scrapie may be the best example of the effect of gene-environment interaction on disease susceptibility; susceptibility genotype and an infectious agent are both required for scrapie to develop.[70] The *PRNP* gene amino acid composition at codons 136 (valine/alanine), 154 (arginine/histidine) and 171 (arginine/glutamine/histidine) largely determines scrapie susceptibility in sheep. Of these codons, 171 contributes most to disease outcome with coding for the amino acid arginine (R) conferring resistance while coding for glutamine (Q) confers susceptibility. While multiple combinations of alleles are possible at these codons, there are five haplotypes predominant in most breeds (Table 5).[71] Furthermore, there are distinct breed susceptibilities to scrapie, and these are due to the breed-specific allele frequencies at the susceptibility codons. For example, scrapie is most common in black-faced European breeds such as Suffolk, Hampshire, Shropshire and Oxford. This outcome is likely due to the fact that approximately 5% of U.S. black-faced sheep are homozygous for the resistance allele (R/R) at codon 171, 40% are heterozygous R/Q and 55% are homozygous for the codon 171 susceptibility allele Q/Q. On the other hand, not all breeds carry the VRQ susceptibility allele at *PRNP*. Thus, genotype information provides a valuable tool for selective breeding of sheep that are resistant to scrapie, and scrapie eradication plans in the U.S. and U.K. are based on the strong association of the ARR allele with disease resistance.

3.3.3. *Bovine Spongiform Encephalopathy*

Bovine spongiform encephalopathy, commonly referred to as "mad cow disease," received wide-spread media attention when it reached epidemic proportions in the U.K. in the late 1980s–early 1990s. It is generally believed that BSE in these cattle was a consequence of consuming feed containing tissue byproducts of scrapie-infected sheep or BSE-infected cattle. The practice of using ruminant-derived tissues as a protein source in feed was banned in 1988 after this link between feed and BSE was identified. Despite this cattle feed practice ban, a novel human prion disease, variant CJD appeared in the U.K. in 1995. Experimental evidence indicted that variant CJD was caused by the same prion causing BSE in cattle and linked the consumption of contaminated meat or nerve tissue from BSE-infected cattle with this new human disease.[51]

Because of the link to human disease, much effort has gone into understanding the molecular mechanisms of BSE. Unlike the relatively polymorphic *PRNP* gene in humans or sheep, less genetic variation has been identified in the bovine *PRNP* gene. Furthermore, the strong genetic association between *PRNP* genotype and susceptibility to prion disease as seen in sheep has not been identified in cattle. For example, the majority of polymorphisms identified in the bovine *PRNP* gene — an octapeptide repeat,[72] silent mutations,[73] and SNPs causing amino acid substitutions,[74] — have little[75] or no[76] evidence of an association to the incidence of prion diseases. In contrast, a 23-bp insertion/deletion polymorphism in the putative *PRNP* promoter found in German cattle breeds was recently reported to have significant differences in distribution between BSE-affected and healthy animals.[77] This exciting result awaits confirmation in more animals and other breeds.

In other species, BSE challenge is associated with the host genotype at *PRNP*. For example, the incubation period was related to the host *PRNP* genotype in goats experimentally exposed to BSE.[76] In mouse models, bovine PrP transgenic mice with one extra octapeptide repeat insertion mutation (7OR-PrP) showed reduced incubation and survival times after BSE prion inoculation compared with transgenic mice expressing bovine PrP.[78]

3.3.4. *Chronic Wasting Disease of Cervidae*

Chronic wasting disease is a TSE affecting specific species of native North American deer, including mule deer and white-tailed deer, as well as Rocky Mountain elk. Both captive (farmed) and free-living populations can be affected.[79] The clinical course of CWD is progressive and includes weight loss, behavioral changes and ataxia. Chronic wasting disease was first identified in the 1960s in Colorado and Wyoming. Its occurrence has spread to several neighboring states in recent years. This spread, along with the links between BSE and variant CJD, has intensified concerns regarding the importance of CWD and its potential transmissibility to humans or domestic livestock.

Like scrapie, CWD appears to be a communicable disease with horizontal transmission among animals of susceptible species. Fortunately, there are currently no confirmed cases of spread of CWD beyond the Cervidae; no disease transmission was observed from affected Cervidae to cattle, despite more than 5 years of co-mingling. As has been found in several other prion diseases, the presence of a molecular barrier based on PrP sequence homology that limits susceptibility of humans, cattle and sheep to CWD has been suggested.[80] However, a report from investigators at the Centers for Disease Control and Prevention describes CJD in three unusually young human subjects who had regularly consumed venison and elk.[81] While this report did not confirm an association between CWD and these human cases, it serves to underscore the importance of understanding this prion disease.

Several polymorphisms have been reported in the coding region of the *PRNP* gene in deer, seven of which were non-synonymous.[82] Additional polymorphisms were present in the *PRNP* promoter. O'Rourke *et al.* found an association between the methionine and leucine alleles at PrP codon 132, with homozygosity for methionine conferring susceptibility to CWD in elk. CWD was absent in the 132leucine/leucine elk population.[83] Thus, CWD, like scrapie, may be due to gene-environment interactions, with host genetic susceptibility playing a role.

3.3.5. *Human Prion Diseases*

Creutzfeldt–Jakob disease, variant CJD, Gerstmann–Straussler–Scheinker disease, kuru, and fatal familial insomnia are the major prion diseases in humans. Like the afore described diseases in animals, TSEs in humans are characterized by incubation periods of years to decades followed by a protracted neurodegenerative clinical course.

Genetic susceptibility to sporadic and acquired forms of CJD are largely determined by a coding polymorphism at codon 129 of *PRNP*, where either methionine or valine can be encoded (M129V). Homozygosity of methionine or valine is correlated with increased susceptibility to the development of sporadic[84,85] and iatrogenic CJD[86] All variant CJD patients have been methionine homozygotes,[87] while heterozygosity has been attributed to a protective effect in inherited prion diseases.[88] The age of onset of disease was delayed or the incubation times were longer in codon 129 heterozygotes compared to homozygotes in some inherited, iatrogenic and sporadic prion diseases.[89-91] It is thought that heterozygosity confers resistance to disease by interfering with the multimerization of identical PrP polypeptides, which are hypothesized to form aggregates, eventually leading to disease.[84,92] Heterozygosity at a different *PRNP* polymorphism, E219K, is also associated with resistance to sporadic CJD in Japanese subjects.[93]

Several other point mutations and insertional mutations in the *PRNP* gene coding region have been implicated in TSEs.[94] Other mutations in *PRNP* for familial CJD (E200K), Gerstmann–Straussler–Scheinker disease (D178N), and fatal familial insomnia (P102L) have been studied in relation to codon 129 composition.[95] Furthermore, along with causative mutations in the *PRNP* gene, disease-modifying QTL have been identified on several chromosomes, implicating the role of other modifier genes in the prion disease process.

4. The Genetics of Noninfectious Inflammatory Diseases

There are a number of diseases with high prevalence in the human population that are due to immune stimulation in the absence of infection. Many of these diseases can be classified as either autoimmune or allergic in nature and include inhalant or food allergies, asthma, inflammatory bowel diseases, multiple sclerosis, and type 1 diabetes, among many others. Both autoimmune and allergic disorders result in chronic inflammation. Along with an inappropriate inflammatory basis, these diseases also share a complex genetic nature; that is, they are caused by the interaction between multiple susceptibility loci and environmental factors. Thus, they are characterized by a genetic predisposition to disease susceptibility. Our understanding of the mechanisms by which the immune system overreacts and the identification of underlying susceptibility loci has been facilitated by the extensive study of murine models.

4.1. *Common Features of Autoimmunity and Allergy*

Autoimmunity and allergy appear to have many parallels, including the dramatic rise in incidence of common autoimmune and allergic diseases in the last half of the 20th century. Both autoimmunity and allergy are the result of inappropriate immune responses. In the case of autoimmunity, the response is self-directed, while allergy is the result of an immune response directed towards innocuous environmental proteins, called allergens. In either case, the body suffers the consequence of ongoing immune activation and subsequent chronic inflammation, with the added insult of tissue-specific injury in autoimmunity. The immune response becomes chronic in both situations; for allergy the inciting allergen may be ubiquitous in the environment or, alternatively, when one allergen is removed, the body often develops responses to new allergens. The adaptive immune response mounted in autoimmunity is usually not able to eliminate the antigen completely; thus, the response is sustained.

A common defect underlying both forms of immune disease may be regulatory T cells. Regulatory T cells serve to moderate T helper cells. T helper cells are upregulated in most autoimmune and allergic diseases; the Th1 subset is most commonly dominant in autoimmunity and Th2 cell activation is associated with allergy. It was previously thought that allergy and autoimmunity were due to an imbalance of Th1 and Th2 cells, and thus autoimmunity and allergy were the results of "opposite" pathologies. However, it is now under consideration that

T helper cells (either Th1 or Th2) may be in a generalized overreactive state due to T regulatory cell dysfunction, which may be common to both disease states.

4.2. *The Genetics of Autoimmunity*

We saw how the MHC complex was crucial to protection from pathogens, but these same molecules can be the source of the problem in autoimmunity. Much of what we know of disease resistance is based on the immune system's accurate recognition of self as distinguished from non-self. When the immune effector cells inappropriately recognize self-antigens as foreign and elicit a response, the result is autoimmunity. Complex autoimmune diseases are chronic conditions initiated by a loss of immunologic tolerance to self-antigens. The most common autoimmune diseases include insulin-dependent diabetes mellitus (IDDM), rheumatoid arthritis, systemic lupus erythematosus, and multiple sclerosis.

4.2.1. *Genetics of Insulin-dependent Diabetes Susceptibility*

Among the autoimmune diseases, IDDM (also referred to as type 1 diabetes) is one of significant medical importance as it generally occurs in children, requires life-long management with injectable insulin, and is characterized by long-term complications. Insulin-dependent diabetes is a complex genetic disease, and, as such, it has its origins in both genetic susceptibility and inciting environmental insults. The order of events in the development of IDDM, as described by Foster,[96] begins with an environmental incident, such as a viral infection, in a genetically susceptible individual. There is a subsequent T cell-mediated inflammatory response in the pancreatic islet cells with resultant alteration of the insulin-producing beta cells such that they are no longer recognized as self. An immune response to the now "non-self" beta cells results in their destruction and subsequent insulin deficiency.

Insulin-dependent diabetes has been extensively studied with mouse models, especially with non-obese diabetic (NOD) mice, which spontaneously develop diabetes and mimic characteristic features of the human disease.[97] As in humans, NOD mice develop spontaneous autoimmune diabetes due to a self-directed pathologic response to the pancreatic islets of Langerhans, with consequent selective destruction of the majority beta cells. This pathology is caused by specific T-lymphocytes and can be alleviated by immune suppressive reagents.

Consistent with the multifactorial nature of IDDM, disease-associated MHC alleles are necessary, but not sufficient to cause diabetes in both humans and

NOD mice. In humans, the highest risk of developing IDDM is observed in heterozygous individuals carrying the HLA genotype DQB1*0302-DRB1*0401/DQB1*0201-DRB1*03. Alternatively, certain HLA genotypes, such as DQB1*0301-DRB1*04, DQB1*0302-DRB1*0406, and DQB1*0602-DRB1*15, protect against IDDM.[98] Some non-MHC susceptibility alleles have also been identified, such as the variable number of tandem repeats in the insulin gene promoter,[99] and its association with the insulin-like growth factor-2 gene.[100] Various additional putative susceptibility loci for type I diabetes are reviewed by Field.[101]

In NOD mice, both the absence of *I-E* and homozygous expression of *I-A*^{g7} MHC class II genes are necessary for diabetes development.[102] The I-E complex is absent because of a deletion in the α chain gene.[103,104] NOD-Eα transgenic mice, which express Eα : Eβ^{g7} together with I-A^{g7}, fail to develop autoimmune diabetes either spontaneously or after treatment with cyclophosphamide, an agent that accelerates diabetes in wild-type NOD mice.[98]

Along with insights into the genetic basis of disease susceptibility, the NOD murine model has been valuable in dissecting the immunopathology of autoimmunity in general, and IDDM in particular. Details of cell types and cellular messengers, such as cytokines, involved in the induction of autoreactive T cells and the failure of autoregulatory mechanisms have been advanced by studies in genetically modified mice. For example, the lack of proliferation of T-lymphocytes into the pancreatic islets with the expression of the *I-E* gene has been suggested to occur via clonal deletion of specific Vβ bearing pathogenic T cells. This suggestion is supported by evidence that this form of MHC-induced protection from diabetes is based on the presentation of an anatomically restricted, non-autoantigenic peptide to highly diabetogenic thymocytes leading to clonal deletion.[105]

4.2.2. *Genetics of Rheumatoid Arthritis Susceptibility*

Rheumatoid arthritis is another autoimmune disease with significant societal impact. It is a crippling disease characterized by chronic inflammation of synovial joints. Collagen induced arthritis is used as an animal model of rheumatoid arthritis. The presentation of type II collagen peptide by the MHC class II molecule H-2A^{q} (equivalent to human HLA-DQ) following immunization with type II collagen is thought to be the most critical event in the induction of collagen induced arthritis,[106] while certain polymorphisms in the *H2-E* gene (equivalent to human HLA-DR) are protective against the disease.[107] This implies that certain DQ alleles predispose to rheumatoid arthritis, whereas DR alleles can modulate the effect of DQ by either enhancing or dominantly protecting against the DQ-associated predisposition to disease. There is a strong linkage disequilibrium between the HLA-DQ and HLA-DR alleles. All the protective DRB1 alleles encode the amino acid sequence motif aspartate–glutamate–arginine–alanine–alanine (DERAA) in the third hypervariable region of the protein. The mechanism of the DERAA related protection may involve the presentation of DRB1-derived peptides by DQ leading either to the deletion of potential autoreactive T cells in the thymus or to the generation of DRB1-specific, DQ-restricted immunoregulatory T helper cells. [108] It has also been demonstrated that DRB1 molecules not only act as restriction elements, but can also act as antigens that are processed and generate DQ-restricted epitopes associated with protection against rheumatoid arthritis.[108] Self-peptides derived from HLA molecules could potentially generate tolerance or autoimmunity depending on their binding affinity with HLA molecules, and a complementation between both DQ and DR molecules is required for susceptibility or protection from disease.[109] Apart from MHC genes, variable number of tandem repeat polymorphisms in the IL-4 gene[110] and an AA genotype in the IL-18 gene promoter[111] have also been reported to have a protective association with rheumatoid arthritis.

4.3. *Genetic Susceptibility to Allergy and Asthma*

Allergic asthma, eczema (also known as atopic dermatitis), rhinitis (hay fever) and food allergy are chronic disorders that have an allergic basis.[112,113] A genetic predisposition to these disorders, along with exposure to environmental factors, are required for development of clinical symptoms. Allergic diseases typically begin in childhood. Children may grow out of their allergies, develop new manifestations, or suffer from persistent disease.

Allergic asthma is generally considered the most severe of the allergic diseases. It is clinically characterized by episodes of airway obstruction,[112,114] which can be so severe as to be fatal. Clinical signs are attributed to underlying chronic airway inflammation, increased mucus production, and bronchial hyperresponsiveness. The inflammation is mediated by Th2 lymphocytes and characterized by eosinophilia and elevated serum IgE.[114]

A familial component to asthma has long been recognized. Twin studies and family studies have demonstrated a heritable predisposition to the expression of asthma and to the primary components of the asthma phenotype.[115,116] Exacerbation by various environmental factors and failure to comply to simple inheritance models led to the conclusion that asthma is a complex genetic disease. In recent years candidate gene, genome screen, and positional cloning approaches have been used to investigate the molecular genetic basis of asthma.

4.3.1. *Mouse Contributions to the Genetics of Asthma*

The murine contribution to understanding the genetics of disease resistance in asthma is great. Inbred mouse models have been successfully used to model the complex genetic susceptibility to asthma and determine the location of QTL that contribute to the disease phenotype. Bronchial hyperresponsiveness is one of the hallmarks of allergic asthma and has been used as a disease marker in both human studies and murine models. Early studies demonstrated a strain-specific variability in bronchial hyperresponsiveness among 9 common inbred mouse strains.[117] Subsequently, other investigators adapted this murine model to more closely mimic the pathology of allergic asthma by sensitizing the mice to allergen. Strain-specific propensity to develop or be protected from Th2-type inflammation and bronchial hyperresponsiveness was noted.[118,119]

A number of QTL that contribute to genetic susceptibility to asthma-related bronchial hyperresponsiveness have been identified in mice. A genome screen for susceptibility loci conducted in a murine model revealed linkage to chromosome 6 for allergen-independent bronchial hyperresponsiveness, near the IL-5 receptor gene (*Il5r*);[120] a functional polymorphism in *Il5r* was subsequently found in an allergic dermatitis mouse model.[121] Comparison of linkage results from similar studies revealed a strong reliance on mouse strain. For example, the genome screen that revealed the QTL on chromosome 6 was performed in a cross between A/J and C3H/HeJ strains;[120] in contrast, similar experiments utilizing crosses between A/J and C57BL/6 mice found QTL on chromosomes 2, 15 and 17.[122] When BP2 and BALB/c strains were tested, bronchial hyperresponsiveness linked to chromosomes 9, 10, 11 and 17.[119] While these results may appear conflicting at first glance, repetition has confirmed strain-specific linkage.[123] Thus, the results reveal the importance of genetic background on gene expression.

Evidence that unique genetic linkages are dependent upon the mouse strains studied supports the hypothesis that genetic variability underlies bronchial hyperresponsiveness and that different mechanisms, or variation at multiple points along a single pathway, may all result in the same endpoint. Evidence for multiple genes affecting bronchial hyperresponsiveness and other asthma-related phenotypes is found in humans as well as mice. Multiple ethnic groups share map locations for several QTL linked to human asthma. For example, linkages to chromosome 11q for IgE and positive skin prick test have been reported in European,[124] Australian,[125] and Japanese[126] populations, while in Amish,[127] Dutch,[128] and African-American[129] families IgE and bronchial hyperresponsiveness link to chromosome 5q. Human chromosome 5q is one of the several genomic locations where the results of human and mouse studies converge; a QTL for atracurium-induced bronchoconstriction was identified on murine chromosome 13, homologous to human chromosome 5q31-q33 in the region containing the gene for IL-9.[130,131] Functional studies produced further evidence in support of *IL9* as a candidate gene for bronchial hyperresponsiveness,[132] and the role of this positional candidate in human asthma and murine models has been actively pursued.[133,134]

Knowledge of the genetic susceptibility to asthma was further enhanced when investigators expanded the utility of the mouse model by simultaneously analyzing DNA polymorphisms and gene expression.[135] A genome screen for allergen-induced bronchial hyperresponsiveness conducted in A/J × C3H/HeJ progeny identified QTL on chromosomes 2 and 7. Simultaneously, expression analysis of lung mRNA conducted via microarray platform revealed an association between allergen-induced bronchial hyperresponsiveness and expression of complement factor 5. These parallel results were linked by the fact that the complement factor 5 gene (*C5*) was located within the QTL interval on chromosome 2, and the C5 receptor gene (*C5r1*) was subsequently found to map to the linked region of chromosome 7.[136] Furthermore, A/J mice carry a deletion mutation in *C5* rendering them deficient in C5 protein. Mechanistic studies supported the need for C5 in cytokine (IL-12) signaling pathways important for protection against allergies.[135] Taken together, these results strongly support a role for the *C5* gene in murine asthma.

In summary, genes identified in mouse models of asthma, such as *C5*, support the importance of immune pathways in the development of asthma-related phenotypes. If defects occur at multiple locations along a pathway, different, multiple or unique candidate genes would be identified in various populations. This may be the case in the murine model in which both *C5* and *C5r1* were linked to the asthma phenotype. Mouse models allow the assessment of the functional significance of various genes and pathways identified in genetic studies.

5. Conclusions

Our understanding of the genetics of disease resistance has matured greatly in the past few decades. Resistance genes are now known to underlie a wide variety of both infectious and non-infectious diseases. Many common medical conditions are due to the complex interactions of susceptibility genes and environmental factors. These disorders are of tremendous public health significance due to the large numbers of individuals affected. Thus, understanding and preventing such conditions will have tremendous societal impact. Furthermore, knowledge of the genetics of disease resistance has direct application in health care. Armed with the growing knowledge of genetic susceptibility loci, health care practitioners may soon be able to look at not only the environmental factors, but also the genetic factors, that determine occurrence, progression, and outcome of many common diseases. As we head into the era of disease prevention by means of incorporating prediction models, we must be able to identify at risk populations and individuals. Thus, it is important to understand the genetics of disease resistance.

The mouse has made significant contributions to this progress. Indeed, the genome of this unassuming animal contains the foundation of our knowledge of the genetics of disease resistance. The mouse has become a valuable research tool, especially in terms of genetics. Using this "tool," researchers have been able to investigate conditions for which there is a genetic component to disease resistance. Furthermore, investigators have begun to decipher the roles of individual genes in their contributions to disease resistance. Finally, as work done in the mouse has applications to the health of human and domestic animal species, the mouse is perhaps the species with most broad-reaching medical impact.

References

1. Morel, L., X.H. Tian, B.P. Croker and E.K. Wakeland. 1999. Epistatic modifiers of autoimmunity in a murine model of lupus nephritis. *Immunity* **11**:131–139.
2. Silver, L.M. 1995. *Mouse Genetics: Concepts and Applications* Oxford University Press, New York.
3. Poltorak, A., X. He, I. Smirnova, M.Y. Liu, C. Van Huffel, X. Du, D. Birdwell, E. Alejos, M. Silva, C. Galanos, M. Freudenberg, P. Ricciardi-Castagnoli, B. Layont and B. Beutler. 1998. Defective LPS signaling in C3H/HeJ and C57BL/10ScCr mice: mutations in Tlr4 gene. *Science* **282**:2085–2088.

4. Blackwell, J.M., T. Goswami, C.W. Evans, D. Sibthorpe, N. Papo, J.K. Whilte, S. Searle, E.N. Miller, C.S. Peacock, H. Mohammed and M. Ibrahim. 2001. SLC11A1 (formerly NRAMP1) and disease resistance. *Cellular Microbiol.* **3**:773–784.
5. Gorer, P.A., S. Lyman and G.D. Snell. 1948. Studies on the genetic and antigenic basis of tumor transplantation: linkage between a histocompatibility gene and 'Fused' in mice. *Proc. R. Soc. London. Ser. B* **135**:499–505.
6. Snell, G.D. and G.F. Higgins. 1951. Alleles at the histocompatibility-2 locus in the mouse as determined by tumor transplantation. *Genetics* **36**:306–310.
7. Hughes, A.L. 2002. Natural selection and the diversification of vertebrate immune effectors. *Immunol. Rev.* **190**:161–168.
8. Janeway, C., Travers, P, Walport, M, Shlomchik, MJ. 2001. *Immunobiology,* Garland Publishing, New York.
9. Kaufman, J., H. Volk and H.J. Wallny. 1995. A "minimal essential Mhc" and an "unrecognized Mhc": two extremes in selection for polymorphism. *Immunol. Rev.* **143**:63–88.
10. Potts, W.K. and P.R. Slev. 1995. Pathogen-based models favoring MHC genetic diversity. *Immunol Rev* **143**:181–197.
11. Messaoudi, I., J.A. Guevara Patino, R. Dyall, J. LeMaoult and J. Nikolich-Zugich. 2002. Direct link between mhc polymorphism, T cell avidity, and diversity in immune defense. *Science* **298**:1797–1800.
12. Dyall, R., I. Messaoudi, S. Janetzki and J. Nikolic-Zugic. 2000. MHC polymorphism can enrich the T cell repertoire of the species by shifts in intrathymic selection. *J. Immunol.* **164**:1695–1698.
13. Cooke, G.S. and A.V. Hill. 2001. Genetics of susceptibility to human genetic disease. *Nat. Rev. Genet.* **2**:967–977.
14. Hendel, H., S. Caillat-Zucman, H. Lebuanec, M. Carrington, S. O'Brien, J.M. Andrieu, F. Schachter, D. Zagury, J. Rappaport, C. Winkler, G.W. Nelson and J.F. Zagury. 1999. New class I and II HLA alleles strongly associated with opposite patterns of progression to AIDS. *J. Immunol.* **162**:6942–6946.
15. Penn, D.J., K. Damjanovich and W.K. Potts. 2002. MHC heterozygosity confers a selective advantage against multiple-strain infections. *Proc. Natl. Acad. Sci. USA* **99**:11260–11264.
16. McClelland, E.E., D.J. Penn and W.K. Potts. 2003. Major histocompatibility complex heterozygote superiority during coinfection. *Infect. Immun.* **71**:2079–2086.
17. Lipsitch, M., C.T. Bergstrom and R. Antia. 2003. Effect of human leukocyte antigen heterozygosity on infectious disease outcome: the need for allele-specific measures. *BMC Med. Genet.* **4**:2.
18. Grundy, J.E., J.S. Mackenzie and N.F. Stanley. 1981. Influence of H-2 and non-H-2 genes on resistance to murine cytomegalovirus infection. *Infect. Immunol.* **32**:277–286.
19. Makino, M., H.C. Morse III, T.N. Fredrickson and J.W. Hartley. 1990. H-2-associated and background genes influence the development of a murine retrovirus-induced immunodeficiency syndrome. *J. Immunol.* **144**:4347–4355.
20. Tucker, H., J. Weens, P. Tsichlis, R.S. Schwartz, R. Khiroya and J. Donnelly. 1977. Influence of H-2 complex on susceptibility to infection by murine leukemia virus. *J. Immunol.* **118**:1239–1243.

21. Vasmel, W.L., M. Zijlstra, T. Radaszkiewicz, C.J. Leupers, R.E. de Goede and C.J. Melief. 1988. Major histocompatibility complex class II-regulated immunity to murine leukemia virus protects against early T- but not late B-cell lymphomas. *J. Virol.* **62**:3156–3166.
22. Chen, C.Y., S.A. Cohen, M.B. Zaleski and B. Albini. 1992. Genetic control of streptococcus-induced hepatic granulomatous lesions in mice. *Immunogenetics* **36**:28–32.
23. Hedrick, P.W. 1998. Balancing selection and MHC. *Genetica* **104**:207–214.
24. Potts, W.K., C.J. Manning and E.K. Wakeland. 1994. The role of infectious disease, inbreeding and mating preferences in maintaining MHC genetic diversity: an experimental test. *Philos. Trans. R. Soc. Lond. B Biol. Sci.* **346**:369–378.
25. Thorp, B.H. and E. Luiting. 2000. Breeding for resistance to production diseases in poultry. In: *Breeding for Disease Resistance in Farm Animals.* ed. R.F.E. Axford, S.C. Bishop, F.W. Nicholas and J.B. Owen. pp. 357–378. CABI Publishing, New York.
26. Wigley, P. 2004. Genetic resistance to Salmonella infection in domestic animals. *Res. Vet. Sci.* **76**:165–169.
27. Edfors-Lilja, I., U. Gustafsson, Y. Duval-Iflah, H. Ellegren, M. Johansson, R.K. Juneja, L. Marklund and L. Andersson. 1995. The porcine intestinal receptor for Escherichia coli K88ab, K88ac: regional localization on chromosome 13 and influence of IgG response to the K88 antigen. *Anim. Genet.* **26**:237–242.
28. Philipsson, J., G. Ral and B. Berglund. 1995. Somatic-cell count as a selection criterion for mastitis resistance in dairy cattle. *Livest. Prod. Sci.* **41**:195–200.
29. Behnke, J.M., F. Iraqi, D. Menge, R.L. Baker, J. Gibson and D. Wakelin. 2003. Chasing the genes that control resistance to gastrointestinal nematodes. *J. Helminthol.* **77**:99–110.
30. Enriquez, F.J., B.O. Brooks, R.H. Cypess, C.S. David and D.L. Wassom. 1988. *Nematospiroides dubius*: two H-2-linked genes influence levels of resistance to infection in mice. *Exper. Parasitol.* **67**:221–226.
31. Else, K.J., D. Wakelin, D.L. Wassom and K.M. Hauda. 1990. The influence of genes mapping within the major histocompatibility complex on resistance to *Trichuris muris* infections in mice. *Parasitology* **101**:61–67.
32. Wassom, D.L., B.O. Brooks, J.G. Babisch and C.S. David. 1983. A gene mapping between the S and D regions of the H-2 complex influences resistance to *Trichinella spiralis* infections in mice. *J. Immunogenetics* **10**:371–378.
33. Wassom, D.L., C.J. Krco and C.S. David. 1987. I-E expression and susceptibility to parasite infection. *Immunol. Today* **8**:39–43.
34. Paterson, K.A., J.C. McEwan, K.G. Dodds and A.M. Crawford. 2001. Fine mapping of a locus affecting host resistance to internal parasites of sheep. *Proc. Assoc. Adv. Anim. Br. Genet.* **14**:91–94.
35. Schwaiger, F.W. and D.S. Wallace. 1996. An ovine lymphocyte antigen is associated with reduced faecal egg counts in four-month-old lambs following natural predominantly *Ostertagia circumcincta* infection. *International J. Parasitol.* **26**:432–428.
36. Coltman, D.W., K. Wilson, J.G. Pilkington, M.J. Stear and J.M. Pemberton. 2001. A microsatellite polymorphism in a gamma interferon gene is associated with resistance to gastrointestinal nematodes in a naturally-parasitized population of Soay sheep. *Parasitology* **122**:571–582.
37. Sacks, D. and N. Noben-Trauth. 2002. The immunology of susceptibility and resistance to *Leishmania major* in mice. *Nat. Rev. Immunol.* **2**:845–858.

38. Gumy, A., J.A. Louis and P. Launois. 2004. The murine model of infection with *Leishmania major* and its importance for the deciphering of mechanisms underlying differences in Th cell differentiation in mice from different genetic backgrounds. *Int. J. Parasitol.* **34**:433–444.
39. Mahida, Y.R. 2003. Host-parasite interactions in rodent nematode infections. *J. Helminthol.* **77**:125–131.
40. Frisch, J.E., C.J. O'Neill and M.J. Kelly. 2000. Using genetics to control cattle parasites - the Rockhampton experience. *Int. J. Parasitol.* **30**:253–264.
41. Courtney, C.H., C.F. Parker, K.E. McClure and R.P. Herd. 1985. Resistance of exotic and domestic lambs to experimental infection with *Hemonchus contortus*. *Int. J. Parasitol.* **15**:101–109.
42. Gruner, L., J. Cabaret, C. Sauve and R. Pailhories. 1986. Comparative susceptibility of Romanov and Lacaune sheep to GI nematodes and small lungworms. *Vet. Parasitol.* **19**:85–93.
43. Claxton, J. and P. Leperre. 1991. Parasite burdens and hosts susceptibility of Zebu and N'Dama cattle in village herds in Gambia. *Vet. Parasitol.* **40**:293–304.
44. Rechav, Y. and M.W. Kostrzewski. 1991. Relative resistance of six cattle breeds to the tick Boophilus decoloratus in South Africa. *Onderstepoort J. Vet. Res.* **58**:181–186.
45. Bock, R.E., T.G. Kingston and A.J. de Vos. 1999. Effect of breed of cattle on transmission rate and innate resistance with Babesia bovis and B bigemina transmitted by Boophilus microplus. *Aust. Vet. J.* **77**:461–464.
46. Spickett, A.M., D. De Klerk, C.B. Enslin and M.M. Scholtz. 1989. Resistance of Nguni, Bonsmara and Hereford cattle to ticks in Bushveld region of South Africa. *Onderstepoort J. Vet. Res.* **56**:245–250.
47. Losson, B.J., J.F. Lonneux and M. Lekimme. 1999. The pathology of Psoroptes ovis infestation in cattle with a special emphasis on breed difference. *Vet. Parasitol.* **83**:219-229.
48. Weissmann, C. and E. Flechsig. 2003. PrP knock-out and PrP transgenic mice in prion research. *Br. Med. Bull.* **66**:43–60.
49. Weissmann, C. 2002. Molecular genetics of transmissible spongiform encephalopathies: an introduction. *J. Toxicol. Sci.* **27**:69–77.
50. Collinge, J. 1997. Human prion diseases and bovine spongiform encephalopathy (BSE). *Hum. Mol. Genet.* **6**:1699–1705.
51. Collinge, J. 2001. Prion diseases of humans and animals: their causes and molecular basis. *Annu. Rev. Neurosci.* **24**:519–550.
52. Bueler, H., A. Aguzzi, A. Sailer, R.A. Greiner, P. Autenried, M. Aguet and C. Weissmann. 1993. Mice devoid of PrP are resistant to scrapie. *Cell* **73**:1339–1347.
53. Bueler, H., M. Fischer, Y. Lang, H. Bluethmann, H.P. Lipp, S.J. DeArmond, S.B. Prusiner, M. Aguet and C. Weissmann. 1992. Normal development and behaviour of mice lacking the neuronal cell-surface PrP protein. *Nature* **356**:577–582.
54. Kocisko, D.A., J.H. Come, S.A. Priola, B. Chesebro, G.J. Raymond, P.T. Lansbury and B. Caughey. 1994. Cell-free formation of protease-resistant prion protein. *Nature* **370**:471–474.
55. Paushkin, S.V., V.V. Kushnirov, V.N. Smirnov and M.D. Ter-Avanesyan. 1997. In vitro propagation of the prion-like state of yeast Sup35 protein. *Science* **277**:381–383.
56. Prusiner, S.B. 1994. Biology and genetics of prion diseases. *Annu Rev Microbiol* **48**:655–686.
57. Prusiner, S.B. 1998. Prions. *Proc. Natl. Acad. Sci. USA* **95**:13363–13383.

58. Brandner, S., S. Isenmann, A. Raeber, M. Fischer, A. Sailor, Y. Kobayashi, S. Marino, C. Weissmann and A. Aguzzi. 1996. Normal host prion protein necessary for scrapie-induced neurotoxicity. *Nature* **379**:339–343.
59. Westaway, D., P.A. Goodman, C.A. Mirenda, M.P. McKinley, G.A. Carlson and S.B. Prusiner. 1987. Distinct prion proteins in short and long scrapie incubation period mice. *Cell* **51**:651–662.
60. Moore, R.C., J. Hope, P.A. McBride, I. McConnell, J. Selfridge, D.W. Melton and J.C. Manson. 1998. Mice with gene targetted prion protein alterations show that Prnp, Sinc and Prni are congruent. *Nat. Genet.* **18**:118–125.
61. Carlson, G.A., C. Ebeling, S.L. Yang, G. Telling, M. Torchia, D. Groth, D. Westaway, S.J. DeArmond and S.B. Prusiner. 1994. Prion isolate specified allotypic interactions between the cellular and scrapie prion proteins in congenic and transgenic mice. *Proc. Natl. Acad. Sci. USA* **91**:5690–5694.
62. Bruce, M.E. 2003. TSE strain variation. *Br. Med. Bul.l* **66**:99–108.
63. Kingsbury, D.T., K.C. Kasper, D.P. Stites, J.D. Watson, R.N. Hogan and S.B. Prusiner. 1983. Genetic control of scrapie and Creutzfeldt-Jakob disease in mice. *J. Immunol.* **131**:491–496.
64. Carlson, G.A., P.A. Goodman, M. Lovett, B.A. Taylor, S.T. Marshall, M. Peterson-Torchia, D. Westaway and S.B. Prusiner. 1988. Genetics and polymorphism of the mouse prion gene complex: control of scrapie incubation time. *Mol. Cell Biol.* **8**:5528–5540.
65. Stephenson, D.A., K. Chiotti, C. Ebeling, D. Groth, S.J. DeArmond, S.B. Prusiner and G.A. Carlson. 2000. Quantitative trait loci affecting prion incubation time in mice. *Genomics* **69**:47–53.
66. Lloyd, S.E., O.N. Onwuazor, J.A. Beck, G. Mallinson, M. Farrall, P. Targonski, J. Collinge and E.M. Fisher. 2001. Identification of multiple quantitative trait loci linked to prion disease incubation period in mice. *Proc. Natl. Acad. Sci. USA* **98**:6279–6283.
67. Moreno, C.R., F. Lantier, I. Lantier, P. Sarradin and J.M. Elsen. 2003. Detection of new quantitative trait loci for susceptibility to transmissible spongiform encephalopathies in mice. *Genetics* **165**:2085–2091.
68. Lloyd, S.E., J.B. Uphill, P.V. Targonski, E.M. Fisher and J. Collinge. 2002. Identification of genetic loci affecting mouse-adapted bovine spongiform encephalopathy incubation time in mice. *Neurogenetics* **4**:77–81.
69. Manolakou, K., J. Beaton, I. McConnell, C. Farquar, J. Manson, N.D. Hastie, M. Bruce and I.J. Jackson. 2001. Genetic and environmental factors modify bovine spongiform encephalopathy incubation period in mice. *Proc. Natl. Acad. Sci. USA* **98**:7402–7407.
70. Hunter, N. 2003. Scrapie and experimental BSE in sheep. *Br. Med. Bull.* **66**:171–183.
71. Belt, P.B., I.H. Muileman, B.E. Schreuder, J. Bos-de Ruijter, A.L. Gielkens and M.A. Smits. 1995. Identification of five allelic variants of the sheep PrP gene and their association with natural scrapie. *J. Gen. Virol.* **76**:509–517.
72. Goldmann, W., N. Hunter, T. Martin, M. Dawson and J. Hope. 1991. Different forms of the bovine PrP gene have five or six copies of a short, G-C-rich element within the protein-coding exon. *J. Gen. Virol.* **72**:201–204.
73. Humeny, A., K. Schiebel, S. Seeber and C.M. Becker. 2002. Identification of polymorphisms within the bovine prion protein gene (Prnp) by DNA sequencing and genotyping by MALDI-TOF-MS. *Neurogenetics* **4**:59–60.

74. Takasuga, A., T. Abe, T. Ito, T. Watanabe, N. Kamatani and Y. Sugimoto. 2003. Novel prion protein polymorphisms in cattle. *Anim. Genet.* **34**:396–397.
75. Neibergs, H.L., A.M. Ryan, J.E. Womack, R.L. Spooner and J.L. Williams. 1994. Polymorphism analysis of the prion gene in BSE-affected and unaffected cattle. *Anim. Genet.* **25**:313–317.
76. Hunter, N., W. Goldmann, G. Smith and J. Hope. 1994. Frequencies of PrP gene variants in healthy cattle and cattle with BSE in Scotland. *Vet. Rec.* **135**:400–403.
77. Sander, P., H. Hamann, I. Pfeiffer, W. Wemheuer, B. Brenig, M.H. Groschup, U. Ziegler, O. Distl and T. Leeb. 2004. Analysis of sequence variability of the bovine prion protein gene (PRNP) in German cattle breeds. *Neurogenetics* **5**:19–25.
78. Castilla, J., A. Gutierrez-Adan, A. Brun, B. Pintado, B. Parra, M.A. Ramirez, F.J. Salguero, F. Diaz San Segundo, A. Rabano, M.J. Cano and J.M. Torres. 2004. Different behavior toward bovine spongiform encephalopathy infection of bovine prion protein transgenic mice with one extra repeat octapeptide insert mutation. *J. Neurosci.* **24**:2156–2164.
79. Salman, M.D. 2003. Chronic wasting disease in deer and elk: scientific facts and findings. *J. Vet. Med. Sci.* **65**:761–768.
80. Raymond, G.J., A. Bossers, L.D. Raymond, K.I. O'Rourke, L.E. McHolland, P.K. Bryant, 3rd, M.W. Miller, E.S. Williams, M. Smits and B. Caughey. 2000. Evidence of a molecular barrier limiting susceptibility of humans, cattle and sheep to chronic wasting disease. *EMBO J.* **19**:4425–4430.
81. Belay, E.D., P. Gambetti, L.B. Schonberger, P. Parchi, D.R. Lyon, S. Capellari, J.H. McQuiston, K. Bradley, G. Dowdle, J.M. Crutcher and C.R. Nichols. 2001. Creutzfeldt-Jakob disease in unusually young patients who consumed venison. *Arch. Neurol.* **58**:1673–1678.
82. Heaton, M.P., K.A. Leymaster, B.A. Freking, D.A. Hawk, T.P.L. Smith, J.W. Keele, W.M. Snelling, J.M. Fox, C.G. Chitko-McKnown and W.W. Laegreid. 2003. Prion gene sequence variation within diverse groups of U.S. sheep, beef cattle, and deer. *Mamm. Genome* **14**:765–777.
83. O'Rourke, K.I., T.E. Besser, M.W. Miller, T.F. Cline, T.R. Spraker, A.L. Jenny, M.A. Wild, G.L. Zebarth and E.S. Williams. 1999. PrP genotypes of captive and free-ranging Rocky Mountain elk (Cervus elaphus nelsoni) with chronic wasting disease. *J. Gen. Virol.* **80**:2765–2769.
84. Palmer, M.S., A.J. Dryden, J.T. Hughes and J. Collinge. 1991. Homozygous prion protein genotype predisposes to sporadic Creutzfeldt-Jakob disease. *Nature* **352**:340–342.
85. Windl, O., M. Dempster, J.P. Estibeiro, R. Lathe, R. de Silva, T. Esmonde, R. Will, A. Springbett, T.A. Campbell, K.C. Sidle, M.S. Palmer and J. Collinge. 1996. Genetic basis of Creutzfeldt-Jakob disease in the United Kingdom: a systematic analysis of predisposing mutations and allelic variation in the PRNP gene. *Hum. Genet.* **98**:259–264.
86. Brown, P., M. Preece, J.P. Brandel, T. Sato, L. McShane, I. Zerr, A. Fletcher, R.G. Will, M. Pocchiari, N.R. Cashman, J.H. d'Aignaux, L. Cervenakova, J. Fradkin, L.B. Schonberger and S.J. Collins. 2000. Iatrogenic Creutzfeldt-Jakob disease at the millennium. *Neurology* **55**:1075–1081.
87. Zeidler, M., G. Stewart, S.N. Cousens, K. Estibeiro and R.G. Will. 1997. Codon 129 genotype and new variant CJD. *Lancet* **350**:668.

88. Baker, H.E., M. Poulter, T.J. Crow, C.D. Frith, R. Lofthouse and R.M. Ridley. 1991. Amino acid polymorphism in human prion protein and age at death in inherited prion disease. *Lancet* **337**:1286.
89. Poulter, M., H.F. Baker, C.D. Frith, M. Leach, R. Lofthouse, R.M. Ridley, T. Shah, F. Owen, J. Collinge, J. Brown. 1992. Inherited prion disease with 144 base pair gene insertion. 1. Genealogical and molecular studies. *Brain* **115**:675–685.
90. Dlouhy, S.R., K. Hsiao, M.R. Farlow, T. Foroud, P.M. Conneally, P. Johnson, S.B. Prusiner, M.E. Hodes and B. Ghetti. 1992. Linkage of the Indiana kindred of Gerstmann-Straussler-Scheinker disease to the prion protein gene. *Nat. Genet.* **1**:64–67.
91. Huillard d'Aignaux, J., D. Costagliola, J. Maccario, T. Billette de Villemeur, J.P. Brandel, J.P. Deslys, J.J. Hauw, J.L. Chaussain, Y. Agid, D. Dormont and A. Alperovitch. 1999. Incubation period of Creutzfeldt-Jakob disease in human growth hormone recipients in France. *Neurology* **53**:1197–1201.
92. Collinge, J. and M.S. Palmer. 1992. Prion diseases. *Curr. Opin. Genet. Dev.* **2**:448–454.
93. Shibuya, S., J. Higuchi, R.W. Shin, J. Tateishi and T. Kitamoto. 1998. Codon 219 Lys allele of PRNP is not found in sporadic Creutzfeldt-Jakob disease. *Ann. Neurol.* **43**:826–828.
94. Kovacs, G.G., G. Trabattoni, J.A. Hainfellner, J.W. Ironside, R.S. Knight and H. Budka. 2002. Mutations of the prion protein gene phenotypic spectrum. *J. Neurol.* **249**:1567–1582.
95. Knight, R., Collins, S. 2001. Human prion diseases: cause, clinical and diagnostic aspects. In: *Prions. A Challenge for Science, Medicine and Public Health Systems.* ed. H.F. Rabenau, J. Cinatl and H.W. Doerr. pp. 68– 92. Karger, Basel.
96. Foster, D.W. 1994. Diabetes mellitus. In: *Harrison's Principles of Internal Medicine.* ed. K.J. Isselbacher, E. Braunwald, J.D. Wilson, J.B. Martin, A.S. Fauci and D.L. Kasper. pp. 1979–2000. McGraw-Hill, Inc., New York.
97. Tochino, Y. 1987. The NOD mouse as a model of type 1 diabetes. *Crit. Rev. Immunol.* **8**:49–81.
98. Adorini, L., S. Gregori and L.C. Harrison. 2002. Understanding autoimmune diabetes: insights from mouse models. *Trends. Mol. Med.* **8**:31–38.
99. Bell, G.I., S. Horita and J.H. Karam. 1984. A polymorphic locus near the human insulin gene is associated with insulin-dependent diabetes mellitus. *Diabetes* **33**:176–183.
100. Paquette, J., N. Giannoukakis, C. Polychronakos, P. Vafiadis and C. Deal. 1998. The INS 5' variable number of tandem repeats is associated with IGF2 expression in humans. *J. Biol. Chem.* **273**:14158–14164.
101. Field, L.L. 2002. Genetic linkage and association studies of Type I diabetes: challenges and rewards. *Diabetologia* **45**:21–35.
102. Wicker, L.S., J.A. Todd and L.B. Peterson. 1995. Genetic control of autoimmune diabetes in the NOD mouse. *Annu. Rev. Immunol.* **13**:179–200.
103. Bohme, J., B. Schuhbaur, O. Kanagawa, C. Benoist and D. Mathis. 1990. MHC-linked protection from diabetes dissociated from clonal deletion of T cells. *Science* **249**:293–295.
104. Hattori, M., J.B. Buse, R.A. Jackson, L. Glimcher, M.E. Dorf, M. Minami, S. Makino, K. Moriwaki, H. Kuzuya, H. Imura , M. Strauss, J.G. Seidman and G.S. Eisenbarth. 1986. The NOD mouse: recessive diabetogenic gene in the major histocompatibility complex. *Science* **231**:733–735.

105. Schmidt, D., A. Amrani, J. Verdaguer, S. Bou and P. Santamaria. 1999. Autoantigen-independent deletion of diabetogenic CD4+ thymocytes by protective MHC class II molecules. *J. Immunol.* **162**:4627–4636.
106. Holmdahl, R., M. Karlsson, M.E. Andersson, L. Rask and L. Andersson. 1989. Localization of a critical restriction site on the I-A beta chain that determines susceptibility to collagen-induced arthritis in mice. *Proc. Natl. Acad. Sci. USA* **86**:9475–9479.
107. Zanelli, E., M.A. Gonzalez-Gay and C.S. David. 1995. Could HLA-DRB1 be the protective locus in rheumatoid arthritis? *Immunol. Today* **16**:274–278.
108. Snijders, A., D.G. Elferink, A. Geluk, A.L. van Der Zanden, K. Vos, G.M. Schreuder, F.C. Breedveld, R.R. de Vries and E.H. Zanelli. 2001. An HLA-DRB1-derived peptide associated with protection against rheumatoid arthritis is naturally processed by human APCs. *J. Immunol.* **166**:4987–4993.
109. Taneja, V. and C.S. David. 2000. Autoimmunity versus tolerance: analysis using transgenic mice. *Hum. Immunol.* **61**:1383–1389.
110. Buchs, N., T. Silvestri, F.S. di Giovine, M. Chabaud, E. Vannier, G.W. Duff and P. Miossec. 2000. IL-4 VNTR gene polymorphism in chronic polyarthritis. The rare allele is associated with protection against destruction. *Rheumatology (Oxford)* **39**:1126–1131.
111. Sivalingam, S.P., K.H. Yoon, D.R. Koh and K.Y. Fong. 2003. Single-nucleotide polymorphisms of the interleukin-18 gene promoter region in rheumatoid arthritis patients: protective effect of AA genotype. *Tissue Antigens* **62**:498–504.
112. Kay, A.B. 2001. Allergy and allergic disease (first of two parts). *N. Engl. J. Med.* **344**:30-37.
113. Holgate, S.T. 1999. The epidemic of allergy and asthma. *Nature* **402**:B2–B4.
114. Busse, W.W. and R.F. Lemanske, Jr. 2001. Asthma. *N. Engl. J. Med.* **344**:350–362.
115. Longo, G., R. Strinati, F. Poli and F. Fumi. 1987. Genetic factors in nonspecific bronchial hyperreactivity. *Am. J. Dis. Child.* **141**:331–334.
116. Hopp, R.J., A.K. Bewtra, G.D. Watt, N.M. Nair and R.G. Townley. 1984. Genetic analysis of allergic disease in twins. *J. Allergy Clin. Immunol.* **73**:265–270.
117. Levitt, R.C. and W. Mitzner. 1989. Autosomal recessive inheritance of airway hyperreactivity to 5-hydroxytryptamine. *J. Appl. Physiol.* **67**:1125–1132.
118. Ewart, S.L., D. Kuperman, E. Schadt, C. Tankersley, A. Grupe, D.M. Shubitowski, G. Peltz and M. Wills-Karp. 2000. Quantitative trait loci controlling allergen-induced airway hyperresponsiveness in inbred mice. *Am. J. Respir. Cell. Mol. Biol.* **23**:537–545.
119. Zhang, Y., J. Lefort, V. Kearsey, J.R. Lapa e Silva, W.O. Cookson and B.B. Vargaftig. 1999. A genome-wide screen for asthma-associated quantitative trait loci in a mouse model of allergic asthma. *Hum. Mol. Genet.* **8**:601–605.
120. Ewart, S.L., W. Mitzner, D.A. DiSilvestre, D.A. Meyers and R.C. Levitt. 1996. Airway hyperresponsiveness to acetylcholine: segregation analysis and evidence for linkage to murine chromosome 6. *Am. J. Respir. Cell. Mol. Biol.* **14**:487–495.
121. Daser, A., K. Koetz, N. Batjer, M. Jung, F. Ruschendorf, M. Goltz, H. Ellerbrok, H. Renz, J. Walter and M. Paulsen. 2000. Genetics of atopy in a mouse model: polymorphism of the IL-5 receptor alpha chain. *Immunogenetics* **51**:632–638.
122. DeSanctis, G.T., M. Merchant, D.R. Beier, R.D. Dredge, J.K. Grobholz, T.R. Martin, E.S. Lander and J.M. Drazen. 1995. Quantitative locus analysis of airway hyperresponsiveness in A/J and C57BL/6J mice. *Nat. Genet.* **11**:150–154.

123. DeSanctis, G.T., J.B. Singer, A. Jiao, C.N. Yandava, Y.H. Lee, T.C. Haynes, E.S. Lander, D.R. Beier and J.M. Drazen. 1999. Quantitative trait locus mapping of airway responsiveness to chromosomes 6 and 7 in inbred mice. *Am. J. Physiol.* **277**:L1118–1123.
124. Cookson, W.O., P.A. Sharp, J.A. Faux and J.M. Hopkin. 1989. Linkage between immunoglobulin E responses underlying asthma and rhinitis and chromosome 11q. *Lancet* **1**:1292–1295.
125. Daniels, S.E., S. Bhattacharrya, A. James, N.I. Leaves, A. Young, M.R. Hill, J.A. Faux, G.F. Ryan, P.N. le Souef, G.M. Lathrup, A.W. Musk and W.O.C.M. Cookson. 1996. A genome-wide search for quantitative trait loci underlying asthma. *Nature* **383**:247–250.
126. Shirakawa, T., T. Hashimoto, J. Furuyama, T. Takeshita and K. Morimoto. 1994. Linkage between severe atopy and chromosome 11q13 in Japanese families. *Clin. Genet.* **46**:228–232.
127. Marsh, D.G., J.D. Neely, D.R. Breazeale, B. Ghosh, L.R. Freidhoff, E. Ehrlich-Kautzky, C. Schou, G. Krishnaswamy and T.H. Beaty. 1994. Linkage analysis of IL4 and other chromosome 5q31.1 markers and total serum immunoglobulin E concentrations. *Science* **264**:1152–1156.
128. Postma, D.S., E.R. Bleecker, P.J. Amelung, K.J. Holroyd, J. Xu, C.I. Panhuysen, D.A. Meyers and R.C. Levitt. 1995. Genetic susceptibility to asthma-bronchial hyperresponsiveness coinherited with a major gene for atopy. *N. Engl. J. Med.* **333**:894–900.
129. Hizawa, N., L.R. Freidhoff, E. Ehrlich, Y.F. Chiu, D.L. Duffy, C. Schou, G.M. Dunston, T.H. Beaty, D.G. Marsh, K.C. Barnes and S.K. Huang. 1998. Genetic influences of chromosomes 5q31-q33 and 11q13 on specific IgE responsiveness to common inhaled allergens among African-American families. *J. Allergy Clin. Immunol.* **102**:449–453.
130. Levitt, R.C. and S.L. Ewart. 1995. Genetic susceptibility to atracurium-induced bronchoconstriction. *Am. J. Respir. Crit. Care Med.* **151**:1537–1542.
131. Levitt, R.C., S.M. Eleff, L.Y. Zhang, S.R. Kleeberger and S.L. Ewart. 1995. Linkage homology for bronchial hyperresponsiveness between DNA markers on human chromosome 5q31-q33 and mouse chromosome 13. *Clin. Exp. Allergy* **25**:61–63.
132. Nicolaides, N.C., K.J. Holroyd, S.L. Ewart, S.M. Eleff, M.B. Kiser, C.R. Dragwa, C.D. Sullivan, L. Grasso, L.Y. Zhang, C.J. Messler, T. Zhou, S.R. Kleeberger, K.H. Buetow and R.C. Levitt. 1997. Interleukin 9: a candidate gene for asthma. *Proc. Natl. Acad. Sci. USA* **94**:13175–13180.
133. Dong, Q., J. Louahed, A. Vink, C.D. Sullivan, C.J. Messler, Y. Zhou, A. Haczku, F. Huaux, M. Arras, K.J. Holroyd, J.C. Renauld, R.C. Levitt and N.C. Nicolaides. 1999. IL-9 induces chemokine expression in lung epithelial cells and baseline airway eosinophilia in trangenic mice. *Eur. J. Immunol.* **29**:2130–2139.
134. Levitt, R.C., M.P. McLane, D. MacDonald, V. Ferrante, C. Weiss, T. Zhou, K.J. Holroyd and N.C. Nicolaides. 1999. IL-9 pathway in asthma: new therapeutic targets for allergic inflammatory disorders. *J. Allergy Clin. Immunol.* **103**:S485–S491.
135. Karp, C.L., A. Grupe, E. Schadt, S.L. Ewart, M. Keane-Moore, P.J. Cuomo, J. Kohl, D. Kuperman, S. Germer, D. Aud, G. Peltz and M. Wills-Karp. 2000. Identification of complement factor 5 as a susceptibility locus for experimental allergic asthma. *Nature Immunol.* **1**:221–226.
136. Shubitowski, D.M., M. Wills-Karp and S.L. Ewart. 2002. The complement factor 5a receptor gene maps to murine chromosome 7. *Cytogenetics Genome Res.* **97**:133–135.

CHAPTER 11

GENOMIC DISSECTION OF COMPLEX TRAIT PREDISPOSITION

Daniel Pomp

Department of Animal Science, University of Nebraska
Lincoln, NE, USA
dpomp@unl.edu

1. Introduction

Most quantitative traits (i.e., those displaying continuous variation) are exceptionally complex, with varying contributions of genetic susceptibility and interacting environmental factors. Predisposition to a phenotypic range for a complex trait such as body weight or body fat results from combinations of relatively small effects of DNA variations within a large number of unidentified polygenes, known as quantitative trait loci (QTL). Well over 200 QTL have been reported for growth and body composition traits in the mouse, likely representing at least 50 to 100 distinct genes (Figure 1). While molecular biology and genetic manipulation have yielded significant gains in understanding complex traits at the physiological and molecular levels, the genetic architecture of predisposition remains essentially undefined. In other words, while we have made great progress in understanding the proteins, pathways and networks that functionally control a complex phenotype, we know very little about the underlying genetic variation that an individual is born with to predispose it to a particular phenotypic range. This large gap between our extensive knowledge of physiological mechanisms underlying complex traits and our embryonic understanding of how genetic predisposition is manifested, greatly impairs identification of genes underlying relevant QTL, inhibiting an important avenue for gene-based development of diagnostic and therapeutic tools with biomedical and/or agricultural relevance.

In this chapter, the current state of knowledge regarding predisposition gene discovery will be discussed, focusing on traits reviewed in Ch. 6 and 7 (body weight, body composition). Strategies to merge analyses of physiology and

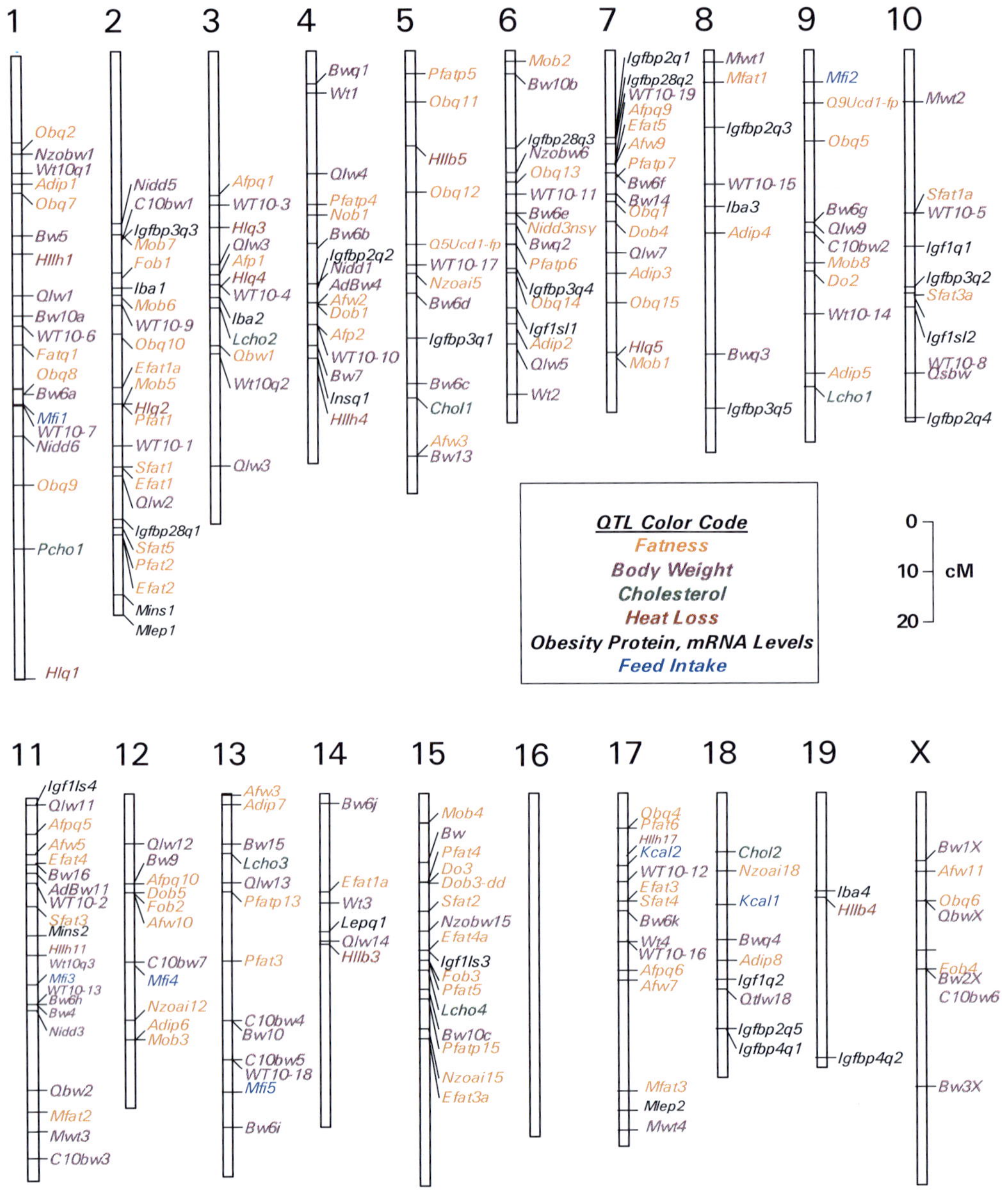
1
Obq2
Nzobw1
Wt10q1
Adip1
Obq7
Bw5
Hllh1
Qlw1
Bw10a
WT10-6
Fatq1
Obq8
Bw6a
Mfi1
WT10-7
Nidd6
Obq9
Pcho1
Hlq1
2
Nidd5
C10bw1
Igfbp3q3
Mob7
Fob1
Iba1
Mob6
WT10-9
Obq10
Efat1a
Mob5
Hlq2
Pfat1
WT10-1
Sfat1
Efat1
Qlw2
Igfbp28q1
Sfat5
Pfat2
Efat2
Mins1
Mlep1
3
Afpq1
WT10-3
Hlq3
Qlw3
Afp1
Hlq4
WT10-4
Iba2
Lcho2
Qbw1
Wt10q2
Qlw3
4
Bwq1
Wt1
Qlw4
Pfatp4
Nob1
Bw6b
Igfbp2q2
Nidd1
AdBw4
Afw2
Dob1
Afp2
WT10-10
Bw7
Insq1
Hllh4
5
Pfatp5
Obq11
Hllb5
Obq12
Q5Ucd1-fp
WT10-17
Nzoai5
Bw6d
Igfbp3q1
Bw6c
Chol1
Afw3
Bw13
6
Mob2
Bw10b
Igfbp28q3
Nzobw6
Obq13
WT10-11
Bw6e
Nidd3nsy
Bwq2
Pfatp6
Igfbp3q4
Obq14
Igf1sl1
Adip2
Qlw5
Wt2
7
Igfbp2q1
Igfbp28q2
WT10-19
Afpq9
Efat5
Afw9
Pfatp7
Bw6f
Bw14
Obq1
Dob4
Qlw7
Adip3
Obq15
Hlq5
Mob1
8
Mwt1
Mfat1
Igfbp2q3
WT10-15
Iba3
Adip4
Bwq3
Igfbp3q5
9
Mfi2
Q9Ucd1-fp
Obq5
Bw6g
Qlw9
C10bw2
Mob8
Do2
Wt10-14
Adip5
Lcho1
10
Mwt2
Sfat1a
WT10-5
Igf1q1
Igfbp3q2
Sfat3a
Igf1sl2
WT10-8
Qsbw
Igfbp2q4
QTL Color Code
Fatness
Body Weight
Cholesterol
Heat Loss
Obesity Protein, mRNA Levels
Feed Intake
0
10
20
cM
11
Igf1ls4
Qlw11
Afpq5
Afw5
Efat4
Bw16
AdBw11
WT10-2
Sfat3
Mins2
Hllh11
Wt10q3
Mfi3
WT10-13
Bw6h
Bw4
Nidd3
Qbw2
Mfat2
Mwt3
C10bw3
12
Qlw12
Bw9
Afpq10
Dob5
Fob2
Afw10
C10bw7
Mfi4
Nzoai12
Adip6
Mob3
13
Afw3
Adip7
Bw15
Lcho3
Qlw13
Pfatp13
Pfat3
C10bw4
Bw10
C10bw5
WT10-18
Mfi5
Bw6i
14
Bw6j
Efat1a
Wt3
Lepq1
Qlw14
Hllb3
15
Mob4
Bw
Pfat4
Do3
Dob3-dd
Sfat2
Nzobw15
Efat4a
Igf1ls3
Fob3
Pfat5
Lcho4
Bw10c
Pfatp15
Nzoai15
Efat3a
16
17
Obq4
Pfat6
Hllh17
Kcal2
WT10-12
Efat3
Sfat4
Bw6k
Wt4
WT10-16
Afpq6
Afw7
Mfat3
Mlep2
Mwt4
18
Chol2
Nzoai18
Kcal1
Bwq4
Adip8
Igf1q2
Qtlw18
Igfbp2q5
Igfbp4q1
19
Iba4
Hllb4
Igfbp4q2
X
Bw1X
Afw11
Obq6
QbwX
Fob4
Bw2X
C10bw6
Bw3X

Figure 1. Mouse Body Weight Predisposition Map. The status (2004) of mapped quantitative trait loci (QTL) for fatness (orange), body weight and weight gain (violet), serum cholesterol (green), heat loss (red) and levels of mRNA or proteins for candidate genes with physiological relevance to obesity (black). All body weight and weight gains are from animals of 6 weeks of age or older. Lengths of chromosomes and QTL map positions are according to the Mouse Genome Database. It is emphasized that QTL map positions may be inherently inaccurate (in some cases 95% confidence intervals for a QTL include a majority of a chromosome. Only QTL mapped with at least genome-wide 5% significance levels were included in the map, but several suggestive QTL were included when their map positions reaffirmed map positions of other QTL from independent experiments. Symbols for QTL are as presented in the literature (see Elo (2003) for full list of references). If authors did not provide symbols, those suggested by Chagnon *et al.*[108] were used. Adapted from Elo.[109]

predisposition will be described, with an immediate focus on moving from QTL to gene to DNA sequence variations. While such information will be highly informative, the overarching goal is to reconstruct genetic pathways, networks and systems, determine how DNA sequence variations work alone and together to impact each, and gain a more holistic understanding of the overall genetic architecture of complex traits. Finally, new frontiers for complex trait genetic analysis will be outlined.

2. Significance

Carcass and body composition traits constitute extremely important considerations of modern livestock production systems where consumer health concerns and marketing perspectives play increasingly prominent roles. The problem of excess fat in livestock and poultry carcasses is ubiquitous and has serious consequences for the animal industry at four levels: health perceptions of consumers; wasteful production of an undesired biological component; labor costs associated with trimming waste fat; and lower biological efficiencies of fatter animals.[1]

An estimated 65% of US adults are overweight, and 31% are obese, with higher percentages in female minority populations.[2] Overweight and obese conditions substantially increase risk of hypertension, dyslipidemia, type 2 diabetes, coronary heart disease, stroke, gall bladder disease, sleep apnea, osteoarthritis and respiratory problems, combining to form the single largest cause of death in developed countries.[3] Extensive epidemiological data indicate that obesity can also significantly alter cancer risk. A recent and very large study[4] determined that being overweight and/or obese potentially accounts for 20% of all cancer deaths in US women and 14% in US men.

Understanding the roles of specific loci in genetic susceptibility to obesity is critical to improving human health and quality of life. As stated by Comuzzie and Allison,[5] "One of the greatest challenges in biomedical research today is the elucidation of the underlying genetic architecture of complex phenotypes such as obesity." Unfortunately, little progress has been made on this front despite significant research attention. Given the ~75 billion dollar annual health costs associated with obesity,[6] finding novel targets for pharmacological intervention and pharmacogenomic management is critical.

Similar pictures can be painted for a wide variety of complex traits with biomedical and/or agricultural relevance. Enhanced understanding of the genetic architecture of complex trait predisposition will be broadly applicable to many important phenotypes.

3. Background

A significant heritable component for disease was formally recognized as long ago as the turn of the 20th century, when Sir Archibald Garrod[7] described the heritable nature of alkaptonuria. Since then, thousands of genetic disorders exhibiting Mendelian inheritance have been described, and the molecular basis for many of these has been identified. These diseases have phenotypes which fall into discrete classes (e.g., afflicted vs. not afflicted). In contrast, most economically relevant traits in animal agriculture and most common human maladies such as obesity exhibit continuous phenotypic variation and a predominantly multifactorial and polygenic basis.[8,9] While certain rare mutations have been identified, accounting for a small minority of extreme phenotypes in animals[10] and humans,[11] respectively, the actual identity of genes segregating and contributing to common phenotypes in populations is essentially unknown.

Location of polygenic factors controlling inheritance of quantitative traits can be established by tracking the segregation of closely linked markers, i.e., QTL mapping. This dates back to early work with *Drosophila*,[12] mice[13] and *Phaseolus vulgaris*[14]. For the subsequent 70 years, analyses continued to use visible phenotypic markers and protein variants. However, DNA marker analysis of quantitative traits has recently gained prominence with development of ubiquitous and polymorphic marker systems[15] and powerful statistical methodologies.[16–19] The first extensive use of DNA markers was in plants, where large suitable families could be generated.[20,21] The feasibility of performing QTL mapping in mammals was first demonstrated using rats,[22,23] but the mouse has subsequently dominated research in polygenic analysis including body weight regulation.[24,25] The prominent role of mice in complex trait analysis mirrors its role in molecular dissection in the post-genome era, building on its general status as one of the major genetic tools responsible for advancing biomedical research (see Ch. 15).

Well over 200 QTL for growth and obesity-related traits have been localized in the mouse. For simplicity and convenience, we have graphically summarized the ~60 papers contributing this information in Figure 1 (see also Table 3[26,27]). This map is a remarkable achievement, representing a comprehensive and exhaustive portrait of the predisposition genomic map for mouse body weight regulation and associated traits. Although it requires a significant effort, QTL mapping has become a relatively routine endeavor, and similar genome maps of predisposition loci can now be built for a variety of complex traits. And while new QTL continue to be revealed for obesity and for many other quantitative phenotypes, this phase of the study of continuous variation is coming to an end.

Many lessons have been learned regarding the nature of complex trait predisposition based on QTL mapping.[26,27] For example, the distribution of polygenic effects clearly does not conform with the uniform distribution proposed in the context of the infinitesimal model, but rather approximates an exponential model[28] that nonetheless maintains an infinitesimal quality.[29] Also, while the primary contributions to genetic control of growth and obesity within a species seem to derive from genome-wide variations (i.e., Figure 1), within a specific population such contributions derive only from localized genomic regions.[26,30]

However, despite the successes of QTL mapping, few of these QTL have been unequivocally resolved to a single gene, let alone the causative source of genetic variation. This immense chasm between the ease of QTL mapping and the difficulty in identifying the underlying genes dramatically limits the ability to harness this critical mass of information for the betterment of human health and/or food production. This scenario is all the more surprising due to the fact that advancements in knowledge on the endocrine, biochemical and molecular underpinnings of obesity represent a significant success story in modern biology, sparked originally by study of spontaneous mutations causing obesity[31,32] and later by targeted single gene mutations in a multitude of genes with relevance to energy balance.[33,34] This frustrating gap in our knowledge of physiological mechanisms underlying obesity, and the nature of genetic predisposition to obesity, was insightfully summarized by Bray and Bouchard[35] who wrote: "Unfortunately, the spectacular gains in understanding the biology of energy balance of the last few years have not yet translated into significant advances on the genetic front. This is particularly striking when one realizes that so far there is not one single obese human being whose excess body fat can be explained by a specific mutation in one of the genes exerting its effects in relevant energy balance pathways."

In recent years there have been a few reports of single gene mutations resulting in human obesity,[11,36] and we have seen a promising trend in the ability to elucidate the underlying identities of QTL,[37] including those underlying obesity-related traits. However, while strategies for identifying genes that determine quantitative traits have been clearly outlined,[38] the painstaking and difficult nature of the process bodes poorly for rapid progress.[39] This dilemma grows in parallel with the continued rapid pace of discovery in understanding energy balance mechanisms.[40,41]

4. Quantitative Genomics: A New Experimental Paradigm for Analysis of Complex Trait Genetics

A central hypothesis has been proposed[25] that substantial genetic variation underlying predisposition to weight regulation and obesity is involved in *trans*-regulation of the primary physiological pathways directly regulating quantitative phenotypes. This hypothesis has been formulated based on several areas of accumulated data. First, few causative mutations have been found within energy balance candidate genes, despite significant detection efforts in humans.[11] Second, studies localizing QTL regulating mRNA or protein levels of such candidate genes have primarily identified *trans*-acting regulation. And third, recent genome-wide evaluations have found that *trans*-acting loci are the primary drivers of variation in gene expression in yeast,[42,43] *Drosophila*[44] and mice[45]. In the next sections of this chapter, it will be argued that the paradigm of "Quantitative Genomics,"[25] whereby large-scale endo-phenotyping at the transcriptional, proteomic and/or metabolomic levels is performed within the context of a QTL mapping population, will be a powerful force in dissecting the genetic architecture of complex trait predisposition. Quantitative Genomics is designed to synergistically harness the powers of recombination and functional analyses.

Within the broad concept of Quantitative Genomics, the combining of QTL mapping with large-scale gene expression analysis currently represents the most powerful mechanism for bridging the gap between our knowledge of the physiology of, and predisposition, to obesity. This has been referred to as "transcriptome mapping"[46] and "genetical genomics"[47,48] and treats gene expression levels of any particular gene measured across different individuals as an expression-level polymorphism that in principle reflects the underlying genetic variation.[47,49,50] This type of analysis was actually pioneered using proteomic evaluation[51,52] and later extended by the same group to the transcriptome in an F_2 population of maize.[53] Transcriptome mapping has been highlighted as a powerful mechanism to dissect complex traits and make more efficient the selection of candidate genes underlying predisposition loci, with recent successful implementation in yeast,[42,43] *Drosophila,*[44,54] *Eucalyptus,*[55] mice,[45,56] and humans.[57]

As discussed earlier, a critically important question that transcriptome mapping promises to help answer regards the underlying nature of obesity QTL. Are primary obesity QTL represented by sequence variation within genes with major roles in energy balance pathways or, as first proposed several years ago,[58] do they regulate such genes in a *trans*-acting manner? Although a plethora of candidate gene analyses in humans has provided mixed results (e.g., *MC3R*;[36] *UCP2*;[59] *LEPR*[60]), the latter speculation is strongly supported when the transcriptome/proteome mapping paradigm has been applied on a modest basis by examining one or a few transcripts/proteins at a time. For example, mapping for determinants of levels of plasma leptin, IGF-1, IGF-1 binding proteins and other individual components of energy balance pathways has identified multiple QTL for each trait, but none coinciding with map positions of the structural genes themselves in mice[61–64] or humans.[65] Several *trans*-acting QTL for induced *Ucp1* mRNA levels have been identified in mice,[66] and we have found similar results for control of hypothalamic mRNA levels of *Rpl3*, *Oxt* and *Timp2*[67] (Wesolowski and Pomp, unpublished data) and for hepatic expression of most of the genes involved in the insulin signaling pathway (Ferrell and Pomp, unpublished data). In pigs, only *trans*-acting QTL influencing plasma FSH levels were identified.[68]

Although these very limited findings provide a tantalizing glimpse into the nature of complex trait predisposition and the underlying QTL, transcriptome mapping offers the potential for several orders-of-magnitude greater power and resolution. In yeast, for example, Brem *et al.*[42] found QTL for 570 expressed genes in a cross between laboratory and wild strains. Of these, 36% were apparently due to polymorphisms within the gene themselves, while the remainder was controlled by a small group of *trans*-acting modulator loci each regulating from 7 to 94 genes of related function. Similar findings were reported in mice,[45] where it was further found that the stronger the evidence for an expression QTL, the more likely it would map within the structural gene itself.

This latter result is expected as DNA mutations within a gene and affecting the expression of that gene should be easier to identify than second-order effects. However, it should be noted that in such cases, the causal polymorphism may be closely linked to, but not part of, the gene whose expression is being evaluated.

Yvert *et al.*[43] determined that most gene expression differences in a cross between laboratory and wild strains of yeast mapped to *trans*-acting loci. Furthermore, *trans*-regulatory variation was broadly dispersed across different classes of genes with a wide variety of functions. Further compelling evidence that obesity QTL may be represented by sequence variation within *trans*-acting loci that regulate genes with major roles in energy balance pathways is provided in a study using *Drosophila.*[44] In that experiment, *trans*-regulatory variation was found in metabolic enzyme activity for each key determinant of metabolism and respiration measured. And in *Eucalyptus*, results suggest that transcription of genes controlling lignin is regulated by *trans*-acting modulation.[55]

5. Expected Outcomes from the Transcriptome Mapping Paradigm

Establishment of a robust system that can genetically map loci that modulate the steady-state levels of any gene showing transcript level variation between populations with divergent phenotypes is expected to provide a wealth of information regarding genetic architecture of complex traits. A hypothetical snap-shot of the types of outcomes this research paradigm can provide is summarized in Figure 2, where the X-axis represents the known genetic map position of each gene represented on a large-scale expression microarray, and the Y-axis represents the estimated genetic map location of the single QTL that explains the most variation in expression levels for each gene on the X-axis. When transcriptome mapping data are plotted in such a manner, a higher-order structure of gene transcript level regulation should be revealed. Based on the early large-scale transcriptome mapping efforts, three patterns of transcript regulation are being revealed (R.W. Williams and D.W. Threadgill, personal communication), with a fourth pattern possible.

First, a large number of genes plotted along the diagonal will suggest that their transcript levels are *cis*-regulated (i.e., the location of their QTL transcript regulators genetically map to the physical location of the genes themselves), likely due to promoter polymorphisms or other variants within or near the genes themselves that affect transcript level. This "*cis*-diagonal"[46] can immediately uncover high-quality candidate genes potentially representing QTL for growth and obesity phenotypes, especially when the map position of the expression QTL

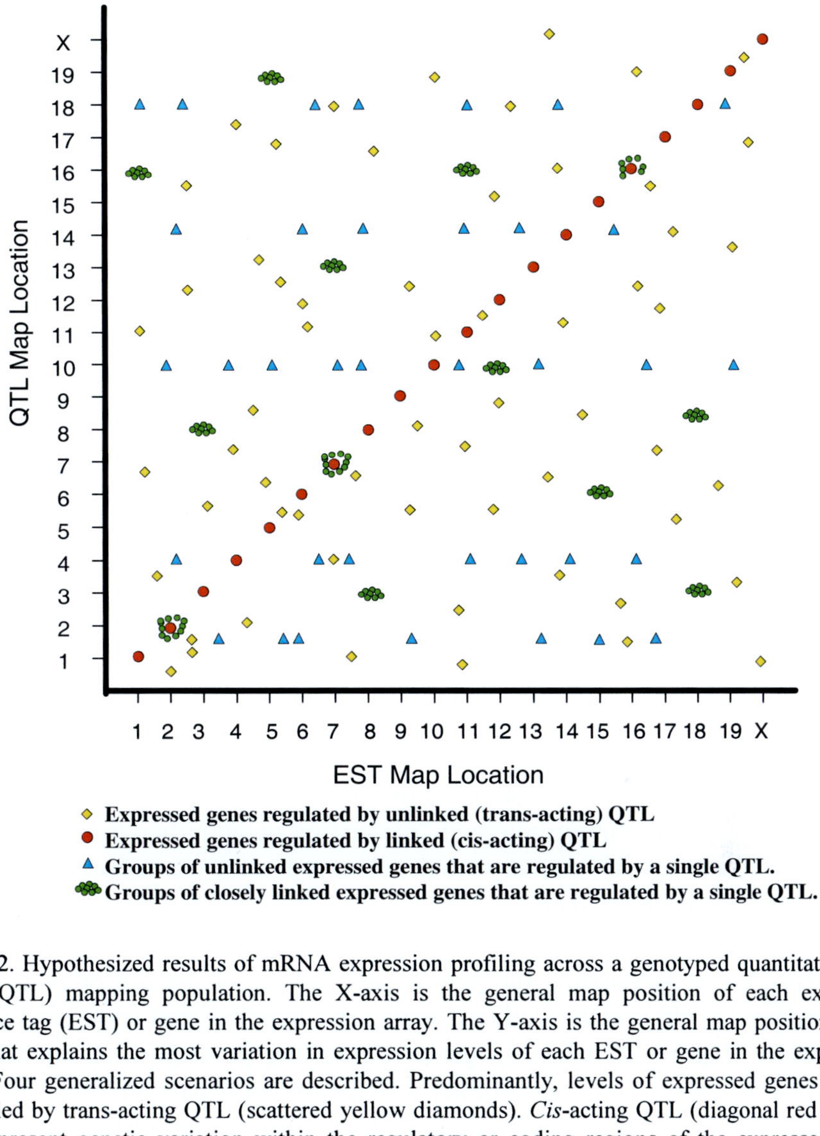

Figure 2. Hypothesized results of mRNA expression profiling across a genotyped quantitative trait locus (QTL) mapping population. The X-axis is the general map position of each expressed sequence tag (EST) or gene in the expression array. The Y-axis is the general map position of the QTL that explains the most variation in expression levels of each EST or gene in the expression array. Four generalized scenarios are described. Predominantly, levels of expressed genes will be controlled by trans-acting QTL (scattered yellow diamonds). *Cis*-acting QTL (diagonal red circles) may represent genetic variation within the regulatory or coding regions of the expressed genes themselves. In addition, a single QTL may result in changes in expression levels of many unlinked genes (horizontal blue triangles), either due to direct pleiotropy or to multiple changes in a regulatory cascade resulting from alternation of expression in a single key gene. Finally, clusters (small green dots) of gene expression changes could result from changes due to linkage of multiple expressed genes to a single regulatory QTL, or alternatively as a result of coordinated expression neighborhoods.

falls under the QTL peak for the end-point phenotype (e.g., weight, fat, feed intake) and the expression and end-point phenotypes exhibit significant genetic correlation. A second class of genes would be those controlled by unlinked *trans*-regulators, as evidenced by genetic locations of controlling QTL being different from the physical locations of the genes they regulate. Evaluation of the *trans*-regulating patterns of expression phenotypes will add immense value to selection of candidate genes representing QTL by implicating pathways and mechanisms underlying the mechanism of action of the QTL.

A third, and highly interesting class of genes, is QTL transcript modulators that would regulate the steady-state transcript levels of large groups of unlinked genes. These master transcript modulators would be identified by the horizontal strips of plotted QTL modulators (Figure 2), indicating the presence of one or a few tightly linked regulatory genes. This class of results will represent two important findings. First, the QTL may be in a key gene within a pathway that, when perturbed by a polymorphism, causes a cascade of effects that are evidenced by multiple expression changes in other genes. Second, the QTL may represent a key genetic control switch such as a transcription factor, within which a polymorphism could cause a multitude of changes in expression of genes throughout the genome. Such expression QTL patterns were found in human immortalized B cells.[57] And studies examining gene expression in mouse brain using recombinant inbred lines have identified such master transcript modulator loci that regulate hundreds to thousands of transcripts (R.W. Williams and D.W. Threadgill, personal communication). Finally, a speculative fourth class of genes would be those representing potential "expression neighborhoods,"[69,70] although evidence for such results has not yet been observed in the transcriptome mapping experiments conducted to date.

We expect these efforts to begin to enable development of an initial framework for understanding the genetic architecture of obesity predisposition. Such studies should greatly facilitate testing our hypothesized structure for such architecture (Figure 3), whereby genes controlling predisposition to complex traits such as body weight and obesity are, for the most part, involved in trans-regulation of the primary physiological pathways directly regulating phenotypes involved in energy balance.

Additional benefits of the transcriptome mapping paradigm are the ability to detect genetic heterogeneity within phenotypic classes at the level of gene expression and to test the effects of such heterogeneity on gene identification for complex traits. For example, Schadt *et al.*[45,56] found one general pattern of hepatic gene expression within those F_2 mice that were relatively resistant to

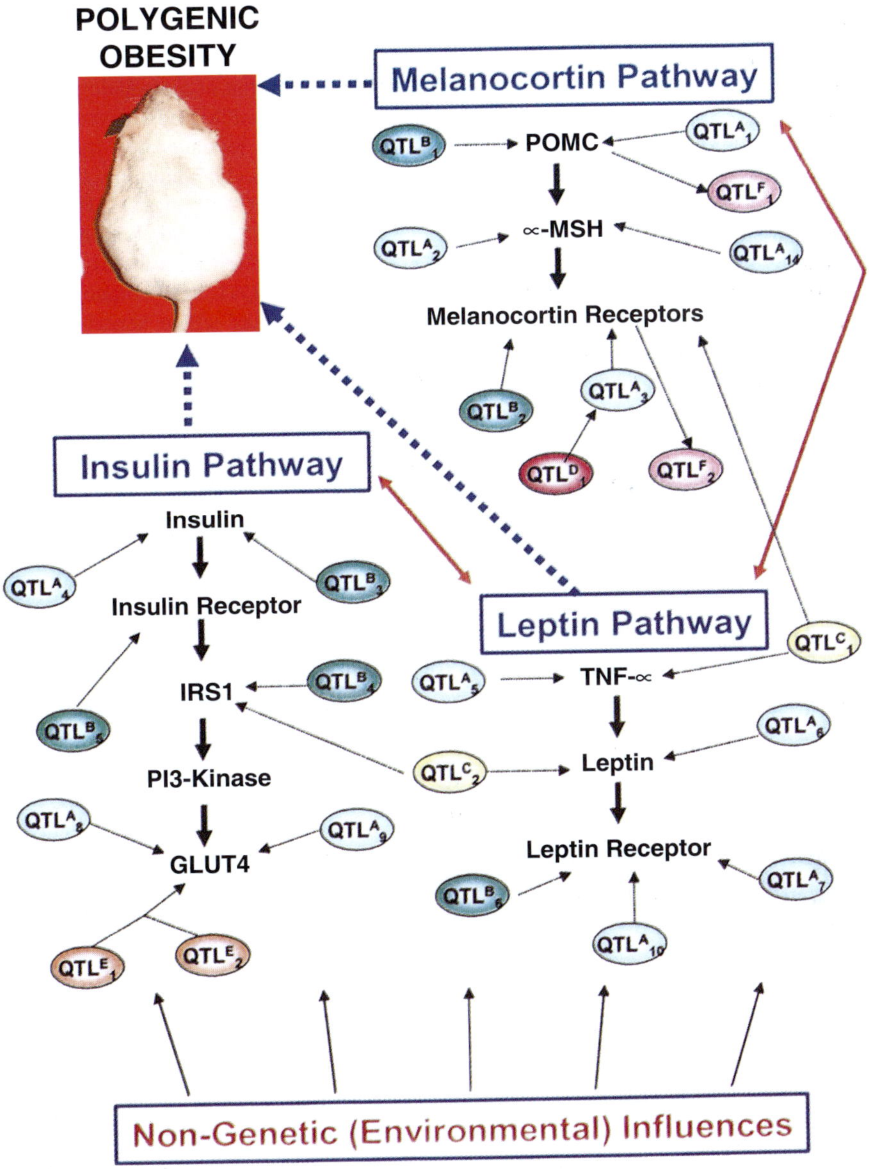

Figure 3. Hypothesized genetic architecture for obesity predisposition. Simplified and generalized components of example pathways (out of many that combine to regulate obesity) are illustrated. We hypothesize that each of these pathways consists of a complex regulatory cascade with the coordinated and interactive expression of genes playing significant roles in the physiology of the pathway, under the regulation of predisposition genes (quantitative trait loci; QTL) and non-genetic (environmental) influences. In other words, key genes in physiological pathways would for the most part not possess relevant

(Figure 3 continued) heritable variation; such variation would instead reside within loci that regulate expression levels of these key physiological genes. There could be several potential modes of regulation and complexity for QTL that control expression or activity of physiological genes. QTL^A (blue): QTL can regulate transcription of a physiological gene. For example, a transcription factor within which genetic variation leads to variability in activity or expression levels. QTL^B (green): QTL can regulate post-translational modification of the physiological gene. QTL^C (yellow): QTL can be pleiotropic factors that regulate multiple physiological genes. QTL^D (red) and QTL^E (orange): QTL can act epistatically to regulate other QTL. For example, certain genetic variation within a QTL is required in order for variation at a second QTL to exert an influence on the expression or activity of a physiological gene (QTL^D), or effects of multiple QTL must combine in order to regulate a physiological gene (QTL^E). QTL^F (pink): cases where heritable genetic variation does exist directly in the physiological genes, either within regulatory or coding regions, contributing to phenotypic variability in expression or activity of that gene (e.g. *POMC, MC4R*).

dietary-induced obesity, while the F_2 mice that were sensitive to dietary-induced obesity could be partitioned into classes represented by two distinct patterns of gene expression. This genetic heterogeneity within the fatter mice had important ramifications for obesity QTL identification within this F_2 population. In one case, a QTL was detectable when all data were included in the analysis; however, the effect was highly significant when the lean mice were compared to one of the classifications of fat mice and was not significant when the lean mice were compared to the other classification of fat mice.[45] In a second case, a QTL was only detected after consideration of the genetic heterogeneity within the mice with high adiposity. Both of these results would clearly have very important ramifications for development and use of diagnostic and therapeutic tools for any complex trait, when such tools are based on gene targets uncovered in QTL analysis.[56]

6. Future Directions for Transcriptome Mapping

No published study to date has evaluated multiple tissues in transcriptome mapping. We expect that evaluation of highly relevant yet diverse tissue types will expose different sets of transcript QTL, even when analyzing the same transcript populations. Discovery of similarities (or differences) across tissues will add power to experimental findings, providing validation especially for *cis*-acting QTL and revealing important underlying biology for the traits of interest.

Extension of the paradigm of transcriptome mapping to the proteome and metabolome would also be unique. Although experimental issues (i.e., lack of genome-wide reagents) would render such extension to be initially on a limited scale, it would represent an important test to determine if QTL profiles underlying protein levels parallel those controlling transcription. The question of whether microarrays are valid indicators of actual protein levels (and hence biological activity) is still of major importance, and such results would test and extend this question in regard to comparison of the underlying polygenic control at each step of the central dogma of biology. In regard to metabolomics, discovery of QTL controlling carbohydrate, glucose, lipid and fatty acid metabolism would have particular relevance to obesity research. Our preliminary efforts in this regard have revealed a large number of QTL underlying *de novo* fatty acid synthesis (Allan, Pomp and Eisen, unpublished data).

7. Statistical Issues in Transcriptome Mapping

Although transcriptome mapping requires significant computing power, the basic aspects of necessary statistical paradigms are already in place based on traditional transcriptome analysis and QTL mapping. However, several layers of analysis should be employed to extract full value from the enormous amount of collected data and gain valuable insight into genetic control of gene expression. As recently noted by Darvasi,[71] "I expect that the combining of genetic information and gene expression will hasten the day when genomics delivers on its promise to improve health care. But we must continue striving to develop and apply sophisticated analytical tools for interpreting the vast, complex data sets that are being produced with modern genomic technologies."

Traditionally, these would include analysis of sex-interactions in genetic control of the transcriptome, determination of the role of genomic imprinting in control of genome-wide gene expression, and evaluation of within- and between-founder line genetic variance. Perhaps more important, transcriptome mapping represents an extremely challenging scenario for thorough implementation of multiple trait analyses. Initially, this may be best implemented for specific situations, such as genes that are part of the same known pathway and genes measured in different tissues. Also, when single-trait expression QTL appear to map to the same region, multi-trait analyses can be used to improve precision and significance.

Most QTL analyses have ignored the potential role of gene interactions in the control of trait variation. However, there is mounting evidence that analyses specifically testing for epistasis can both identify QTL that are not otherwise found and explain a greater proportion of the genetic variation.[72–76] In the context of understanding the genetic control of the transcriptome and proteome, it is critical that we extend analyses to include epistasis, as this is likely to play an important role in interpretation of the network of gene interactions that contribute to obesity. Transcriptome mapping data sets can also be applied to Bayesian model selection approaches for genome wide epistatic QTL mapping, methods for fine mapping that attempt to separate epistasis from haplotype effects, and methods using empirical Bayes and other methodology to combine QTL analysis and microarray analysis in single statistical models where each type of information borrows strength from the other. An excellent example of development of novel statistical methods to use transcriptome mapping data to reconstruct gene networks was recently presented by Zhu *et al.*[77]

When transcriptome mapping is implemented within a very large structured pedigree, the opportunity exists to merge traditional quantitative genetic analyses, such as genetic parameter estimation, with QTL analysis. This would enable estimation of genetic correlations among, and heritabilities of, the sub- (e.g., transcriptional, proteomic, endocrine) and end-point phenotypic traits. Such an analysis was recently implemented on a relatively small scale using expression data for lymphoblastoid cell lines across members of 15 Centre d'Etude du Polymorphisme Humain (CEPH) families.[78] As with multi-trait analysis, data reduction is likely required to make this effort feasible and enable extraction of meaningful information when larger data sets are considered. For example, heritabilities can be measured for all endpoint phenotypes and for all endo-phenotypes for which at least one significant QTL is identified. Genetic correlations can then be estimated for the following sets of traits: A) between each pair of endpoint phenotypes; B) between each sub-phenotype for which heritability is estimated significantly different from zero; and between traits in categories A and B.

Genetic parameter estimation will add unique value to this research paradigm. A strong heritability of a sub-phenotype should coincide with presence of strong evidence for QTL, providing validation of the process. More important, genetic correlation between a sub-phenotype and body fat levels will be critical to differentiate among and rank multiple linked transcriptional and/or proteomic QTL that could represent positional candidate genes for obesity predisposition. Finally, as we first proposed,[25] it is interesting to speculate that genetic correlation analysis can be a useful method for clustering of array results and identification of biologically relevant pathways. This clustering would be an extension of the stratification of obesity phenotypic classes based on combined expression phenotypes[45] and the use of principal components analysis recently proposed by Lan *et al.*[79]

Both QTL analysis and genetic parameter estimation, in the context of transcriptome mapping, have immense computational requirements. It is likely that such efforts, when carried out on a large scale, will require recoding of existing programs to run on multiple parallel processors.

8. Implementation of Transcriptome Mapping to Study Obesity

Analysis of complex trait genetic architecture using "Quantitative Genomics," manifested through the transcriptome (and/or proteome and metabolome) mapping paradigm can be applied to essentially any QTL mapping experiment where samples with spatial and temporal biological relevance have been stored in a manner that maintains their integrity. In regard to obesity, the study by Schadt *et al.*[45] specifically targeted dietary-induced obesity in mice and identified two promising candidates potentially representing the MMU2 QTL described earlier.[80] This distal region of mouse chromosome 2 is one of the most relevant to obesity predisposition in the mouse genome.[11,81] Not only is this region well populated with multiple body weight and fatness QTL, from crosses employing different approaches and genetic backgrounds,[27,61,80] QTL harbored in this region have among the largest effects of any body weight and obesity polygenes ever localized.[26,27,81,82]

We are applying this approach using the polygenic obese M16 line of mice,[64,83,84] and its non-obese ICR control line (Figure 4). Having identified differences between the lines for a wide variety of traits including transcriptional, proteomic and metabolomic phenotypes with relevance to energy balance,[64,85] we have established a large F_2 QTL mapping population and have phenotyped ~1,200 mice for growth, fatness, feed consumption and an endocrine panel (Allan, Pomp and Eisen, unpublished data). Tissues with relevance to energy balance were stored for endo-phenotyping, while DNA was extracted from all animals. Currently, in a large collaboration with the pharmaceutical industry, liver, adipose, muscle and hypothalamus samples from most F_2 individuals are being profiled using a microarray containing the majority of mouse transcripts, and DNA is being densely genotyped for single nucleotide polymorphism (SNP) markers. Furthermore, the F_2 was designed to provide a family structure suitable for genetic parameter estimation for all phenotypes measured across the population. By applying large-scale transcriptional phenotyping to the M16 × ICR F_2 QTL mapping population, our goal is to advance the positional cloning of obesity polygenes and begin to obtain significant understanding of the genetic architecture of obesity predisposition.

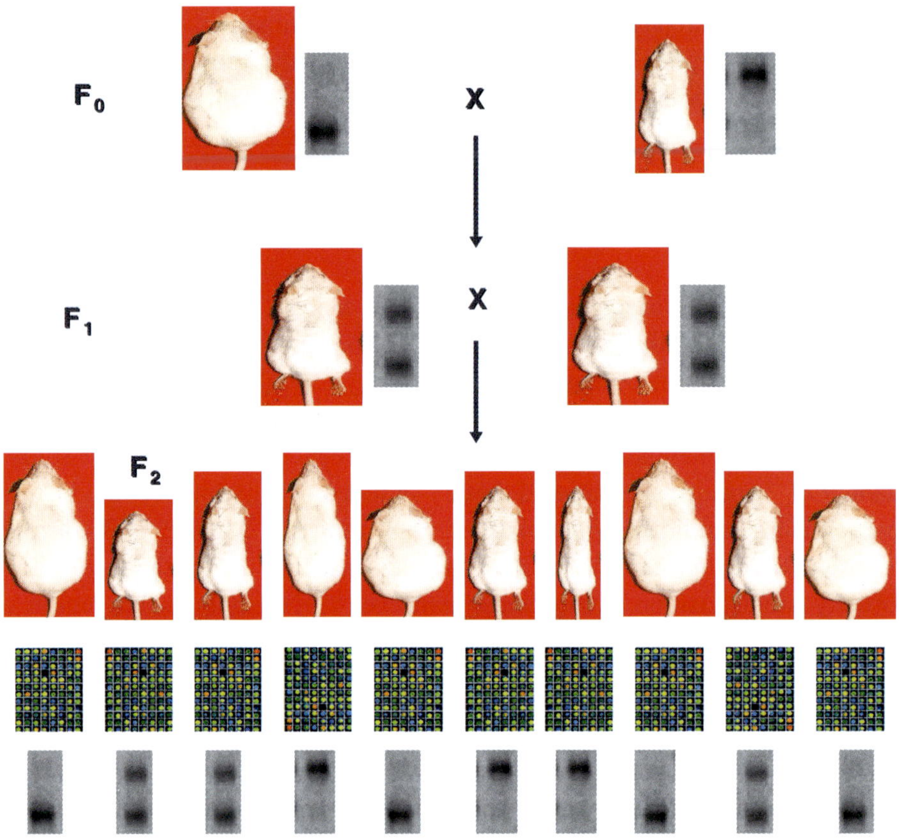

Figure 4. Application of the transcriptome mapping approach to body weight and obesity using an F_2 cross between a line selected for rapid weight gain (M16, top left) and its unselected control line (ICR, top right). Individuals in the segregating F_2 population (images here are graphical representations created from pictures of the two F_0 mice) will be phenotyped for body weight, body composition and feed intake. Additionally, tissue samples from F_2 individuals will be assayed using a microarray containing most expressed murine genes. Alternative parental line forms of DNA markers will segregate and can be tracked in the F_2 population, facilitating a QTL analysis for both weight and composition end-point phenotypes and gene expression sub-phenotypes. This project is in collaboration with Mark Allan and Gene Eisen.

A much broader and powerful platform would be provided by development of a large cohort of recombinant inbred lines (RIL) developed from a multi-way cross of strains representing the majority of phenotypic diversity available in mice.[86–88] A set of 1000 RIL originating from a cross of 8 inbred lines would theoretically achieve 0.1 cM precision when mapping QTL with additive effects of > 0.25 SD. Applying transcriptome mapping to this organized assortment of defined recombinational breakpoints would dramatically increase the rate of positional cloning of genes underlying QTL, and would significantly enhance the understanding of the genetic architecture of complex trait predisposition for obesity and for a wide variety of other agriculturally and biomedically relevant phenotypes.

9. Limitations of Transcriptome Mapping

It is prudent to acknowledge that, despite the potential power and breadth of the transcriptome mapping approach, it has important drawbacks and limitations that can and will restrict its utility. We will use examples from recent and well characterized gene discoveries in livestock species to generally illustrate the limitations of transcriptome mapping, assuming hypothetically that this approach were able to have been applied in each case.

One important issue is that some QTL may not be manifested by changes in steady-state levels of mRNA. Not only would transcriptome mapping fail to identify correct candidate genes underlying such QTL, it may in fact mislead the investigator into examining the wrong candidates. For example, the double muscling phenotype in cattle is known to be caused by mutations in the *myostatin* gene. However, these mutations are not manifested by changes in mRNA levels, but rather by alterations in protein function.[89] Transcriptome mapping would not have identified *myostatin* as a candidate gene in a resource population segregating the double muscling phenotype, while other expression QTL falling on the cis-diagonal (see Figure 2) in the chromosomal region where double muscling had been mapped to may have been falsely identified as candidate genes. The approach would still, however, provide important information on transcriptional changes that are downstream from the QTL's effect and which are important in the context of understanding the overall genetic architecture of the trait.

Since the transcriptome mapping approach relies on gene expression phenotypes, it is critical that selection of both spatial (what tissue) and temporal (when the tissue is collected) coordinates captures as much significant biology as possible. An example of this is clearly demonstrated in the recent finding of a QTL represented by a regulatory mutation in IGF-2 causing a major effect on muscle growth in pigs.[90] Given that this mutation is manifested by gene expression changes in postnatal skeletal and cardiac muscle but not in fetal muscle or postnatal liver,[90,91] transcriptome mapping would have been of immediate assistance in finding this mutation only if postnatal muscle was evaluated. In cases such as obesity, multiple tissues are implicated in control of specific pathways that contribute to the end-phenotype, and expression of many genes will vary over time and across environments (e.g., diet). Thus, a thorough transcriptome mapping effort would constitute a massive undertaking involving multiple tissues, time points, environments and genetic backgrounds that, combined with the high cost of microarrays, is likely beyond the scope of most research budgets.

A third example of the potential limitations of the transcriptome mapping approach is demonstrated by the possible complexity underlying gene expression changes manifested by a QTL. An excellent example of this would be the callipyge (*CLPG*) locus causing muscle hypertrophy in sheep, which is characterized by a unique polar overdominance mode of inheritance. The *CLPG* mutation was identified as an A to G transition in a highly conserved dodecamer motif located between the imprinted *DLK1* and *GTL2* genes.[92] The *CLPG* mutation was shown, in postnatal skeletal muscle, to enhance the transcript levels of the *DLK1*, *PEG11*, *GTL2*, and *MEG8* genes in *cis* without altering their imprinting status.[93,94] If transcriptome mapping had been applied in postnatal muscle to the callipyge model, the expression changes in these multiple genes that are tightly linked to each other and to the mutation underlying the QTL, would all have appeared as a cluster of genes on the *cis*-diagonal (see Figure 2). Although we present such complexity as a potential limitation, in reality it may also be seen as evidence of the potential power of the approach.

Given that significant effort and expense are invested in population development and data collection using transcriptome mapping, it is imperative that data and results be made broadly available to the research community. One such powerful and useful environment is provided by WebQTL.[95,96] WebQTL (http:/www.webqtl.org/) is a web-based package for complex trait analysis and a tool for multi-dimensional searches among large data sets derived from high-throughput analysis techniques. Furthermore, exploring these data sets in a systematic way will be a challenge. This challenge has already been partially addressed by those describing molecular biology and genetics of simpler eukaryotes. For example, the GRID is a database for genetic or molecular interactions among products of yeast, fly and worm genes,[97] and OSPREY[98] is a software platform for visualization of interaction networks. Such tools then will be directly applicable to exploration of obesity-related interactions uncovered in transcriptome mapping.

10. New Frontiers for Analysis of Complex Trait Genetics

Transcriptome mapping holds significant promise for bridging the knowledge gap between physiology and predisposition of complex traits, and thus selecting and prioritizing candidate genes that may underlie a QTL. However, we are still faced with the tremendous challenge of finding the correct gene and determining the identity and function of the responsible genetic variation. Although the use of inbred strains represents the vast majority of complex trait analyses in mice, several recent QTL cloning successes in livestock species highlight the power of analysis within outbred populations.[99] This strategy was also recently employed to achieve one of the first successes in going from a QTL with modest effect to the underlying gene in mice, when Yalein *et al.*[100] showed that *Rgs2* harbors genetic variation modulating anxiety using the MF1 outbred strain. Furthermore, they showed that the method of quantitative complementation testing (see, e.g., De Luca *et al.*[101]) could be powerfully applied to identify small-effect genes underlying QTL. This further demonstrates the power inherent to the combining of genetic and functional approaches for complex trait analysis.

As if complex traits were not complex enough, as we learn more about the genome and how it works and is structured, new complexities arise that pose challenging yet exciting frontiers for future avenues of quantitative trait analysis. Among a growing list, three examples will be highlighted here. First, epigenetic (non-Mendelian) modifications appear to be heritable.[102,103] This trans-generational epigenetic inheritance, which can be influenced by genetic background,[102] adds another layer of complexity that may be an important contributor to continuous variation and affect complex trait predisposition.[104] Second, the emerging importance of non-coding RNA and RNAi adds significant complexity to the regulation of complex traits,[105] and vastly expands the candidate genomic regions where quantitative trait variation may be manifested by specific polymorphisms. Finally, the recent reports of large-scale copy number polymorphisms in the human genome[106,107] may represent novel forms of genetic variation that could have important implications for analysis of susceptibility to disease.

Acknowledgments

This chapter is a contribution of the University of Nebraska Agricultural Research Division and was supported in part by funds provided through the Hatch Act. The authors are grateful to Mark Allan, Stephanie Wesolowski, Dale

Van Vleck, Joao Rocha, Kari Elo, Rob Williams, David Threadgill, Ken Manly and Chris Haley for useful discussions. Some research in progress discussed in this paper is part of a fruitful and ongoing collaboration with Dr. Gene Eisen at North Carolina State University. The source of the figures and significant sections of the text in this Chapter is from the *Journal of Animal Science*, and copyright is owned by the American Society of Animal Science. The author is grateful to the American Society of Animal Science for permission to reproduce this information.

References

1. Eisen, E.J. 1989. Selection experiments for body composition in mice and rats: a review. *Livest. Prod. Sci.* **23**:17–32.
2. Hedley, A.A., C.L. Ogden, M.D. Carroll, L.R. Curtin and K.M. Flegal. 2004. Prevalence of overweight and obesity among US children, adolescents, and adults. 1999–2002. *JAMA.* **291**:2847–2850.
3. National Institute of Health (NIH). National Heart, Lung and Blood Institute (NHLBI). 1998. Clinical guidelines in the identification, evaluation, and treatment of overweight and obesity in adults. HHS, Public Health Service (PHS).
4. Calle, E.E., C. Rodriguez, K. Walker-Thurmond and M.J. Thun. 2003. Overweight, obesity, and mortality from cancer in a prospectively studied cohort of U.S. adults. *N. Engl. J. Med.* **348**:1625–2638.
5. Comuzzie, A. G. and D. B. Allison. 1998. The search for human obesity genes. Science **280**:1374–1377.
6. Finkelstein, E.A., I.C. Fiebelkorn and G. Wang. 2004. State-level estimates of annual medical expenditures attributable to obesity. *Obes.Res.* **12**:18–24.
7. Garrod, A. 1902. The incidence of alkaptonuria: A study in chemical individuality. *Lancet* **2**:1616–1620.
8. Festing, M. F. W. 1979. The inheritance of obesity in animal models of obesity. In: *Animal Models of Obesity*. ed. M.F.W. Festing. pp.15–37. MacMillan Press, London.
9. Rich, S. S. 1990. Mapping genes in diabetes. *Diabetes* **39**:1315–1319.
10. McPherron, A. C. and S. J. Lee. 1996. The transforming growth factor β superfamily. In: *Growth Factors and Cytokines in Health and Disease.* ed. D. LeRoith and C. Bondy. pp. 357–393. JAI Press, Greenwich, CT.
11. Snyder, E.E., B. Walts, L. Perusse, Y.C. Chagnon, S.J. Weisnagel, T. Rankinen and C. Bouchard. 2004. The human obesity gene map: the 2003 update. *Obes. Res.* **12**:369–439.
12. Sturtevant, A. H. 1913. The linear arrangement of six sex-linked factors in Drosophila, as shown by their mode of association. *J. Exp. Zool.* **14**:43–59.
13. Haldane, J. B., A. D. Sprunt and N. M. Haldane. 1915. Reduplication in mice. *Science* **5**:133–135.
14. Sax, K. 1923. The association of size differences with seed-coat pattern and pigmentation in *Phaseolus vulgaris*. *Genetics* **8**:552–560.

15. Dietrich, W. F. *et al.* 1996. A comprehensive genetic map of the mouse genome. *Nature* **380**: 149–152.
16. Lander, E. S. and D. Botstein. 1989. Mapping Mendelian factors underlying quantitative traits using RFLP linkage maps. *Genetics* **121**:185–199.
17. Zeng, Z-B. 1993. Theoretical basis for separation of multiple linked gene effects in mapping quantitative trait loci. *Proc. Natl. Acad. Sci. USA* **90**:10972–10976.
18. Zeng, Z-B. 1994. Precision mapping of quantitative trait loci. *Genetics* **136**:1457–1468.
19. Haley, C. S., S. A. Knott and J. M. Elsen. 1994. Mapping Quantitative trait loci in crosses between outbred lines using least squares. *Genetics* **136**:1195–1207.
20. Paterson, A. H., E. S. Lander, J. D. Hewitt, S. Peterson, S. E. Lincoln and S. D. Tanksley. 1988. Resolution of quantitative traits into Mendelian factors by using a complete linkage map of restriction fragment length polymorphisms. *Nature*. **335**:721–726.
21. Martin, B., J. Nienhuis, G. King and A. Schaefer. 1989. Restriction fragment length polymorphisms associated with water use efficiency in tomato. *Science*. **243**:1725–1728.
22. Hilbert, P., K. Lindpainter, J. S. Beckmann, T. Serikawa, F. Soubrier, C. Dubay, P. Cartwright, B. De Gouyon, C. Julie, S. Takahasi, M. Vincent, D. Ganten, M. Georges and G. M. Lathrop. 1991. Chromosomal mapping of two genetic loci associated with blood-pressure regulation in hereditary hypertensive rats. *Nature* **353**:521–529.
23. Jacob, H. J., K. Lindpainter, S. E. Lincoln, K. Kusumi, R. K. Bunker, Y. – P. Mao, D. Ganten, V. J. Dzau and E. S. Lander. 1991. Genetic mapping of a gene causing hypertension in the stroke-prone rat. *Cell* **67**:213–224.
24. Brockmann, G. A. and M. R. Bevova. 2002. Using mouse models to dissect the genetics of obesity. *Trends Genet.* **18**:367–376.
25. Pomp, D., Allan, M.F. and Wesolowski, W.R. 2004. Quantitative genomics: exploring the genetic architecture of complex trait predisposition. *J. Anim. Sci.* **82E-Suppl.**: E300–E312.
26. Rocha, J. L., L. D. Van Vleck, E. J. Eisen and D. Pomp. 2004. A large-sample QTL study in mice: I. Growth. *Mamm. Genome* **15**:83–89.
27. Rocha, J. L., L. D. Van Vleck, E. J. Eisen and D. Pomp. 2004. A large-sample QTL study in mice: II. Organ and body composition traits. *Mamm. Genome* **15**:100–113.
28. Robertson, A. 1967 The nature of quantitative genetic variation. In: *Heritage from Mendel.* ed. A. Brink. pp. 265–280. University of Wisconsin Press, Madison.
29. Jinks J.L. 1977 Discussion of Dr. Eaves' paper. *J.Roy. Statist. Soc., Series* A. **140:**352–353.
30. Knott SA, L. Marklund, C.S. Haley , K. Andersson K, W. Davies, H. Ellegren, M. Fredholm, I. Hansson, B. Hoyheim, K. Lundstrom, M. Moller and L. Andersson. 1998 Multiple marker mapping of quantitative trait loci in a cross between outbred wild boar and Large White pigs. *Genetics* **149**:1069–1080.
31. Zhang, Y., R. Proenca, M. Maffei, M. Barone, L. Leopold and J. Friedman. 1994. Positional cloning of the mouse obese gene and its human homologue. *Nature* **372**:425–432.
32. Michuad, E. J., S. J. Bultman, M. L. Klebig, M. J. van Vugt, L. S. Stubbs, L. B. Russell and R. P. Woychik. 1994. A molecular model for the genetic and phenotypic characteristics of the mouse lethal yellow (A^y) mutation. *Proc. Natl. Acad. Sci. USA* **91**:2562–2566.
33. Bray, G. A. and L. A. Tartaglia. 2000. Medicinal strategies in the treatment of obesity. *Nature* **404**:672–677.
34. Barsh, G. S. and M. W. Schwartz. 2002. Genetic approaches to studying energy balance: perception and integration. *Nat. Rev. Genet.* **3:**589–600.

35. Bray, G. and C. Bouchard. 1997. Genetics of human obesity: research directions. *FASEB J* **11**: 937–945.
36. Farooqi, I. S., J. M. Keogh, G. S. Yeo, E. J. Lank, T. Cheetham and S. O'Rahilly. 2003. Clinical spectrum of obesity and mutations in the melanocortin 4 receptor gene. *N. Engl. J. Med.* **348**:1085–1095.
37. Korstanje, R. and B. Paigen. 2002. From QTL to gene: the harvest begins. *Nat. Genet.* **31**:235–236.
38. Abiola, O. *et al.* 2003. The nature and identification of quantitative trait loci; a community's view. *Nat. Rev. Genet.* **11:**911–916.
39. Nadeau, J. H. and W. N. Frankel. 2000. The roads from phenotypic variation to gene discovery: mutagenesis versus QTLs. *Nat. Genet.* **25**:381–384.
40. Schwartz, M. W., S. C. Woods, R. J. Seeley, G. S. Barsh, D. G. Baskin and R. L. Leibel. 2003. Is the energy homeostasis system inherently biased toward weight gain? *Diabetes* **52**:232–238.
41. Wang, Y. X., C. H. Lee, S. Tiep, R. T. Yu, J. Ham, H. Kang and R. M. Evans. 2003. Peroxisome- proliferator-activated receptor delta activates fat metabolism to prevent obesity. *Cell* **113**:159–170.
42. Brem, R. B., G. Yvert, R. Clinton and L. Kruglyak. 2002. Genetic dissection of transcriptional regulation in budding yeast. *Science* **296**:752–755.
43. Yvert, G., R. B. Brem, J. Whittle, J. M. Akey, E. Foss, E. N. Smith, R. Mackelprang and L. Kruglyak. 2003. Trans-acting regulatory variation in Saccharomyces cerevisiae and the role of transcription factors. *Nat. Genet.* **35**:57–64.
44. Montooth, K. L., J. H. Marden and A. G. Clark. 2003. Mapping determinants of variation in energy metabolism, respiration and flight in Drosophila. *Genetics* **165**:623–635.
45. Schadt, E. E., S. A. Monks, T. A. Drake, A. J. Lusis, N. Che, V. Colinayo, T. G. Ruff, S. B. Milligan, J. R. Lamb, G. Cavet, P. S. Linsley, M. Mao, R. B. Stoughton and S. H. Friend. 2003. Genetics of gene expression surveyed in maize, mouse and man. *Nature* **422**:397–402.
46. Williams, R. W., S. Shou, L. Lu, Y. Qu, J. Wang, K. F. Manly, E. J. Chesler, H. C. Hsu, J. D. Mountz and D. W. Threadgill. 2002. Genomic Analysis of transcriptional networks: combining microarrays with complex trait analysis. *Proceedings 1st Annual Complex Trait Consortium Meeting*. Memphis, TN. http//www.complextrait.org/ctc2002/williams.html
47. Jansen, R. C. and J. P. Nap. 2001. Genetical genomics: the added value from segregation. *Trends Genet.* **17**:388–391.
48. Jansen, R.C. 2003. Studying complex biological systems using multifactorial perturbation. *Nat. Rev. Genet.* **4**:145–151.
49. Dumas, P., Y. Sun, G. Corbell, S. Trenblay, Z. Pausova, V. Kren, D. Krenova, M. Pravence, P. Hamet and J. Treblay. 2000. Mapping of quantitative trait loci (QTL) of differential stress gene expression in rat recombinant inbred strains. *J. of Hypertension* **18**:545–551.
50. Doerge, R. W. 2002. Mapping and analysis of quantitative trait loci in experimental populations. *Nat. Rev. Genet.* **3**:43–52.
51. Damerval, C., A. Maurice, J. M. Josse and D. de Vienne.1994.Quantitative trait loci underlying gene product variation: a novel perspective for analyzing regulation of genome expression. *Genetics* **37**:289–301.
52. de Vienne, D., A. Maurice, J. M. Josse, A. Leonardi and C. Damerval. 1994. Mapping factors controlling genetic expression. *Cell. Mol. Biol.* (*Noisy-le-grand*) **40**:29–39.

53. Consoli, L, A. Lefevre, M. Zivy, D. de Vienne and C. Damerval. 2002. QTL analysis of proteomeand transcriptome variations for dissecting the genetic architecture of complex traits in maize. *Plant Mol. Biol.* **48**:575–581.
54. Wayne, M. L. and L. M. McIntyre. 2002. Combining mapping and arraying: An approach to candidate gene identification. *Proc. Natl. Acad. Sci.USA* **99**:14903–14906.
55. Kirst, M., A.A. Myburg, J.P. De Leon, M.E. Kirst, J. Scott and R. Sederoff. 2004. Coordinated genetic regulation of growth and lignin revealed by quantitative trait locus analysis of cDNA microarray data in an interspecific backcross of *Eucalyptus*. *Plant Physiol.* **4**:2368–2378.
56. Schadt, E.E., S.A. Monks and S.H. Friend. 2003. A new paradigm for drug discovery: Integrating clinical genetic, genomic and molecular phenotypic data to identify drug targets. *Biochem. Soc. Trans.* **31**:437–443.
57. Morley, M., C.M. Molony, T.M. Weber, J.L. Devlin, K.G. Ewens, R.S. Spielman and V.G. Cheung. 2004. Genetic analysis of genome-wide variation in human gene expression. *Nature.* **430**:743–747.
58. Pomp, D. 1999. Animal models of obesity. *Molecular Medicine Today* **5**:459–460.
59. Esterbauer, H., C. Schneitler, H. Oberkofler, C. Ebenbichler, B. Paulweber, F. Sandhofer, G. Ladurner, E. Hell, A. D. Strosberg, J. R. Patsch, F. Krempler and W. Patsch. 2001. A common polymorphism in the promoter of *UCP2* is associated with decreased risk of obesity in middle-aged humans. *Nat. Genet.* **28**:178–183.
60. Heo, M. *et al.* 2001. Pooling analysis of genetic data: the association of leptin receptor (LEPR) polymorphisms with variables related to human adiposity. *Genetics* **159**:1163–1178.
61. Mehrabian, M., P. Z. Wen, J. Fisler, R. C. Davis and A. J. Lusis. 1998. Genetic loci controlling body fat, lipoprotein metabolism, and insulin levels in a multifactorial mouse model. *J. Clin. Invest.* **101**:2485–2496.
62. Brockmann, G. A., J. Kratzsch, C. S. Haley, U. Renne and M. Schwerin. 2000. Single QTL effects, epistasis, and pleiotropy account for two-thirds of the phenotypic F_2 variance of growth and obesity in DU6i x DBA/2 mice. *Genome Res.* **10**:1941–1957.
63. Rosen, C. J., G. A. Churchill, L. R. Donahue, K. L. Shultz, J. K. Burgess, D. R. Powell, C. Ackert and W. G. Beamer. 2000. Mapping quantitative trait loci for serum insulin-like growth factor-1 levels in mice. *Bone* **27**:521–528.
64. Allan, M.F., Eisen, E.J. and D. Pomp. 2004. The M16 mouse: an outbred animal model of early onset polygenic obesity and diabesity. *Obes. Res.* **12:**1397–407.
65. Hixson, J. E., L. Almasy, S. Cole, S. Birnbaum, B. D. Mitchell, M. C. Mahaney, M. P. Stern, J. W. MacCluer, J. Blangero and A. G. Comuzzie. 1999. Normal variation in leptin levels is associated with polymorphisms in the proopiomelanocortin gene, POMC. *J. Clin. Endocrinol. Metab.* **84**:3187–3191.
66. Koza, R. A., S. M. Hohmann, C. Guerra, M. Rossmeisl and L. P. Kozak. 2000. Synergistic gene interactions control the induction of the mitochondrial uncoupling protein (Ucp1) gene in white fat tissue. *J. Biol. Chem.* **275:**34486–34492.
67. Wesolowski, S. R., M. F. Allan, M. K. Nielsen and D. Pomp. 2003. Evaluation of hypothalamic gene expression in mice divergently selected for heat loss. *Physiol. Genomics* **13**:129–137.
68. Rohrer, G. A., T. H. Wise, D. D. Lunstra and J. J. Ford. 2001. Identification of genomic regions controlling plasma FSH concentrations in Meishan-White Composite boars. *Physiol. Genomics* **6**:145–151.
69. Oliver, B., M. Parisi and D. Clark. 2002. Gene expression neighborhoods. *J. Biol.* **1**:4.

70. Spellman, P. T. and G. M. Rubin. 2002. Evidence for large domains of similarly expressed genes in the Drosophila genome. *J. Biol.* **1**:5.
71. Darvasi, A. 2003. Genomics: Gene expression meets genetics. *Nature* **422:**269–270.
72. Shimomura, K., S. S. Low-Zeddues, D. P. King, T. D. L. Steeves, A. Whiteley, J. Kushla, P. D. Zemenides, A. Lin, M. H. Vitaterna, G. A. Churchill and J. S. Takahashi. 2001. Genome-wide epistatic interaction analysis reveals complex genetic determinants of circadian behavior in mice. *Genome Res.* **11**:959–980.
73. Leamy, L. J., E. J. Routman and J. M. Cheverud. 2002 An epistatic genetic basis for fluctuating asymmetry of mandible size in mice. *Evolution Int. J. Org. Evolution.* **56**:642–653.
74. Carlborg, Ö., S. Kerje, K. Schutz, L. Jacobsson, P. Jensen and L. Andersson. 2003 A global search reveals epistatic interaction between QTLs for early growth in the chicken. *Genome Res.* **13**:413–421.
75. Carlborg, O. and C.S. Haley. 2004. Epistasis: too often neglected in complex trait studies? *Nat. Rev. Genet.* **5:**618–625.
76. Yi, N., A. Diament, S. Chiu, K. Kim, D.B. Allison, J.S. Fisler and C.H. Warden. 2004. Characterization of epistasis influencing complex spontaneous obesity in the BSB model. *Genetics.* **167:**399–409.
77. Zhu, J., P.Y. Lum, J. Lamb, D. GuhaThakurta, S.W. Edwards, R. Thieringer, J.P. Berger, M.S. Wu, J. Thompson, A.B. Sachs and E.E. Schadt. 2004. An integrative genomics approach to the reconstruction of gene networks in segregating populations. *Cytogenet. Genome Res.* **105**:363–374.
78. Monks, S.A., A. Leonardson, H. Zhu, P. Cundiff, P. Pietrusiak, S. Edwards, J.W. Phillips, A. Sachs and E.E. Schadt. 2004. Genetic inheritance of gene expression in human cell lines. *Am. J. Hum. Genet.* **75**:1094–1095.
79. Lan, H., J. P. Stoehr, S. T. Nadler, K. L. Schueler, B. S. Yandell and A. D. Attie. 2003. Dimension reduction for mapping mRNA abundance as quantitative traits. *Genetics* **164**:1607–1614.
80. Lembertas, A. V., L. Perusse, Y. C. Chagnon, J. S. Fisler and C. H. Warden. 1997. Identification of an obesity quantitative trait locus on mouse chromosome 2 and evidence of linkage to body fat and insulin on the human homologous region 20q. *J. Clin. Invest.* **100**:1240–1247.
81. Jerez-Timaure, N., D. Pomp and E.J. Eisen. 2004. Characterization of quantitative trait loci with major effects on fatness and growth on mouse chromosome 2. *Obes. Res.* **12**:1408–1420.
82. Pomp, D. 1997. Genetic dissection of obesity in polygenic animal models. *Behav. Genet.* **27**:285–306.
83. Hanrahan, J. P., E. J. Eisen and J. E. Legates. 1973. Effects of population size and selection intensity on short-term response to selection for post-weaning gain in mice. *Genetics* **73**:513–530.
84. Eisen, E. J. 1986. Maturing patterns of organ weights in mice selected for rapid postweaning gain. *Theor. Appl. Genet.* **73**:148–157.
85. Pomp, D., N. Jerez, M. F. Allan and E. J. Eisen. 2002. Integrated genomic, proteomic and metabolomic dissection of polygenic selection response for murine growth and fatness. *Proceedings 7th World Congress on Genetics Applied to Livestock Production*, Montpellier, France.

86. Williams, R. W., K. W. Broman, J. M. Cheverud, G. A. Churchill, R. W. Hitzemann, K. W. Hunter, J. D. Mountz, D. Pomp, R. H. Reeves, L. C. Schalkwyk and D. W. Threadgill. 2002. A collaborative cross for high-precision complex trait analysis. *1st Workshop Report of the Complex Trait Consortium*: www.complextrait.org/Workshop1.pdf
87. Vogel, G. 2003. Scientists dream of 1001 complex mice. *Science* **301**:456–457.
88. Churchill, G.A. *et al.* 2004. The Collaborative Cross, a community resource for the genetic analysis of complex traits. *Nat Genet.* **36**:1133–1137.
89. Kambadur R., M. Sharma, T. P. Smith and J. J. Bass. 1997. Mutations in myostatin (GDF8) in double-muscled Belgian Blue and Piedmontese cattle. *Genome Res.* **7**:910–916.
90. Van Laere, A. S., M. Nguyen, M. Braunschweig, C. Nezer, C. Collette, L. Moreau, A. L. Archibald, C. S. Haley, N. Buys, M. Tally, G. Andersson, M. Georges and L. A. Andersson. 2003. A regulatory mutation in IGF2 causes a major QTL effect on muscle growth in the pig. *Nature* **425**:832–836.
91. Braunschweig, M.H., A.S. Van Laere, N. Buys, L. Andersson, and G. Andersson. 2004. IGF2 antisense transcript expression in porcine IGF2 antisense transcript expression in porcine postnatal muscle is affected by a quantitative trait nucleotide in intron 3. *Genomics.* **6**:1021–1029.
92. Freking, B.A., S.K. Murphy, A.A. Wylie, S.J. Rhodes, J.W. Keele, K.A. Leymaster, R.L. Jirtle and T.P. Smith. 2002. Identification of the single base change causing the callipyge muscle hypertrophy phenotype: the only known example of polar overdominance in mammals. *Genome Res.* **12**:1496–1506.
93. Charlier, C., K. Segers, L. Karim, T. Shay, G. Gyapay, N. Cockett and M. Georges. 2001. The callipyge mutation enhances the expression of coregulated imprinted genes in *cis* without affecting their imprinting status. *Nat. Genet.* **27**:367–369.
94. Davis, E., C.H. Jensen, H.D. Schroder, F. Farnir, T. Shay-Hadfield, A. Kliem, N. Cockett, M. Georges and C. Charlier. 2004. Ectopic expression of DKL1 protein in skeletal muscle of padumnal heterozygotes causes the callipyge phenotype. *Curr. Biol.* **20**:1858–1862.
95. Wang, J., R. W. Williams, E. J. Chesler and K. F. Manly. 2003. WebQTL: Web-based complex trait analysis. *Neuroinformatics* **1**:299–308.
96. Chesler, E.J., L. Lu, J. Wang, R.W. Williams and K.F. Manly. 2004. WebQTL: rapid exploratory analysis of gene expression and genetic networks for brain and behavior. *Nat. Neurosci.* **5**:485–486.
97. Breitkreutz, B.J., C. Stark and M. Tyers. 2003. Osprey: a network visualization system. *Genome Biol.* **3**:R22.
98. Breitkreutz, B. J., C. Stark and M. Tyers. 2003. The GRID: the General Repository for Interaction Datasets. *Genome Biol.* **4**:R23.
99. Andersson, L. and M. Georges. 2004. Domestic-animal genomics: deciphering the genetics of complex traits. *Nat. Rev. Genet.* **3**:202–212.
100. Yalcin, B., S.A. Willis-Owen, J. Fullerton, A. Meesaq, R.M. Deacon, J.N. Rawlins, R.R. Copley, A.P. Morris, J. Flint and R. Mott. 2004. Genetic dissection of a behavioral quantitative trait locus shows that Rgs2 modulates anxiety in mice. *Nat. Genet.* **36**:1197–1202.
101. De Luca, M., N.V. Roshina, G.L. Geiger-Thornsberry, R.F. Lyman, E.G. Pasyjkova and T.F. Mackay. 2003. Dopa decarboxylase (*Ddc*) affects variation in *Drosophila* longevity. *Nat. Genet.* **34**:429–433.

102. Rakyan, V.K., S. Chong, M.E. Champ, P.C. Cuthbert, H.D. Morgan, K.V. Luu and E. Whitelaw. 2003. Transgenerational inheritance of epigenetic states at the murine Axin(Fu) allele occurs after maternal and paternal transmission. *Proc. Natl. Acad. Sci. USA* **100:**2538–2543.

103. Lin, Q., Q. Chen, L. Lin and J. Zhou. 2004. The Promoter Targeting Sequence mediates epigenetically heritable transcription memory. *Genes Dev.* **18**:2639–2651.

104. Bjornsson, H.T., M.D. Fallin and A.P. Feinberg. 2004. An integrated epigenetic and genetic approach to common human disease. *Trends Genet.* **20**:350–358.

105. Mattick, J.S. 2004. RNA regulation: a new genetics? *Nat. Rev. Genet.* **5**:316 –323 .

106. Iafrate, A.J., L. Feuk, M.N. Rivera, M.L. Listewnik, P.K. Donahoe, K. Qi, S.W. Scherer and C. Lee. 2004. Detection of large- scale variation in the human genome. *Nat. Genet.* **36**:949–951.

107. Sebat, J. *et al.* 2004. Large-scale copy number polymorphism in the human genome. *Science.* **305**:525–528.

108. Chagnon, Y.C., T. Rankinen, E.E. Snyder, S.J. Weisnagel, L. Perusse and C. Bouchard. 2003. The human obesity gene maps: The 2002 update. *Obes. Res.* **11**:313–367.

109. Elo, K. 2003. Towards integrated genomic approaches – lessons from energy metabolism studies. *Proceedings NJF's 22nd Congress.* pp. 222–227. Turku, Finland.

CHAPTER 12

MOUSE MUTAGENESIS

D. R. Beier

Genetics Division
Brigham and Women's Hospital/Harvard Medical School, Boston, MA, USA
beier@receptor.med.harvard.edu

1. Introduction

Utilization of mutagenesis for genetic analysis has been extremely productive in invertebrate systems such as *S. cerevesiae, C. elegans*, and *D. melanogaster*, and has been recently used for the analysis of the zebrafish *D. rerio*. Yet even this vertebrate is 400 million years divergent in evolution from humans, which may qualify the inferences one can draw from its analysis with respect to mammalian organogenesis and physiology. Fortunately, a variety of efficient mutagenesis techniques have been developed for the mouse. With the completion of the characterization of the mouse genome sequence and the development of powerful molecular and computational tools for positional cloning, there has been a recent revival of interest in the utilization of mutagenesis for the study of mammalian biology. This review focuses on strategies that employ the chemical N-ethyl-N-nitrosourea (ENU) for the induction of mutations, as this is the most widely used protocol for phenotype-driven studies; other approaches will be discussed more briefly.

2. ENU Mutagenesis: History

The identification of ENU as a potent mutagen grew out of a larger effort by Russell and colleagues to study the effects of radiation and chemicals on the mammalian germ-line.[1] The key to this analysis was the development of the specific-locus test, which allowed these investigators to accurately quantitate relatively rare mutagenic events.[2] This approach entailed treating a mouse with a mutagen and then crossing it with a strain of mice that carried visible recessive mutations at seven "specific loci." Progeny of this cross would appear wild-type unless the mutagenic treatment created a new mutation at one of the seven loci (or created a visible dominant mutation). Because these could be easily scored by simple inspection, it enabled the analysis of the thousands of mice necessary to establish accurate quantitation. Use of seven independent loci helped assess the randomness of the effects. Now all of the specific loci have been cloned: *agouti* (*agouti*), *tyrosinase related protein 1* (*brown*), *tyrosinase* (*albino*), *myosin 5a* (*dilute*), *Bone morphogenetic protein 5* (*short ear*), *pink-eyed dilution* (*pink-eyed dilution*), *endothelin-B receptor* (*piebald*). The average open reading frame of 2049 base pairs is representative of most transcribed loci, which supports the validity of the specific-locus test as a general means to determine per-locus mutation frequencies.

Using this assay, Russell and colleagues found that treatment with ENU could efficiently generate germ cell mutations without causing systemic morbidity.[3] Analysis of the timing of the effects suggested that it is the spermatogonial stem cells that are sensitive to this agent. Additional studies showed that the efficiency of mutagenesis could be optimized by using a fractionated dose.[4] In this fashion, frequencies of 6–15 $\times$ 10^{-4} mutations per locus per gamete can be obtained, greater than that obtained using radiation mutagenesis and 1000-fold higher than the spontaneous rate of mutation of 0.5 $\times$ 10^{-6}. Characterization of ENU mutations in *Drosophila* revealed that the molecular basis of these mutants were single-base changes; many recent studies have confirmed that this is the case in mammalian cells as well, e.g.[5]

The first application of ENU mutagenesis was not for the creation of novel mutations but rather to generate allelic series at specific loci, notably those in the T haplotypes.[6,7] This goal was facilitated by the simplicity of analysis, as it required only a single generation breeding scheme. Similarly, novel mutation discovery focused on the analysis of dominant phenotypes[8] or those that could be readily scored as heterozygotes, such as protein isomorphisms.[9–14] Many of these mutations were in clinically relevant proteins and served as models of human diseases. However, the true power of mutagenesis was demonstrated in Bode's elegant use of a 3-generation breeding strategy for uncovering recessive mutations (Figure 1) to generate mouse models of hyperphenylalaninemia.[15,16] This strategy was paradigmatic not only in its significance with respect to mutation discovery relevant to human disease, but in the application of a screening test (i.e., the Guthrie test on blood spots) that was amenable to a high throughput analysis.

However, despite the very clear demonstration of the efficacy of ENU, it was not widely embraced for mammalian investigation, most likely due to the difficulty at the time of identifying loci that were mutated as a consequence of a single base change. Fortunately, the ENU mutagenesis torch was carried by a few laboratories, notably, those of Dove, who used it to generate the widely used *Min* mouse model of the familial cancer syndrome adenomatosis polyposis coli,[17] and Favor, who studied primarily eye phenotypes.[18] As development of the tools for genomic analysis made it feasible to contemplate the positional cloning of ENU-induced mutant loci, there was an increased appreciation of the potential of this approach for phenotype-driven analysis of mouse biology. Its utility was dramatically shown by Takahashi and colleagues in their studies of the circadian rhythm defect in the *clock* mutant,[19] which they showed was due to a mutation in a transcriptional regulator.[20] Bucan and colleagues have also identified mice with circadian rhythm defects. The positional cloning of one of these mutants demonstrates the utility of phenotype-driven analysis as the causal gene was found to be *Rab3a*; a null allele of this gene had been characterized, but the circadian rhythm defect had not been previously appreciated.[21]

ENU mutagenesis was also shown to be effective for other applications; e.g., the generation of allelic series (originally done as previously noted by Bode and colleagues for the T locus,[6,7] and more recently by others.[22,23] Mutagenesis was also shown to be effective for the comprehensive region-specific generation of recessive mutations using deletion strains.[24] Alleles obtained in this fashion proved useful for the characterization of embryonic ectoderm development *(eed),* which was eventually shown to be a mutation in a Polycomb-related protein that is required for X-inactivation and genomic imprinting.[25,26]

The relative simplicity and efficiency of the ENU mutagenesis regimen, combined with the prospect of obtaining full length genome sequence, encouraged several investigators to initiate large scale programs that involve the use of large numbers of mice and a wide variety of phenotype assays. The efforts by Brown at Harwell[27] and Balling and Hrabe De Angelis at the GSF National Research Center for Environment and Health[28] were the first of this type. In the Harwell study, over 26,000 mice were screened and 1,089 mice with visible or other anomalies were identified. In the subset of lines that was further tested, 50% were found to have a heritable phenonotype; thus the investigators extrapolate that they have recovered approximately 500 mutants from the screening program. In the GSF program over 14,000 mice were screened for a large number of clinically relevant parameters, and 182 heritable mutants for a variety of phenotypes were recovered, with a large additional number of phenotypic variants still in testing. Similar large-scale programs have been implemented in Canada[29] and Japan.[30] Many of these efforts focused initially on characterization of dominant and semi-dominant phenotypes in first generation (G1) mice; both the GSF and Harwell studies reported that the rate of recovery of dominant mutations was approximately 2%. Most of these programs have now incorporated screens for recessive mutations as well.

The first large scale mutagenesis effort in the United States was initiated by Justice and colleagues at Baylor. This program included a novel strategy employing balancer chromosomes that facilitated the identification of recessive lethal mutations by simple examination.[31] Specifically, the mutagenized mice are crossed to a line that was engineered using gene targeting in embryonic stem (ES) cells to carry an inversion with defined endpoints. Loci within the inversion will not recombine with loci on the mutagenized chromosome, and, by using visible markers for the respective chromosomes, it is possible to score for both haplotypes. Absence of mice homozygous for alleles derived from the mutagenized parent indicates the existance of a recessive lethal mutation. Note that phenotypes due to "off-target" mutations will also be ascertained in this screen; these mutations can be characterized using standard mapping strategies.

Additional large-scale programs include a consortium of investigators that have developed mutagenesis efforts that focus on neurological phenotypes,[32] and a program at the Jackson Laboratory designed to examine physiological traits.[33]

Concurrently with these efforts, several investigators demonstrated that ENU mutagenesis could be effectively used in small laboratories in more focused studies. Screens for mutations affecting developmental process were particularly productive, as they could be efficiently evaluated in large numbers of progeny by inspection. Anderson showed that screening of mid-gestation third generation (G3) mice could be used to uncover mutations that perturb embryonic patterning; several of these mutations have since been shown to affect the Sonic hedgehog (*Shh*) signaling pathway.[34–36] Of particular note is the fact that this phenotype-driven analysis uncovered a role for the *polaris* gene in *Shh* signaling, which was not appreciated despite the extensive analysis of both null and hypomorphic mutants.[37]

Studies by Peterson demonstrated how mutagenesis could be used to specifically investigate brain development and discovered the developmental role of FK506 binding protein 12-rapamycin associated protein 1 (Frap1, also known as mammalian target of rapamycin (mTOR)).[38,39] Cordes and Barsh used ENU mutagenesis to generate a second allele of the *kreisler* mutation, which affects hind-brain patterning.[22] This result proved crucial to the discovery of the underlying mutant gene, as the original allele was a genomic rearrangement that affected multiple loci.

Our own research targeted late embryonic phenotypes as models of organogenesis and birth defects.[40] In one example, we found that a mutant line with a defect in diaphragmatic and pulmonary development carried a hypomorphic mutation of the transcription cofactor *Fog2 (Zfpm2)* due to a sequence change in a splice donor site (Ackerman and Beier, unpublished) . The mutant phenotype is similar to the abnormalities found in the often devastating but poorly understood human disorder congenital diaphragmatic hernia (CDH). A role for this gene in CDH was suggested by the association of the disorder with chromosomal translocations at 8q22.3, where human FOG2 is located. To test this, sequencing of FOG2 was done on 30 autopsy samples, and a heterozygous base change resulting in a premature stop codon was identified in one case. Analysis of the parents revealed that this was a *de novo* mutation, which strongly implicates FOG2 as causal for CDH in the affected child. This result clearly demonstrates the utility of ENU mutagenesis as a tool for finding candidate genes for human developmental defects.

A number of investigators have shown that *in vitro* assays can be employed to query specific cellular processes. Schimenti and colleagues used this approach to identify mutations affecting chromosome stability.[41] They used flow cytometry of peripheral blood to measure micronucleation, which was used as a quantitative indicator of chromosome damage *in vivo*. Using this approach they found a mutation in *Polq*, encoding DNA polymerase theta. Beutler and colleagues assayed macrophages isolated from mutagenized mice for their response to inducers of the Toll-like receptor microbial sensor system. Using this cellular analysis, they identified a mutant line, which lacked cytokine responses to double-stranded RNA and had an impaired response to the endotoxin lipopolysaccharide. The mutation proved to be in a Toll/interleukin-1 receptor/resistance (TIR) adaptor protein.[42] In a variation on this analysis of cellular phenotypes as a means to query immune function, Goodnow and co-workers have used flow cytometry analysis of peripheral blood to identify mice that lack specific subsets of lymphoid cells.[43]

The efficiency of mutagenesis has emboldened a number of investigators to use this approach in the analysis of very narrowly defined phenotypes. For example, Cordes has applied mutagenesis to uncover mutations affecting response to a serotonergic agonist, a phenotype relevant to human psychiatric disorders.[44] Carlson and colleagues are attempting to generate modifiers of a transgenic model of Alzheimer's disease (personal communication). There is considerable enthusiasm for using mutagenesis to identifying enhancer and suppressors of mutant phenotypes, and the first report describing success using this approach has been published.[45]

3. ENU Mutagenesis: Practical Aspects

There are a number of components of an ENU mutagenesis experiment that require attention. These include treatment protocols, husbandry, screening assays, establishing heritability, locus mapping, mutation identification, and establishing causality.

3.1. *Treatment Protocols*

In their original studies, Russell and colleagues demonstrated that a fractionated dose treatment (i.e., treating mice with 3–4 moderate doses of ENU over an extended period) was more effective than treatment with a single large dose.[46] That this protocol is highly effective has been corroborated in multiple reports,[47] and a treatment regimen of 90 mg/kg mouse weight given once a week for 3 weeks is now a commonly employed protocol. In our own studies we have used an empiric approach in which we employ a range of doses (e.g., 80–100 mg/kg), and mice treated with the highest doses that recover fertility are used for breeding. Single dose protocols can also be effective, as is evident from the success of the Medical Research Council program.[27] In either case it is advisable to generate and test multiple males. This is because the spermatogonial cell population that recovers after ENU treatment may be small, and consequently the repertoire of mutations in progeny of a single treated male may be limited.

There is no particular concensus with respect to strain selection, although it has been shown that FVB/N mice are much more sensitive to ENU than most strains.[47] The BTBR strain has been considered particularly efficient for mutagenesis; however, this is based primarily on a single study[48] and has not been systematically examined. It is generally advisable to use an inbred strain for mutagenesis treatment in order to facilitate genetic mapping and reduce phenotypic variability. With the development of highly informative single nucleotide polymorphisms (SNP) panels (as discussed below), mutagenesis of an inbred strain is no longer an absolute requirement, although mutation mapping in very mixed genetic backgrounds remains challenging. However, crossing the mutagenized mouse to a different strain introduces genetic informativeness early in the experiment, which can facilitate mapping, as discussed below.

3.2. *Husbandry*

Treatment with a mutagenic dose of ENU results in a period of infertility. Recovery from this occurs after 8–12 weeks, and the treated mice (designated G0 for Generation 0) are then bred. Dominantly or semi-dominantly inherited phenotypes can be assayed in their first generation (G1) progeny. A number of screens have focused on this population, as it is logistically feasible to screen large numbers of mice. However, because most heterozygous mutations do not result in readily scored phenotypes, it is frequently necessary to employ a multi-generation breeding scheme to uncover mutations with recessive effects. The general strategy described by Bode is the most efficient means to this end.[15] In this approach third-generation progeny of the original mutagenized male are examined, as is illustrated in Figure 1. Because the G0 male may not remain fertile for sufficiently long to be completely tested, it is crossed to a wild-type female, and its G1 male progeny are screened. These G1 males are mated to wild-type mice, generating G2 progeny. If the G1 male carries a recessive mutation, one half of these progeny will be heterozygous for the mutation. The G1 male is then mated to progeny G2 females and the G3 offspring are examined

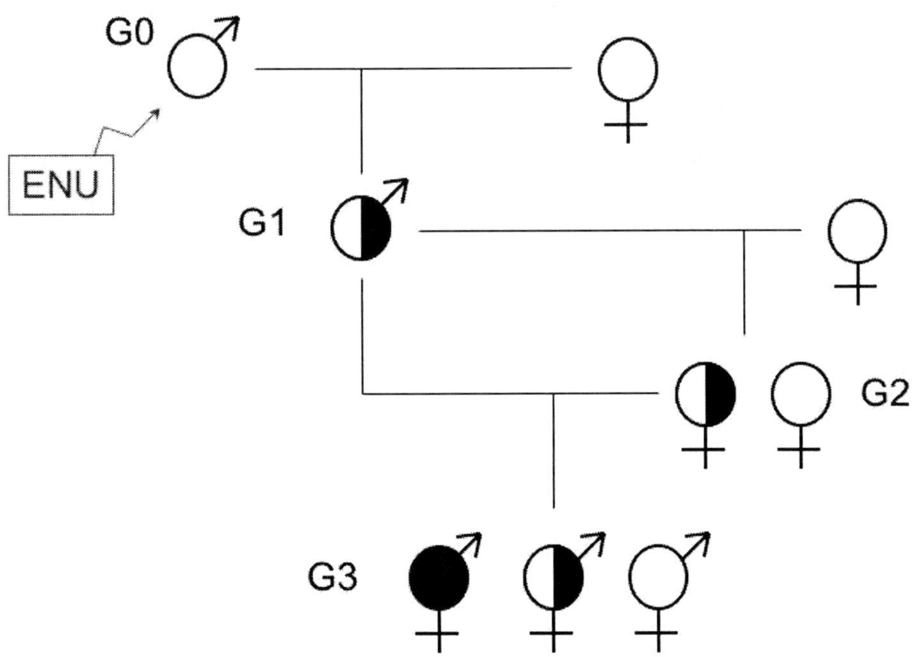

Figure 1. Three generation strategy for generating recessive mutations.

for the phenotype of interest. If one tests sufficient G2 females, the likelihood of observing a mutation carried by the G1 male is high.

Screens for recessive mutations are logistically more complex than analysis of first-generation progeny but also have a number of significant advantages. Mutant mice in a recessive screen can generally be more readily ascertained, due to the fact that most mutations in the mouse have no or only subtle effects in heterozygotes. Additionally, recessive screens can identify mutations that result in lethality. Also, establishing that an observed phenotype is heritable and monogenic is facilitated in a screen for recessive mutations, as it would be expected that multiple affected mice will be seen in the analysis of any G1 "family." The husbandry demands make it less feasible to test large numbers of G1 mice than in a screen for dominant mutations; however, this disadvantage is mitigated by the fact that each G1 male carries many mutations.

3.3. *Screening Assays*

While it is possible to imagine almost any assay as part of a mutagenesis experiment, several general principles should be considered. First, these assays must be amenable to "high-throughput," i.e., to the efficient analysis of large numbers of mice. While ENU is a highly effective mutagen, it should be anticipated that it will be necessary to screen through hundreds to thousands of animals to find a true phenodeviant. In the same vein, it is necessary that the threshold for considering a mouse as possibly mutant be set high. In an assay of a normally distributed quantitative trait, five percent of mice will be greater than two standard deviations from the mean. Distinguishing these from a true mutant (which is likely to be less frequent) will be inefficient. To facilitate high throughput analysis, many studies employ a heirarchical approach in which a simplified screening assay is followed by more detailed characterization.

3.4. *Heritability Testing*

When a mouse with an abnormal phenotype is identified, it is necessary to further evaluate it to assess whether this abnormality is a consequence of a monogenic mutation. For dominant or semi-dominant traits this evaluation can be done by progeny testing. This procedure can employ an outcross to a different strain, in which case the affected offspring can be used for genetic mapping. One caution here is that a phenotype may be altered or suppressed in a mixed genetic background. If the mutant mice are rare or phenotypically ambiguous, the mutation will be difficult to map and positionally clone, and further investment of effort will depend on the aims of the experiment.

3.5. *Locus Mapping*

A crucial step in an analysis of a novel mutant line is the mapping of the mutation. The development of highly informative, easily assayed simple sequence length polymorphism (SSLP) markers has made this task relatively straightforward.[49] The general approach is to test 4–5 markers distributed across every chromosome (not including the X chromosome, as these mutations will not be derived from the mutagenized G0 male) in 10–20 affected offspring. If attention is given to haplotype structure, even fewer markers can be employed with high sensitivity.[50] Analysis of this modest number of mice will yield only low to moderate resolution localization; however, this can be highly informative. First, it confirms the presumption that the phenotype is monogenic in origin. Second, it provides a means to follow the mutation by genotype (i.e., analysis of flanking markers), as opposed to the more onerous method of progeny testing. Third, this degree of localization may be sufficient to suggest the mutation is allelic with a known gene, which may influence enthusiasm for further analysis. Fourth and finally, the astute (or lucky) investigator may correctly identify candidate loci based on knowledge of the potential mechanism of the observed defect and the presumptive function of genes that reside in the recombinant interval.

Recent studies have characterized large numbers of SNP that can be used for linkage analysis.[51] The utility of these biallelic markers resides primarily in the fact that they are amenable to automated multiplex analysis, expediting the genome-wide analysis that is necessary for mapping. This technology promises to markedly facilitate genetic studies, as it can reduce the time required to localize a mutation to a matter of days. In our own research we have developed a panel of over 400 SNP, which are analyzed as mutliplexes using Sequenom mass spectrometry technology. As not all SNP are polymorphic between any two strains, the number of informative SNP will vary depending on the genetic distance between strains; this number has proven to be about 200 for most strain combinations. We have successfully mapped multiple mutations with this panel, using as few as four affected intercross progeny.

3.6. *Mutation Identification*

The characterization of full-length mouse genome sequence has dramatically simplified the task of positional cloning. The laborious tasks of contig generation and gene discovery are no longer necessary, at least for most of the genome. However, the effort necessary to identify the single base change that is the basis of an ENU-induced mutation remains substantial. The general strategy continues to employ high-resolution recombination analysis that defines the recombinant interval in which the causal gene must reside, followed by the characterization of the genes in the interval for sequence changes. A number of factors have made it possible to pursue mutational analysis for intervals, which may have been resolved only to 1–2 cM resolution. These factors include the decreasing costs of sequencing or other mutation detection technologies, the increasingly reliable annotation of the genome, and the generation of gene expression databases. The integration of these considerations with the availability of genomic sequence is of particular relevance to positional cloning, as it allows inspection of the recombinant interval for genes that are candidates based on their expression in tissues relevant to the observed phenotype.

Presently, most mutation analysis is either gene or exon-directed. That is, it is either cDNAs or exons (generally including adjacent splice sites) that are analyzed. This approach is reasonable given that there is no report as yet documenting an ENU-induced mutation occurring in a non-transcribed regulatory region. However, there is no reason to presume that these mutations cannot occur, and there have been several reports of cases in which mutations of known loci (proven by complementation) show no changes in coding sequence. Furthermore, gene-directed analysis relies on the quality of their annotation, which may be less dependable in 5′ untranslated regions or for alternative splice forms.

It would thus be useful to consider alternative methods, such as transformation-associated recombination (TAR) cloning[52] which could be employed for cloning large segments of genomic DNA for complete resequencing. This would be highly desirable, as it would not be necessary to rely on gene annotations for exon structure, and, if amenable to "shotgun" sequencing, would facilitate high-throughput analysis and would not require expensive primer syntheses. At this point such techniques remain time- and labor-intensive, and, although theoretically attractive, it is not yet clear that they will be more efficient than a gene-based resequencing approach for mutational analysis.

3.7. *Proof of Causality*

A significant issue with respect to chemical mutagenesis is the fact that for many mutant loci only the single ENU-induced allele may exist. While causality can be suggested by functional studies, a formal proof requires either correcting the defect by complementation or identifying independent alleles. The former can be achieved by bacterial artifical chromosome (BAC) transgenesis; however, it will generally be more useful to generate a conditional or reporter-tagged allele by homologous recombination, despite the additional work this may entail. This is particularly the case for hypomorphic mutations, since it will be of interest to determine the null phenotype.

Recent developments in gene targeting technologies have made it practical to contemplate pursuing this even for significant numbers of mutant lines. The ability to modify genomic sequences using homologous recombination in bacteria facilitates the rapid generation of targeting vectors. Alternatively, it has been proven that transgenic expression of short double-stranded RNA can be used to disrupt gene expression in mice. These transgenes can be introduced either via ES cells[53] or by direct injection into fertilized oocytes.[54] The latter approach is attractive as it permits a relatively rapid confirmation since tested clones can be assayed in the first generation. A potential complicating factor is the variability of expression attenuation that can be expected; however, this may prove to be informative, since it will effectively generate an allelic series.

4. Alternative Mutagenesis Techniques

While ENU is the most frequently used mutagen in mice, many other chemicals have been studied as well.[55] One of particular interest is the chemotherapeutic agent chlorambucil, which has been shown to be efficient, and to generate deletions and rearrangements.[56] These can be more readily detected than single base changes, and in many cases are less ambiguous with respect to their potential effect on gene function.[57] However, it is post-spermatogonial cells that are sensitive to the mutagen, which has the practical consequence that there is only a limited window of breeding that can be done, as this population will be replaced by normal cells.

Another mutagenesis technology that has seen a recent revival is transgenic insertional mutagenesis. In this approach exogenous DNA of known sequence composition is introduced into the genome, in some cases disrupting a functional transcription unit and resulting in a phenotype. The sequences flanking the insertion site can be identified using techniques such as hybridization-based cloning or inverse PCR. These can then be used as a probe for further cloning or with the more recent generation of full-length sequence, simply compared to a sequence database to identify the disrupted gene. Insertional mutagenesis was of particular importance prior to the completion of the mouse genome sequence as it facilitated the rapid identification of the disrupted locus. This approach resulted in the characterization of a number of mutant loci, including *microphthalmia* and *fused.*[58,59]

However, transgenic insertions could frequently be associated with deletions or rearrangements, which complicated both molecular cloning and gene identification. In addition, the introduction of exogenous DNA, which frequently inserts as multi-copy concatamers could itself cause effects at a distance. Thus, the *formin* gene was identified as the disrupted locus in two independent *limb deformity* (*ld*) mutants caused by transgenic insertions (as well as a third allele carrying a deletion in the same region).[60] However, recent data has shown that *ld* is due to a defect in *gremlin*, a transcription factor adjacent to *formin*, and it appears likely that transgenic insertion affected regulatory sequences that control *gremlin* expression.[61,62]

With the development of robust tools for high-resolution genetic mapping, as well as full-length genomic sequence, transgenic insertional mutagenesis has proven less popular. However, the task of positional cloning can still be substantial, and there is an advantage to having a clonable tag at the insertion site. For this reason several investigators have been developing an insertional mutagenesis system in which a quiescent transposable element is mobilized by introduction of a transposase. An example of this method utilizes the Sleeping Beauty transposon and transposases, which have been modified to enhance their activity.[63] While the utility of these for generating readily-cloned mutant lines has been shown, they have some potential limitations, such as the evidence that transposition events tend to be local and the requirement for the generation and characterization of novel transgenic lines, and their efficiency relative to other systems is as yet untested.

5. Genotype-Driven Mutagenesis

The strategies described above represent phenotype-driven approaches. These have particular utility since they facilitate unbiased examination of the molecular basis of mammalian traits. However, the complementary approach of genotype-driven analysis is of considerable importance, as is evident by the remarkable productivity of experiments employing targeted gene modification using embryonic stem cell technology. While this analysis is usually done by homologous recombination of specific genes of interest, random mutagenesis techniques have proven useful for genotype-driven analysis as well. The best example of this is the generation of libraries of "gene-trap" ES cell clones carrying selectable markers that only function when they have been inserted into a transcription unit. The insertion site can be identified using inverse-PCR or similar techniques, and the presumptive disrupted gene determined by examination of the genome sequence databases. Gene-trap approaches have proven quite robust, and a number of large scale efforts are in progress.[64] Fittingly, one of the clear applications of this method has been to generate a resource for validation of candidate genes discovered using phenotype-driven mutagenesis (K.V. Anderson, personal communication).

In a related strategy, investigators have used mutational detection techniques such as heteroduplex analysis or direct sequencing to screen for sequence changes in ENU mutagenized populations. This screening has been done effectively using either ES cell clones[65,66] or archived sperm from G1 male mice.[5] An advantage of this approach is that it enables one to potentially obtain an allelic series for any gene of interest; i.e., mutations that cause both nonsense and missense mutations, which may potentially allow functional inferences based on the observed resultant phenotype. However, the repertoire of mutations obtained is random, and the efficiency of this method relative to direct engineering has not been clearly established, especially given the development of novel technologies for plasmid modification in bacteria.

6. Conclusions

In recent years there has been remarkable progress in the development of means to modify the germline of the mouse, and the insight provided by examining the consequences of the perturbation of specific genes has been phenomenal. Potent mutagenesis techniques existed well prior to the era of germ-line modification, but the practical difficulty in characterizing mouse mutations at the molecular level limited their application, except in the laboratories of a few resolute investigators. The characterization of the mouse genome sequence and the development of robust computational and biological resources for genomic analysis have greatly facilitated the task of characterizing even single-base sequence changes. As a consequence, mouse mutagenesis is enjoying a revival that encompasses both large programs with broad aims and smaller efforts targeting specific developmental or physiological processes. The success of these efforts demonstrates clearly how well phenotype-driven investigation can complement and synergize with gene-driven analysis.

In their seminal paper describing the use of a three-generation screen to uncover novel mutations affecting metabolic processes, McDonald and Bode wrote: "In addition to the inherent value of the mutant and the information that may be obtained from it, we hope that this study will alert mammalian biologists working in a variety of fields to the availability of this method for obtaining mutants, and will stimulate consideration of which specific mouse mutant phenotypes would be informative in particular systems, and lead to the design of new ways for detecting and using such mutants."[15] It is fair to say this hope has been realized, and the potential for continued discovery remains great.

References

1. Russell, W.L. 1972. Radiation and chemical mutagenesis and repair in mice. *Johns Hopkins Med. J. Suppl.* **1**:239–247.
2. Russell, W.L. 1951. X-ray-induced mutations in mice. *Cold Spring Harb. Symp. Quant. Biol.* **16**:327–336.
3. Russell, W.L., E.M. Kelly, P.R. Hunsicker, J.W. Bangham, S.C. Maddux, and E.L. Phipps. 1979. Specific locus test shows ethylnitrosurea to be the most potent mutagen in the mouse. *Proc. Natl. Acad. Sci. USA* **76**:5818–5819.
4. Russell, W., P. Hunsicker, D. Carpenter, C. Cornett, and G. Guinn. 1982. Effect of dose-fractionation on the ethylnitrosurea induction of specific-locus mutations in mouse spermatogonia. *Proc. Natl. Acad. Sci. USA* **79**:3592–3593.
5. Coghill, E.L., A. Hugill, N. Parkinson, C. Davison, P. Glenister, S. Clements, J. Hunter, R.D. Cox, and S.D. Brown. 2002. A gene-driven approach to the identification of ENU mutants in the mouse. *Nat. Genet.* **30**:255–256.
6. Bode, V.C. 1984. Ethylnitrosourea mutagenesis and the isolation of mutant alleles for specific genes located in the T region of mouse chromosome 17. *Genetics* **108**:457–470.
7. Justice, M.J. and V.C. Bode. 1986. Induction of new mutations in a mouse t-haplotype using ethylnitrosourea mutagenesis. *Genet. Res.* **47**:187–192.
8. Ehling, U.H., J. Favor, J. Kratochvilova, and A. Neuhauser-Klaus. 1982. Dominant cataract mutations and specific-locus mutations in mice induced by radiation or ethylnitrosourea. *Mutat. Res.* **92**:181–192.
9. Johnson, F.M. and S.E. Lewis. 1981. Electrophoretically detected germinal mutations induced in the mouse by ethylnitrosourea. *Proc. Natl. Acad. Sci. USA* **78**:3138–3141.
10. Murota, T., T. Shibuya, and K. Tutikawa. 1982. Genetic analysis of an N-ethyl-N-nitrosourea-induced mutation at the hemoglobin beta-chain locus in mice. *Mutat. Res.* **104**:317–321.
11. Popp, R.A., E.G. Bailiff, L.C. Skow, F.M. Johnson, and S.E. Lewis. 1983. Analysis of a mouse alpha-globin gene mutation induced by ethylnitrosourea. *Genetics* **105**:157–167.

12. Pretsch, W. and D.J. Charles. 1984. An inherited variant of mouse sn-glycerol-3-phosphate dehydrogenase detected by isoelectric focusing: genetical and biochemical analyses. *Biochem. Genet.* **22**:419–428.
13. Peters, J., S.J. Andrews, J.F. Loutit, and J.B. Clegg. 1985. A mouse beta-globin mutant that is an exact model of hemoglobin Rainier in man. *Genetics* **110**:709–721.
14. Lewis, S.E., R.P. Erickson, L.B. Barnett, P.J. Venta, and R.E. Tashian. 1988. N-ethyl-N-nitrosourea-induced null mutation at the mouse Car-2 locus: an animal model for human carbonic anhydrase II deficiency syndrome. *Proc. Natl. Acad. Sci. USA* **85**:1962–1966.
15. McDonald, J.D. and V.C. Bode. 1988. Hyperphenylalaninemia in the hph-1 mouse mutant. *Pediatr. Res.* **23**:63–67.
16. McDonald, J., A. Shedlovsky, and W. Dove. 1990. Investigating inborn errors of phenylalanine metabolism by efficient mutagenesis of the mouse germline. In *Banbury Report 34: Biology of Mammalian Germ Cell Mutagenesis.* Cold Spring Harbor Laboratory Press, Cold Spring Harbor.
17. Moser, A.R., H.C. Pitot, and W.F. Dove. 1990. A dominant mutation that predisposes to multiple intestinal neoplasia in the mouse. *Science* **247**:322–324.
18. Favor, J., A. Neuhauser-Klaus, and U.H. Ehling. 1991. The induction of forward and reverse specific-locus mutations and dominant cataract mutations in spermatogonia of treated strain DBA/2 mice by ethylnitrosourea. *Mutat. Res.* **249**:293–300.
19. Vitaterna, M., D. King, A. Chang, J. Kornhauser, P. Lowrey, J. McDonald, W. Dove, L. Pinto, F. Turck, and J. Takahashi. 1994. Mutagenesis and mapping of a mouse gene, *clock*, essential for cricadian behavior. *Science* **264**:716–719.
20. King, D., Y. Zhao, A. Sangoram, L. Wilsbacher, M. Tanaka, M. Antoch, T. Steeves, M. Vitaterna, J. Kornhauser, P. Lowrey, F. Turek, and J. Takahashi. 1997. Positional cloning of the mouse circadian *clock* gene. *Cell* **89**:641–653.
21. Kapfhamer, D., O. Valladares, Y. Sun, P.M. Nolan, J.J. Rux, S.E. Arnold, S.C. Veasey, and M. Bucan. 2002. Mutations in Rab3a alter circadian period and homeostatic response to sleep loss in the mouse. *Nat. Genet.* **32**:290–295.
22. Cordes, S.P. and G.S. Barsh. 1994. The mouse segmentation gene kr encodes a novel basic domain-leucine zipper transcription factor. *Cell* **79**:1025–1034.
23. Rajaraman, S., W.S. Davis, A. Mahakali-Zama, H.K. Evans, L.B. Russell, and M.A. Bedell. 2002. An allelic series of mutations in the kit ligand gene of mice. I. Identification of point mutations in seven ethylnitrosourea-induced Kitl(Steel) alleles. *Genetics* **162**: 31–340.
24. Rinchik, E.M. and D.A. Carpenter. 1999. N-ethyl-N-nitrosourea mutagenesis of a 6- to 11-cM subregion of the Fah-Hbb interval of mouse chromosome 7: Completed testing of 4557 gametes and deletion mapping and complementation analysis of 31 mutations. *Genetics* **152**: 373–383.
25. Wang, J., J. Mager, Y. Chen, E. Schneider, J.C. Cross, A. Nagy, and T. Magnuson. 2001. Imprinted X inactivation maintained by a mouse Polycomb group gene. *Nat. Genet.* **28**:371–375.
26. Mager, J., N.D. Montgomery, F.P. de Villena, and T. Magnuson. 2003. Genome imprinting regulated by the mouse Polycomb group protein Eed. *Nat. Genet.* **33**:502–507.
27. Nolan, P.M. *et al.* 2000. A systematic, genome-wide, phenotype-driven mutagenesis programme for gene function studies in the mouse. *Nat. Genet.* **25**:440–443.

28. Hrabe de Angelis, M.H. *et al.* 2000. Genome-wide, large-scale production of mutant mice by ENU mutagenesis. *Nat. Genet.* **25**:444–447.
29. Rossant, J. and C. McKerlie. 2001. Mouse-based phenogenomics for modelling human disease. *Trends. Mol. Med.* **7**:502–507.
30. Masuya, H., Y. *et al.* 2004. Development and implementation of a database system to manage a large-scale mouse ENU-mutagenesis program. *Mamm. Genome.* **15**: 404–411.
31. Kile, B.T., K.E. Hentges, A.T. Clark, H. Nakamura, A.P. Salinger, B. Liu, N. Box, D.W. Stockton, R.L. Johnson, R.R. Behringer, A. Bradley, and M.J. Justice. 2003. Functional genetic analysis of mouse chromosome 11. *Nature* **425**:81–86.
32. Bult, C., W.A. Kibbe, J. Snoddy, M. Vitaterna, D. Swanson, S. Pretel, Y. Li, M.M. Hohman, E. Rinchik, J.S. Takahashi, W.N. Frankel, and D. Goldowitz. 2004. A genome end-game: understanding gene function in the nervous system. *Nat. Neurosci.* **7**:484–485.
33. Svenson, K.L., M.A. Bogue, and L.L. Peters. 2003. Invited review: Identifying new mouse models of cardiovascular disease: a review of high-throughput screens of mutagenized and inbred strains. *J. Appl. Physiol.* **94**:1650-1659; discussion 1673.
34. Kasarskis, A., K. Manova, and K.V. Anderson. 1998. A phenotype-based screen for embryonic lethal mutations in the mouse. *Proc. Natl. Acad. Sci. USA* **95**:7485–7490.
35. Eggenschwiler, J.T., E. Espinoza, and K.V. Anderson. 2001. Rab23 is an essential negative regulator of the mouse Sonic hedgehog signalling pathway. *Nature* **412**:194–198.
36. Caspary, T., M.J. Garcia-Garcia, D. Huangfu, J.T. Eggenschwiler, M.R. Wyler, A.S. Rakeman, H.L. Alcorn, and K.V. Anderson. 2002. Mouse dispatched homolog1 is required for long-range, but not juxtacrine, Hh signaling. *Curr. Biol.* **12**:1628–1632.
37. Huangfu, D., A. Liu, A.S. Rakeman, N.S. Murcia, L. Niswander, and K.V. Anderson. 2003. Hedgehog signalling in the mouse requires intraflagellar transport proteins. *Nature* **426**:83–87.
38. Hentges, K., K. Thompson, and A. Peterson. 1999. The flat-top gene is required for the expansion and regionalization of the telencephalic primordium. *Development* **126**:1601–1609.
39. Hentges, K.E., B. Sirry, A.C. Gingeras, D. Sarbassov, N. Sonenberg, D. Sabatini, and A.S. Peterson. 2001. FRAP/mTOR is required for proliferation and patterning during embryonic development in the mouse. *Proc. Natl. Acad. Sci. USA* **98**:13796–13801.
40. Herron, B.J., W. Lu, C. Rao, S. Liu, H. Peters, R.T. Bronson, M.J. Justice, J.D. McDonald, and D.R. Beier. 2002. Efficient generation and mapping of recessive developmental mutations using ENU mutagenesis. *Nat. Genet.* **30**:185–189.
41. Shima, N., S.A. Hartford, T. Duffy, L.A. Wilson, K.J. Schimenti, and J.C. Schimenti. 2003. Phenotype-based identification of mouse chromosome instability mutants. *Genetics* **163**:1031–1040.
42. Hoebe, K., X. Du, J. Goode, N. Mann, and B. Beutler. 2003. Lps2: a new locus required for responses to lipopolysaccharide, revealed by germline mutagenesis and phenotypic screening. *J. Endotoxin. Res,* **9**: 250–255.
43. Nelms, K.A. and C.C. Goodnow. 2001. Genome-wide ENU mutagenesis to reveal immune regulators. *Immunity* **15**:409–418.
44. Weiss, K.C., D.Y. Kim, C.T. Pawson, and S.P. Cordes. 2003. A genetic screen for mouse mutations with defects in serotonin responsiveness. *Brain Res. Mol. Brain Res.* **115**:162–172.
45. Carpinelli, M.R., D.J. Hilton, D. Metcalf, J.L. Antonchuk, C.D. Hyland, S.L. Mifsud, L. Di Rago, A.A. Hilton, T.A. Willson, A.W. Roberts, R.G. Ramsay, N.A. Nicola, and W.S. Alexander. 2004. Suppressor screen in Mpl-/- mice: c-Myb mutation causes

supraphysiological production of platelets in the absence of thrombopoietin signaling. *Proc. Natl. Acad. Sci. USA* **101**:6553–6558.

46. Russell, W.L., P.R. Hunsicker, D.A. Carpenter, C.V. Cornett, and G.M. Guinn. 1982. Effect of dose-fractionation on the ethylnitrosurea induction of specific-locus mutations in mouse spermatogonia. *Proc. Natl. Acad. Sci. USA* **79**:3592–3593.
47. Justice, M.J., D.A. Carpenter, J. Favor, A. Neuhauser-Klaus, M. Hrabe de Angelis, D. Soewarto, A. Moser, S. Cordes, D. Miller, V. Chapman, J.S. Weber, E.M. Rinchik, P.R. Hunsicker, W.L. Russell, and V.C. Bode. 2000. Effects of ENU dosage on mouse strains. *Mamm. Genome* **11**:484–488.
48. Shedlovsky, A., J.D. McDonald, D. Symula, and W.F. Dove. 1993. Mouse models of human phenylketonuria. *Genetics* **134**:1205–1210.
49. Dietrich, W., J.C. Miller, R.G. Steen, M. Merchant, D. Damron, R. Nahf, A. Gross, D.C. Joyce, M. Wessel, R.D. Dredge, A. Marquis, L.D. Stein, N. Goodman, D.C. Page, and E.S. Lander. 1994. A genetic map of the mouse with 4,006 simple sequence length polymorphism. *Nat. Genet.* **7**:220–245.
50. Neuhaus, I.M. and D.R. Beier. 1998. Efficient localization of mutations by interval haplotype analysis. *Mamm. Genome* **9**:150–154.
51. Lindblad-Toh, K., E. Winchester, M.J. Daly, D.G. Wang, J.N. Hirschhorn, J.P. Laviolette, K. Ardlie, D.E. Reich, E. Robinson, P. Sklar, N. Shah, D. Thomas, J.B. Fan, T. Gingeras, J. Warrington, N. Patil, T.J. Hudson, and E.S. Lander. 2000. Large-scale discovery and genotyping of single-nucleotide polymorphisms in the mouse. *Nat. Genet.* **24**:381–386.
52. Noskov, V.N., N. Kouprina, S.H. Leem, I. Ouspenski, J.C. Barrett, and V. Larionov. 2003. A general cloning system to selectively isolate any eukaryotic or prokaryotic genomic region in yeast. *BMC Genomics* **4**:16.
53. Kunath, T., G. Gish, H. Lickert, N. Jones, T. Pawson, and J. Rossant. 2003. Transgenic RNA interference in ES cell-derived embryos recapitulates a genetic null phenotype. *Nat. Biotechnol.* **21**:59–561.
54. Rubinson, D.A., C.P. Dillon, A.V. Kwiatkowski, C. Sievers, L. Yang, J. Kopinja, D.L. Rooney, M.M. Ihrig, M.T. McManus, F.B. Gertler, M.L. Scott, and L. Van Parijs. 2003. A lentivirus-based system to functionally silence genes in primary mammalian cells, stem cells and transgenic mice by RNA interference. *Nat. Genet.* **33**:401–406.
55. Favor, J. 1999. Mechanisms of mutation induction in germ cells of the mouse as assessed by the specific locus test. *Mutat. Res.* **428**:227–236.
56. Rinchik, E., J. Bangham, P. Hunsicker, N. Cacheiro, B. Kwon, I. Jackson, and L. Russell. 1990. Genetic and molecular analysis of chlorambucil-induced germ-line mutations in the mouse. *Proc. Natl. Acad. Sci. USA* **87**:1416–1420.
57. Flaherty, L., A. Messer, L.B. Russell, and E.M. Rinchik. 1992. Chlorambucil-induced mutations in mice recovered in homozygotes. *Proc. Nat. Acad. Sci. USA* **89**:2859–2863.
58. Hodgkinson, C.A., K.J. Moore, A. Nakayama, E. Steingrimsson, N.G. Copeland, N.A. Jenkins, and H. Arnheiter. 1993. Mutations at the mouse micropthalmia locus are associated with defects in a gene encoding a novel basic-helix-loop-helix-zipper protein. *Cell* **74**:395–404.
59. Perry, W.L. 3rd, T.J. Vasicek, J.J. Lee, J.M. Rossi, L. Zeng, J. Zhang, S.M. Tilghman and F. Costantini. 1995. Phenotypic and molecular analysis of a transgenic insertional allele of the mouse fused locus. *Genetics* **141**:321-332.

60. Vogt, T.F., L. Jackson-Grusby, A.J. Wynshaw-Boris, D.C. Chan, and P. Leder. 1992. The same genomic region is disrupted in two transgene-induced limb deformity alleles. *Mamm. Genome.* **3**:431–437.
61. Khokha, M.K., D. Hsu, L.J. Brunet, M.S. Dionne, and R.M. Harland. 2003. Gremlin is the BMP antagonist required for maintenance of Shh and Fgf signals during limb patterning. *Nat. Genet.* **34**:303–307.
62. Zuniga, A., O. Michos, F. Spitz, A.P. Haramis, L. Panman, A. Galli, K. Vintersten, C. Klasen, W. Mansfield, S. Kuc, D. Duboule, R. Dono, and R. Zeller. 2004. Mouse limb deformity mutations disrupt a global control region within the large regulatory landscape required for Gremlin expression. *Genes Dev.* **18**:1553–1564.
63. Horie, K., K. Yusa, K. Yae, J. Odajima, S.E. Fischer, V.W. Keng, T. Hayakawa, S. Mizuno, G. Kondoh, T. Ijiri, Y. Matsuda, R.H. Plasterk, and J. Takeda. 2003. Characterization of Sleeping Beauty transposition and its application to genetic screening in mice. *Mol. Cell. Biol.* **23**:9189–9207.
64. Skarnes, W.C., H. von Melchner, W. Wurst, G. Hicks, A.S. Nord, T. Cox, S.G. Young, P. Ruiz, P. Soriano, M. Tessier-Lavigne, B.R. Conklin, W.L. Stanford, and J. Rossant. 2004. A public gene trap resource for mouse functional genomics. *Nat. Genet.* **36**:543–544.
65. Munroe, R.J., R.A. Bergstrom, Q.Y. Zheng, B. Libby, R. Smith, S.W. John, K.J. Schimenti, V.L. Browning, and J.C. Schimenti. 2000. Mouse mutants from chemically mutagenized embryonic stem cells. *Nat. Genet.* **24**:318–321.
66. Chen, Y., D. Yee, K. Dains, A. Chatterjee, J. Cavalcoli, E. Schneider, J. Om, R.P. Woychik, and T. Magnuson. 2000. Genotype-based screen for ENU-induced mutations in mouse embryonic stem cells. *Nat. Genet.* **24**:314–317.

CHAPTER 13

EMBRYO BIOTECHNOLOGIES

Carl A. Pinkert[1] and Michael J. Martin[2]

[1]*Department of Pathology and Laboratory Medicine*
Center for Aging and Developmental Biology
University of Rochester Medical Center, Rochester, NY, USA
carl_pinkert@urmc.rochester.edu
and
[2]*Functional Genetics, Inc.*
Albany, OH, USA
cy81martin@earthlink.net

1. Introduction

In vitro embryo culture and associated assisted reproductive techniques have come to represent integral components across a wide range of mammalian research efforts. Embryo biotechnologies, with particular emphasis on the mouse model, have evolved considerably over the last century. They have become useful and important tools for studies in animal genetics and breeding research, ranging from the efficient generation, propagation, and rescue of valuable lineages, to a variety of *in vitro* and *in vivo* manipulations related to genetic engineering technologies.

Embryo biotechnologies have developed from the late 1800s to become a critical component and necessity for a host of mouse modeling efforts today (see Table 1). One cannot readily contemplate where genetic engineering techniques would stand today without the efforts of early mammalian embryologists and the advent of successful embryo manipulations both *in vitro* and *in vivo*. It has been over 100 years since the first successful embryo transfer experiments — dating back to the efforts first published in the 1880s by Schenk,[1] followed by Heape's success at transferring mammalian ova in 1891.[2] By the time Hammond's mouse embryo culture studies were reported in the late 1940s and early 1950s, culture systems were developed that sustained ova through several cleavage divisions.[3] Such methods provided a means to systematically identify and test procedures for a variety of egg manipulations. These early studies led to experiments that included the production of chimeric animals, transfer of inner cell mass cells and teratocarcinoma cells, the injection of nucleic acids into mouse embryos and nuclear transfer technologies. Without the ability to culture or maintain mouse embryos, the necessary insights needed to establish such technologies would not have been possible (see Table 1).[4,5]

Table 1. Seminal events in the development of embryo biotechnologies. With rare exceptions, the mouse has been a mammalian model of choice as embryo biotechnologies have evolved (adapted from ref. 6).

Year	Event
1880	Mammalian embryo cultivation attempted
1891	First successful mammalian embryo transfer
early 1900s	*In vitro* embryo culture develops
1937	Successful mouse embryo transfer
1949	*In vitro* embryo culture and cleavage divisions achieved
1961	Mouse embryo aggregation to produce chimeras
1966	Mouse zygote microinjection technology established
1974	Development of teratocarcinoma cell transfer into mouse ova
1977	mRNA and DNA transferred into Xenopus eggs
1980	mRNA transferred into mouse ova
1980–1981	Transgenic mice first documented
1981	Transfer of ES cells derived from mouse embryos
1982	Transgenic mice and a growth (GH) phenotype
1987	ES cell transfer into mouse blastocysts: Chimeric "knock-out" mice
1989	ES cell transfer into mouse blastocysts: Targeted DNA integration and germline chimeras
1993	Live mice derived from vitrified ova
1993	ES cell co-culture: germline chimeric mice produced
1997	Mitochondrial transfer into mouse ova: Transmitochondrial "mitomice"
1997	"Cloning" via nuclear transfer in sheep
1998	"Cloning" via nuclear transfer in mice
2003	ES cells differentiated into gametes
2004	Genetically modified mouse parthenogenotes viable and fertile
2005	????

2. Embryo Techniques

2.1. *Superovulation and Generation of Ova*

For a variety of embryo biotechnologies, generation and harvest of oocytes or preimplantation ova are the initial steps in most protocols. While there is some debate regarding the quality and subsequent viability of ova derived via natural mating (no hormonal stimulus) *vs.* superovulatory regimes, considerable efforts have gone into identification of suitable *in vitro* culture and handling criteria to optimize yields and survival.

Hormonal induction of ovulation and superovulation is routinely employed for generation of mouse ova.[7–9] Advantages of hormonal induction include: a) an increase in numbers of viable ova suitable for various manipulations, b) synchronization of estrus in cohorts of mice, c) utilization of prepuberal or peripuberal mice, and d) collection of ova from subfertile and mildly reproductively impaired animals. Using various mouse strains, 20 to 40 oocytes or preimplantation ova may be obtained from females, in significant contrast to the yield of 6 to 15 oocytes obtained from normally cycling females.[7–10]

Superovulation of mature and peripuberal female mice by exogenous administration of gonadotropins has been performed for nearly half a century, allowing one to obtain large numbers of fertilizable eggs for various manipulations.[11] Pregnant Mare's Serum Gonadotropin (PMSG; mimicking endogenous Follicle Stimulating Hormone (FSH) oocyte stimulation) and human Chorionic Gonadotropin (hCG; mimicking endogenous Luteinizing Hormone (LH) induction of ovulation) are routinely used to superovulate mice.[7–9] Alternatively, FSH and LH can be administered by osmotic pump and injection, respectively.[8] This latter method, while more tedious and at greater expense, has been useful in inducing superovulation in obese and older mice that did not respond to previous PMSG/hCG induction.

Treatment with PMSG/hCG has proven to be a most cost-effective technique and is, therefore, routinely used for generation of oocytes, and zygote to blastocyst stage ova. The specific time intervals between a) injection of PMSG and hCG and b) from hCG to recovery of mouse ova have been well documented.[8–17] The effects of multiple- and high-dose gonadotropin stimulation on fertilization and preimplantation development after IVF have been described, as have the comparative efficiencies of various mouse strains.[7–9,12,17–20] As anticipated in these studies, relative performance and yields varied by strain (efficiencies of hybrid strains > outbred strains > inbred strains), specific ages of donors, and a number of husbandry practices. In our experience, it should be noted that there are dramatic differences in fertilization results between individual repetitions for the outbred and inbred strains.[17,20,21] Yet, this was predictable based on earlier studies using various representative strains for natural matings to derive fertilized ova.

2.2. *Embryo Culture, IVF and Zona Disruption*

The ability to manipulate the mouse genome can be traced back to pioneering studies in which murine embryo culture and transfer procedures were first developed (see Table 1). In 1949, a physiological saline based-medium supplemented with egg yolk and white allowed 8-cell ova (but not 2-cell ova) to develop to the blastocyst stage.[3] Seven years later, Whitten's Medium was published (a Krebs-Ringer bicarbonate based medium supplemented with glucose, penicillin, streptomycin and 1% fresh, thin egg white) that also supported development of 8-cell mouse embryos to the blastocyst stage.[22] It was the use of Whitten's medium with pioneering embryo transfer techniques that resulted in the birth of the first live mice from *in vitro* cultured blastocysts.[23–25]

Further developments in mouse embryo culture media occurred in the mid-1960s when it was observed that the addition of both pyruvate and lactate to Whitten's medium improved the development of 2-cell mouse ova to the blastocyst stage.[26,27] These findings resulted in the creation of Brinster's Modified Ova Culture medium (BMOC), a modified version of which is still commercially available and widely used.[28] Today, several culture media (e.g., CZB, MEM and M16), can support development of mouse zygotes and reconstructed or nuclear transfer (NT) embryos to the blastocyst stage.[8,9,29,30] It is important to note that supplementation of a chemically defined medium, such as M16, with fetal calf serum has been shown to significantly reduce pre- and post-implantation development of mouse embryos. Though the exact mechanism is not known, culture of mouse embryos in the presence of serum appears to result in an alteration in the expression and methylation of some growth related genes.[31]

In vitro fertilization (IVF) of mouse ova has been utilized as a model system for developing and testing methodologies for human IVF and as an assisted reproduction technique for the propagation of transgenic mice that have been made infertile by mutations.[32] Early studies that examined the relationship between sperm concentration, normal fertilization and polyspermy led to the development of a protocol(s) that generated live IVF mice using very low sperm:egg ratios.[33,34] These results were subsequently used to help reduce polyspermy in human IVF protocols.

Mouse ova have been fertilized *in vitro* in bicarbonate buffered media such as M199 and Whitten's medium or HEPES-buffered modified Ham's medium.[35–37] A typical IVF protocol includes recovery of ovulated ova and their co-culture with sperm obtained from the cauda epididymis. During co-culture, the cumulus mass surrounding each ovum was deemed critical, as a number of studies have shown that its absence can result in increased polyspermy and reduced development of monospermic embryos to the blastocyst stage.[32,37] Indeed, conflicting reports regarding the requirement for cumulus-intact oocytes in IVF protocols do exist.[14,38–40] Yet, the cumulus cell mass does disaggregate during co-culture with sperm, and the effect of cumulus cell removal did not affect IVF outcome in inbred, and hybrid strains in one study.[17] Using outlined IVF procedures, one could expect ~40 to 90% of cumulus-intact and 30 to 90% of cumulus-free ova to be fertilized following IVF procedures.[17,32] The wide variation again was dependent upon specific strain and optimization of specific collection regimens.[17,20,32]

Mouse oocyte IVF in the absence of cumulus cells was reported for inbred C57BL/6,[14] hybrid [C57BL10ScSn/Ph × A/Ph] F5,[38] and outbred Swiss mice.[40] However, others observed a more restricted range of sperm concentration in which cumulus-free oocytes were fertilized *in vitro* compared to cumulus-intact oocytes.[39] Using a fixed concentration of 2×10^6 spermatozoa/ml for IVF with representative inbred, outbred and hybrid strains (C57BL/6, ICR, and B6SJLF1), no within-strain fertilization rate differences, with or without intact cumulus cells, were observed.[17] Yet, fertilization rates were not consistent between individual repetitions of the protocol. Since a confounding variable was the use of different sperm donors, it is possible that the relative fertility of individual males may have contributed to the disparity in fertilization results observed. As such, a means of enriching or characterizing relative fertility of males, or using semen from a pool of males may be warranted.

In an attempt to further reduce the number of sperm required for *in vitro* fertilization of mouse ova, the effect of disruption of the zona pellucida on the incidence of sperm penetration and subsequent embryo development was examined. Zonae were first disrupted by a procedure that involved the use of a microneedle filled with acid Tyrode's solution. After bringing the needle into contact with the zona surface, a localized disruption (10 to 12 μm puncture) of the zona was created by expulsion of the acidified solution. It was initially found that the zona drilling procedure resulted in a 100-fold reduction in the number of sperm required for *in vitro* fertilization of mouse ova.[41] Furthermore, neither the polyspermy rate nor the proportion of transplanted embryos that developed to term differed between control and zona drilled oocytes. However, in a later study, a proportion of zona drilled embryos lost their zonae entirely and exhibited what were described as "unique" cleavage patterns during preimplantation development.[42] Complete loss of zonae pellucidae resulted in blastomere separation and random aggregation of embryos to form giant "composite" morulae and blastocysts. Based on these observations, zona drilling of murine ova can increase the proportion of live births when compared to intact ova. It may, however, cause abnormal development of some embryos.[42]

Zona disruption can also be achieved by cutting the zona pellucida with a solid glass microneedle. In comparison to zona drilling, zona cutting creates a slit that is much larger (40×2 µm) and more variable in size.[43] When these two zona disruption techniques were compared, zona drilling resulted in more fertilized oocytes but a higher oocyte loss rate. Yet, neither the proportion of embryos that developed into blastocysts nor the proportion that implanted after transfer to pseudopregnant recipient mice differed.[44] In addition, to zona disruption, zonae can be completely removed by culture in acid Tyrode's solution prior to morula aggregation as mentioned in the following section.

2.3. *Chimera Development*

2.3.1. *Embryo Aggregation*

With the advent of improved embryo culture systems in the mouse, it became possible to create viable multigenotypic individuals composed of two or more types of cells, known as allophenic chimeric mice. During the past 40 years, chimeric mice have been used to study several aspects of development ranging from cell to cell interaction, embryo remodeling and sex determination to the function of the immune system. The first chimeras were created by aggregating two or more zona-free 2- to 8-cell embryos.[45,46] In 1973, the addition of a plant mucoprotein, phytohemagglutin (PHA), to medium was observed to cause the cells from mouse preimplantation embryos to adhere "rapidly and firmly" after only a brief period of exposure.[47] The subsequent birth of live chimeras led these researchers to conclude that PHA was a simple and effective substitute for temperature induced (37°C) aggregation of mouse blastomeres *in vitro*.

In the late 1970s and early 1980s, the development of hexa and octaparental aggregation mouse chimeras was examined.[48,49] The number of inner cell mass (ICM) cells that gave rise to the embryo proper was estimated to be fairly small at 3 to 5 cells. Interestingly, female hexaparental chimeras were also capable of producing ova from all three contributing genotypes.[49]

When preimplantation mouse embryos are randomly aggregated, half of the resulting chimeras, on average, will be sex mosaics (XX-XY) or sex chimeras. Though some sex chimeras become hermaphrodites, the majority (3:1) develop into normal fertile males.[50,51] Studies in the 1970s concluded that sex mosaic chimeras can only develop into females when the embryos contain a "great preponderance" of XX cells.[52,53] Since then, others have found that the gonads in a high proportion (50%) of sex mosaic chimeras initially differentiate as ovotestes between days 10 and 12 of gestation. As development continues, however, the ovarian portion, which is typically confined to the poles of the gonad, undergoes regression while the testicular component continues to grow and develop.[54]

Surprisingly, when parthenogenetic embryos or those that develop from teratocarcinoma cells were aggregated with normal fertilized embryos, both cell types contributed to full term development of chimeric mice. Offspring produced by these female chimeras exhibited markers (alleles) from parthenogenetic or teratocarcinoma cells. These results demonstrated that both cell types can yield fully functional totipotent ova when combined with normal cells in a chimeric environment.[55,56] Most recently, genetic engineering technology has identified regulatory genes, that when modified, allowed development of parthenogenic mice to surmount an imprinting hurdle and develop normally to adulthood.[57]

Lastly, genetically-modified chimeras were derived from the aggregation of morulae and embryonic stem (ES) cells, providing an additional methodology for creating gene targeted animals (see Ch. 14).

2.3.2. *Blastocoel Injection Chimeras*

Chimeric mice can also be produced by the injection of ES cells[58] or teratocarcinoma cells[59–62] into the blastocoel cavity of recipient embryos. A major advantage in the use of ES cells, is that they can be readily maintained and manipulated in culture (e.g., for genetic modification or gene targeting studies) prior to their introduction into embryos.[63–64] Injection of genetically altered ES cells into murine blastocysts has been used to introduce a variety of genetic modifications into mice and ultimately through the germline.[58,64] A sampling of these modifications include specific mutations in mitochondrial genes,[65,66] correction of a deletion mutation in a hypoxanthine phosphoribosyltransferase gene (HPRT)[67] and disruption of a β2 microglobulin gene.[68]

Inner cell mass cells isolated by immunosurgery have also been injected into the blastocoel of recipient embryos to generate chimeras within and between species. Intra- and inter-specific chimeras in laboratory and other species have been used to study various aspects of physiology, behavior and developmental biology.[51–53,69–71]

2.4. *Intracytoplasmic Sperm Injection (ICSI)*

In addition to the development of human ICSI protocols, ICSI in the mouse has been used to help characterize a variety of events that are associated with fertilization and embryo development. These include determining when genomic imprinting is completed in the sperm genome, what developmental stages of sperm can be used to create viable offspring and what physical portion(s) of the spermatozoan is required for normal embryonic development.

It is now clear from a variety of studies in the mouse that injection of primary spermatocytes, secondary spermatocytes, round spermatids or testicular spermatozoa into ova can yield viable offspring.[72,73] While the proportion of embryos that developed to term (~3.8%) was lowest following injection of murine primary spermatocytes, the birth of normal pups suggests that genomic imprinting had been completed in some primary spermatocytes.[73] Injections of sperm that had been decapitated by sonication alone,[74] sonication in Triton X-100, or light sonication followed by cryopreservation[75] were also found to result in normal development of embryos to the blastocyst stage and, in the last instance, live offspring. These findings demonstrated that spermatozoa did not need to be structurally intact and that the plasma and acrosomal membranes were not required for normal embryo development in the mouse.[75]

Other important applications of ICSI include the restoration of fertility in transgenic mice that have been rendered reproductively defective by a genetic modification and the enhancement of fertility associated with cryopreserved semen. Intracytoplasmic sperm injection has been used to successfully circumvent infertility caused by oligospermia, low sperm motility[75,76] and abnormal sperm morphology.[77] While murine sperm can be cryopreserved, there are instances; e.g., with use of inbred mouse strains, where the post-thaw viability is low. Not surprisingly, one can significantly improve the fertilization rate of frozen-thawed sperm by directly injecting them into the cytoplasm of ova. In fact, ICSI was recently found more effective than *in vitro* fertilization for producing viable embryos from cryopreserved sperm.[78]

While intracytoplasmic sperm injection has been used for human oocytes not fertilized after 24 to 48 h postculture (failed fertilization),[79] the use of unfertilized ova obtained from standard superovulatory and natural breeding protocols may be readily fertilized with reasonable efficiency.[9] In mouse experiments, particularly when inbred strains are used, high proportions of ova that are collected after mating (and identification of copulatory plug formation) are not fertilized. However, the unfertilized ova are not identified until after cumulus removal. Routinely, these ova are discarded, but may indeed be a source of fertilizable oocytes using modified IVF protocols.[17]

While the majority of transgenic animals to date have been produced by pronuclear microinjection of DNA or cloned from transformed cell lines, transgenic mice have also been generated by ICSI. Between 60 and 90% of murine ova co-injected with membrane disrupted sperm and DNA that encoded either a green fluorescent protein (GFP) or β-galactosidase reporter gene expressed the transgene product at the blastocyst stage of development.[80] When these embryos were transferred to recipients, 20% of the pups expressed the integrated transgene. ICSI mediated delivery of a transgene has been attempted in other species including the pig and rhesus monkey with minimal success. Though GFP positive embryos were produced in both instances, the subsequent transfer of primate embryos into recipients yielded offspring that failed to express the transgene.[81–83]

2.5. *Embryo Transfer*

Important considerations related to successful embryo transfer include: a) viability and placement of ova into the reproductive tract of recipient females, b) identification of the number of transferred ova required to establish and maintain pregnancy, and c) determination of the upper limit of transferred ova that uteri of recipient mice can sustain to term.[7–9] Factors involved in successful implantation of mouse ova transferred to oviducts or uteri of foster mothers have built upon the accomplishments of earlier workers, who reported the first successful mouse ova transfers[84] and first cultured mouse ova *in vitro* to the blastocyst stage.[3] To verify the existence of a "law of diminishing returns," ova were transferred in increasing numbers to recipient females, but early efforts failed to identify an absolute upper limit to the number of eggs that would implant.[85] However, a detectable limit was indeed identified when the effects of hormonally-induced superpregnancy on subsequent implantation, embryonic mortality and loss at birth were studied.[20,86] In the end, the results of a number of embryo transfer studies were suggestive, but not conclusive evidence, of a "uterine capacity" phenomenon, in which it is thought that beyond a certain finite number, the uterus does not sustain the development of additional embryos to term.[20,23,85–87] When transferring up to 15 zygotes to a single uterine horn of a pregnant recipient mouse, evidence that would indicate the existence of a uterine capacity phenomenon was lacking.[85] However, the number of fetuses developing in the uterine horn not receiving transferred ova (i.e., containing native ova) decreased when large numbers of ova were transferred to the opposite horn. In a separate study utilizing hormonally induced "superpregnant" females, it appeared that embryonic death after Day 9 was dependent on uterine crowding.[86] It was therefore suggested that since there is a defined threshold, above which the death rate (Day 10 to Day 12) increases abruptly, that the limiting factor may be the amount of uterine surface available for the placentation of a linear series of implantations rather than the overall distension of the uterus. Lastly, after transferring up to 25 zygotes into a single oviduct of pseudopregnant mice, the law of diminishing returns was apparent as embryo and fetal survival began to plateau. Interestingly, increasing fetal mortality between day 12 and term were also observed.[20] Speculation as to how the uterus or maternal system might enforce such a limitation has not brought about definitive answers to date.

The groundwork established by these studies and many others since, made possible the development of many enabling technologies (for example, maintenance and rederivation of valuable strains, and transgenic animal production), which were dependent on the efficient transfer of early preimplantation ova. In this regard, several groups have successfully obtained pathogen-free offspring from stocks of mice contaminated with murine pathogens.[9,88] The classical method of rederiving pathogen-free stocks of mice involves the aseptic collection of fetuses by Caesarian-section with transfer to pathogen-free foster mothers. However, while many infections can be vertically transmitted, the zona pellucida can provide a barrier to infection of preimplantation ova, making embryo transfer techniques most useful in maintaining and preserving mouse lineages.[9,88]

3. Assisted Reproductive Technologies and Embryo Engineering

3.1. *Nuclear Transfer* (*NT*)

A considerable amount of work in recent years has focused on using nuclear transfer for developmental studies and genetic engineering efforts. The genetic uniformity of offspring, elimination of a generation interval in developing useful models, time and cost-constraints associated with confirmation of genetic identity, and chromosomal segregation from parental lineages to offspring are illustrative of why nuclear transfer procedures have been of critical importance for *in vitro* and *in vivo* experimentation to date. Nuclear transfer studies can also be used to test totipotency of specific cell lineages. Significant progress has been made in the areas of oocyte *in vitro* maturation, fertilization, and activation as well as embryo culture techniques.[89-91]

The first NT or cloned mice were produced from somatic (cumulus) cells in 1998.[92] Since then, mice have been cloned from additional somatic cell types, e.g., Sertoli and adult tail-tip cells,[93,94] and ES cells.[95,96] As with other species, the cloning of mice entails several steps. These include coordination of donor cell cycles, enucleation of recipient cytoplasts, injection of donor nuclei into cytoplasts, activation of reconstructed ova, and culture and transfer to suitable recipients. A review of the above methodology was recently published.[30]

The production of NT mice in a majority of laboratories continues to be problematic, especially when donor cells are derived from inbred ES cell lines.[30, 97] In an effort to address this difficulty, the influences of various media components on cell development and overall NT efficiency were explored.[98] The diploid NT mouse embryos exhibited "unusual responses" to the culture environment when compared to normal fertilized embryos, parthenogenetic embryos and tetraploid embryos that had been produced by the injection of cumulus cell nuclei into ova with intact spindles. Furthermore, the differences were most pronounced prior to the 8-cell stage. These observations led the authors to propose that the variation in cellular physiology or metabolism observed between cloned and normal embryos may have been the result of incomplete nuclear reprogramming before the 8-cell stage. Others observed that NT embryos were also more sensitive than parthenogenetic embryos to *in vitro* culture conditions and concluded that modifications of culture conditions "not noticed" by control embryos have a significant effect on the *in vitro* development of cloned embryos.[99]

In a situation analogous to that seen in ruminant clones, fetal overgrowth and postnatal death have also been observed in mice cloned from both somatic and ES cells.[30,90,100] In an attempt to address losses, two culture components (genetic background and culture confluence of the donor cell line) were evaluated in regard to the development of murine embryos cloned from ES cells.[96] The proportion of reconstructed embryos that cleaved, developed into blastocysts, and yielded live pups was significantly higher when the donor cells were 80 to 90% as opposed to 60 to 70% confluent. This result held true regardless of the donor cell's genetic background. While the overall cloning efficiency was lower than that observed for most heterozygous ES cell lines,[97] live NT mice were produced from both inbred HM-1 ES cells and intercrossed R1 ES cells. This finding led to the conclusion that cloned mice can be derived from certain strains of inbred ES cells provided these cells possess a sufficient "developmental potential." The components studied perhaps reflected cell cycle control with greater confluence potentially arresting cells in G1. Alternatively, nutrient availability may have been severely restricted, thus influencing experimental outcome.[101]

Though some somatic and ES cell lines can yield live normal appearing clones, microarray analysis of global gene expression in placentae and liver has revealed significant differences in expression patterns between cloned and control mice.[102] Of the 4% of abnormally expressed genes identified in the placentae of cloned mice, a majority were shared by mice cloned from both somatic and ES cells. This finding refutes the previously held notion that somatic donor nuclei are more "normal" than ES cell nuclei.[103] Based on these and other observations, it was suggested that the normalcy of cloned animals that survive to term should be assessed by "detailed molecular analyses of tissues from adult cloned animals" and not "superficial clinical examinations."[102]

3.2. *Microinjection*

The production of transgenic mice has been hailed as a seminal event in the development of animal biotechnology (see Ch. 14). While some progress seems extremely rapid, it is still difficult to believe that following the first published report of a microinjection method in 1966,[104] it took 15 years before transgenic animals were created. The pioneering laboratories that reported success at gene transfer would not have been able to do so, were it not for development of two different embryo technologies — DNA microinjection and retroviral-mediated gene transfer (along with the attendant recombinant DNA technologies).[5,9,105, Ch.14]

3.3. *Retrovirus-Mediated Gene Transfer*

Transfer of foreign genes into animal genomes has also been accomplished using viral transfection procedures. Although embryos can be infected with retroviruses up to midgestation, embryos up to the 16-cell stage are used for infection with one or more recombinant retroviruses harboring a gene construct of interest.[106,107] Immediately following infection, reverse transcriptase regulates retrovirus production of a DNA copy of its RNA genome. Retroviruses effectively transfect mitotically active cells; thus, completion of this process requires that the host cell undergoes the S phase of the cell cycle. Modifications to the retrovirus frequently consist of removal of structural genes that support viral particle formation. Additionally, most retroviruses and complementary lines are ecotropic in that they infect only rodents and rodent cell lines rather than humans.

3.4. *ES Cell Transfer*

From the mid-1980s technologies involving ES cells have been used to produce a host of mouse models.[108–111] Pluripotential ES cells are generally derived from early pre-implantation embryos and maintained in culture for a sufficient period for one to perform various *in vitro* manipulations. The cells may then be injected directly into a host blastocyst or incubated in association with earlier stage ova.[5,9,109] Following injection, embryos are then transferred into intermediate hosts or surrogate females for continued development. Currently, the efficiency of chimeric mouse production results in about 30 percent of the live-born animals containing tissue derived from the injected ES cells.

Pluripotent ES cells have also been extremely valuable in elucidating genes and enhancers involved in directing tissue differentiation and organogenesis. Enhancer and gene trapping are two strategies that permit the isolation of developmentally regulated genes and will enhance our understanding of their roles in development.[5,9,64,109] In both of these strategies a reporter gene is introduced into the genome of the ES cells. These reporter genes are partially or totally void of transcriptional control elements. For example, enhancer traps only have a minimal promoter preceding the marker gene. Thus, in order for the reporter gene to be expressed, *cis* regulatory elements of the "trapped" endogenous gene are required. The expression pattern of the reporter gene in the ES cell chimeras is used to determine the temporal and spatial patterns of gene expression in the developing embryos and fetuses. A gene or promoter trap differs from the enhancer trap in that the promoter sequences have also been removed from the reporter gene. In this case, expression of the reporter gene depends on the integration into an endogenous gene in correct orientation and in frame. The enhancer and promoter trap strategies are yet more examples of the important role of ES cell technology in genetic modification strategies.

Using various methods, the production of "chimeric" animals using ES cells has given researchers another valuable tool to produce transgenic animals. In such techniques, the power of gene transfer technology has been catapulted forward because such processes can allow for targeted insertions into the genome. Such targeting is extremely important, particularly in areas of gene therapy and correction, wherein previous technologies (microinjection and retroviral mediated transfer) allowed only for random integration events.[6,9,64,110,111]

3.5. *Primordial Germ Cell Technologies*

Epiblast-derived primordial germ (PG) cells may prove to be yet another source of cells to be used in the derivation of non-murine pluripotent cell lines. This strategy was also first developed in the mouse. Using certain growth factors and *in vitro* cell culture conditions, mouse PG cells can form ES cell colonies.[112] In addition, PG cell-derived ES cells have been shown to form germ-line chimeras.[113] However, in order for this strategy to work, the PG cells must first undergo a transformation into ES cell lines, since attempts to produce chimeras from freshly isolated PG cells were largely unsuccessful.

3.6. *Germ Cells as Vectors*

ICM cells, the source of many embryonic cell lines, were used as nuclear donors in the nuclear transfer procedure, resulting in live offspring in bovine species.[114,115] In contrast, similar nuclear transfer experiments have had limited success in the mouse. Donor nuclei from eight-cell stage embryos are the most advanced stage to give rise to nuclear transfer offspring,[116] and mouse ES cell donor nuclei fail to direct development past the preimplantation developmental stages.[89,117] Thus, there are also species dissimilarities in the totipotency of early embryonic nuclei used in nuclear transfer procedures.

In 1989, sperm-mediated gene transfer as a vehicle to introduce transgenes into mouse oocytes was reported but remained quite controversial for some time.[118–120] The initial sperm-mediated gene transfer story generated sufficient interest that also led to the development of spermatogonial cell transplantation procedures in mice as an alternative methodology for *in vivo* gene transfer.[121,122]

3.7. *Mitochondrial Transfer into Preimplantation Ova*

While cloning and nuclear transfer have obvious genomic implications, a parallel genetic mechanism involves mitochondrial genetics and techniques for mitochondrial transfer into embryos.[66] With the advancement of various gene transfer technologies, the establishment of mitochondrial transfer as a viable technique to genetically engineer the small 16.5 kb mitochondrial genome in mice has paradoxically lagged behind other genetic technologies. The lack of demonstrable recombination in mitochondrial DNA (mtDNA) necessitated different approaches to conventional transgenesis-based techniques. Initially, heteroplasmic mice were created to explore disease pathogenesis and mitochondrial dynamics in an *in vivo* system. Ultimately, transmitochondrial mouse models will be used to explore the role of the mitochondrial genome in human disease processes and in the development of novel human gene therapies. Specific modeling and the procedures for mitochondrial transfer will be of considerable importance toward our understanding of specific mitochondrial mutations, as well as leading to the development of novel strategies and therapies for human diseases influenced by mitochondrial DNA mutations.[66,123–125]

Curiously, early reports on development of cloned animals by nuclear transfer resulted in conflicting consequences when retrospective studies on mitochondrial transmission were reported.[126–130] Indeed, dependent upon the specific methodology employed for nuclear transfer and cytoplasm/ooplasm transfer to rescue low-quality embryos, additional models of heteroplasmy (where two or more populations of mitochondrial genetics may exist) may or may not have been created.[131–136] In the aggregate, such techniques may illustrate mechanisms underlying the dynamics related to persistence of foreign mitochondria and maintenance of heteroplasmy in various experimental protocols.

3.8. *Embryo Cryopreservation*

Successful cryopreservation of preimplantation-stage mouse ova was first described in 1972.[8,137,138] Procedures for cryopreservation of murine eggs have been greatly simplified over the last few years.[139,140] In conjunction with IVF protocols, advantages of various cryopreservation and vitrification technologies include: a) maintenance of genetic lineages and protection from random mutation (i.e., normal genetic drift), b) safeguarding a colony from potential disease, line contamination, or catastrophic events, c) significantly reducing or eliminating the costs associated with housing and maintaining lines of mice that may not be of immediate need, and d) use in conjunction with rederivation of pathogen-free mouse stocks.

In mouse experiments, both embryo transfers of frozen and non-frozen eggs have been used to rederive important animals and eliminate a variety of rodent pathogens. Survival rates greater than 80% were reported for a number of mouse strains after vitrification and storage in liquid nitrogen (LN_2). While the reagents can be individually obtained and the procedure performed in the laboratory, a one-step freezing method for cryopreservation of murine eggs has been most efficient and economical.[139] We have obtained a 50% or greater survival rate of 2- to 8-cell eggs; however, our results with morula stage eggs were significantly lower.

Efforts on technologies including vitrification, focus on eliminating inter- and intra-cellular ice crystal formation as well as a reduction in the time and expense associated with conventional slow-freezing protocols.[141] Initial attempts to cryopreserve mouse embryos by vitrification were not successful,[142] but paved the way for the production of the first live mice derived from vitrified embryos.[143] Mice have since been derived from embryos vitrified at the pronuclear, 2-cell and blastocyst stages of development.[144–146] Recent comparisons illustrate that post-thaw survival and *in vitro* hatching rates of cryopreserved blastocysts were comparable when using vitrification and conventional slow-freezing procedures.[146] Several factors can influence the efficacy of vitrification, including the pre-vitrification equilibration conditions (duration and temperature)[145] and the specific embryo handling techniques (using solid surface vitrification or straw systems).[144,147]

4. Future of Embryo Biotechnologies

From studies conducted over the last century, it is amazing to see how mouse embryo technologies have continued to evolve. Recent breakthrough studies demonstrating that ES cells can be differentiated in culture to produce oocyte-like morphology and developmentally-competent germ cells brings us perhaps perilously closer to the ***Brave New World*** described by Huxley in 1932.[148–151] Recent discoveries, coupled with the identification of specific genes that allow putatively normal development of parthenogenetically-derived mice,[57] provide dramatic evidence of the magnitude of current reproductive biotechnologies and the near term biological, biomedical and agricultural applications that await us.

Whole animal and somatic-cell techniques will continue to evolve in development of efficient and effective embryo biotechnologies. While beyond the scope of this chapter, various diagnostic techniques employing blastomere and single-cell analyses, take advantage of techniques from PCR to hybridization analyses that are amenable to high throughput genomic analyses as described for microarray and proteomic technologies.[152] Undeniably, the use of high throughput methods provides an economy-of-scale and subsequent rapid completion of multiple analyses — facilitating characterization of genetic and genomic analyses as well as animal health status. Lastly, but of consequence in any studies related to mouse embryology, from a humane standpoint, the welfare of animals used in such efforts is of paramount importance, as is efficient development of new technologies and representative data collection and analysis. In the end, it is clear that various techniques applied appropriately will be of important societal impact. The challenges that we face relate to how and in what light the technologies are perceived, coupled by the added benefits to the animals used and experimental outcomes.

Acknowledgements

We gratefully acknowledge our mentors and colleagues whose invaluable guidance and assistance through the years were very much appreciated. We also thank R.L. Howell and C. Adams in the review of this manuscript. Recent efforts have been funded by NCRR, NIDCR, The University of Rochester Medical Center, and Functional Genomics, Inc.

References

1. Schenk, S.L. 1880. Das Saugethierei kunstlick befruchtet ausserhalb des mutterthieres. *Mitt. Embr. Inst. K. K. Univ. Wien* **1**:107.
2. Heape, W. 1891. Preliminary note on the transplantation and growth of mammalian ova within a uterine foster mother. *Proc. Royal Soc. London* **48**:457–458.
3. Hammond, J. 1949. Recovery and culture of tubal mouse ova. *Nature* **163**:28–29.
4. Brinster, R.L. and R.D. Palmiter. 1986. Introduction of genes into the germ line of animals. *The Harvey Lectures, Series* **80**:1–38.
5. Pinkert, C.A. 2002. Introduction to transgenic animals. In: *Transgenic Animal Technology: A Laboratory Handbook*. ed. C.A. Pinkert. pp. 3–12. Academic Press, San Diego.
6. Pinkert, C.A. 1997. The history and theory of transgenic animals. *Lab. Anim.* **26**:29–34.
7. Overbeek, P.A. 2002. Factors affecting transgenic animal production. In: *Transgenic Animal Technology: A Laboratory Handbook.* ed. C.A. Pinkert. pp. 71–112. Academic Press, San Diego.
8. Polites, H.G. and C.A. Pinkert. 2002. DNA microinjection and transgenic animal production. In: *Transgenic Animal Technology: A Laboratory Handbook.* ed. C.A. Pinkert. pp. 15–70. Academic Press, San Diego.
9. Nagy A., M. Gertsenstein K. Vintersten and R. Behringer. 2003. *Manipulating the Mouse Embryo: A Laboratory Manual.* 3rd ed., Cold Spring Harbor Laboratory Press, Cold Spring Harbor, New York.
10. Rafferty, K.A. 1970. *Methods in Experimental Embryology of the Mouse.* Johns Hopkins Press, Baltimore.
11. Fowler, R.E. and R.G. Edwards. 1957. Induction of superovulation and pregnancy in mature mice by gonadotrophins. *J. Endocrinol.* **15**:374–384.

12. Iwamatsu, T. and M.C. Chang. 1971. Factors involved in the fertilization of mouse eggs in vitro. *J. Reprod. Fertil.* **26**:197–208.
13. Fraser, L.R. 1979. Rate of fertilization in vitro and subsequent nuclear development as a function of the post-ovulatory age of the mouse egg. *J. Reprod. Fertil.* **55**:153–160.
14. Hashlamoun, L.A. and G.J. Killian. 1985. Effects of timing of ovum recovery, cumulus cells, sperm preincubation time, and pH on in vitro fertilization in C57BL/6 mice. *Arch. Androl.* **15**:159–171.
15. Hillier, S.G., A.K.S. Siddiquey and R.M.L. Winston. 1985. Fertilization in vitro of cumulus-enclosed mouse oocytes: Effect of timing of the ovulatory HCG injection. *Intl. J. Fertil.* **30**:34–38.
16. Edgar, D.H., K.M. Whalley and J.A. Mills. 1987. Preimplantation development following in vitro fertilization of mouse oocytes: effects of timing of superovulation and preincubation in vitro. *J. In Vitro Fertil. Embryo Transf.* **4**:111–115.
17. Vergara, G.J., M.H. Irwin, R.J. Moffatt and C.A. Pinkert. 1997. *In vitro* fertilization in mice: strain differences in response to superovulation protocols and effect of cumulus cell removal. *Theriogenology* **47**:1245–1252.
18. Niwa, K., M. Araki and A. Iritani. 1980. Fertilization in vitro of eggs and first cleavage of embryos in different strains of mice. *Biol. Reprod.* **22**:1155–1159.
19. Edgar, D.H., K.M. Whalley and J.A. Mills. 1987. Effects of high-dose and multiple-dose gonadotropin stimulation on mouse oocyte quality as assessed by preimplantation development following in vitro fertilization. *J. In Vitro Fertil. Embryo Transf.* **4**:273–276.
20. Johnson, L.W., R.J. Moffatt, F.F. Bartol and C.A. Pinkert. 1996. Optimization of embryo transfer protocols for mice. *Theriogenology* **46**:1267–76.
21. Pinkert, C.A. 1990. A rapid procedure to evaluate foreign DNA transfer into mammals. *BioTechniques* **9**:38–39.
22. Whitten, W.K. 1956. Culture of tubal ova. *Nature* **177**:96.
23. McLaren, A. and D. Michie. 1956. Studies in the transfer of fertilized mouse eggs to uterine foster mothers. I. Factors affecting the implantation and survival of native transferred eggs. *J. Exp. Biol.* **33**:394–416.
24. Biggers, J.D. and A. McLaren. 1958. "Test tube" animals. The culture and transfer of early mammalian embryos. *Discovery (London)* **19**:423–426.
25. McLaren, A. and J.D. Biggers. 1958. Successful development and birth of mice cultivated in vitro as early embryos. *Nature* **182**:877–878.
26. Brinster, R.L. 1965. Studies on the development of mouse embryos in vitro. II. The effect of energy sources. *J. Exp. Zool.* **158**:59–68.
27. Brinster, R.L. 1965. Studies on the development of mouse embryos in vitro. IV. Interaction of energy sources. *J. Reprod. Fertil.* **10**:227–240.
28. Brinster, R.L. 1971. In vitro culture of embryos. In: *Pathways to Conception*. ed. A.I. Sherman. pp. 245–277. Charles C. Thomas, Springfield IL.
29. Lawitts, J.A. and J.D. Biggers. 1991. Optimization of mouse embryo culture media using simplex methods. *J. Reprod. Fertil.* **91**:543–556.
30. Gao, S., M. McGarry, K.E. Latham and I. Wilmut. 2003. Cloning of mice by nuclear transfer. *Cloning & Stem Cells* **5**:287–294.

31. Khosla, S., W. Dean, D. Brown, W. Reik and R. Feil. 2001. Culture of preimplantation mouse embryos affects fetal development and the expression of imprinted genes. *Biol. Reprod.* **64**:918–926.
32. Bleil, J. 1993. In vitro fertilization. In: *Guide to Techniques in Mouse Development. Methods in Enzymology.* ed. P.M. Wassarman and M.L. DePamphilis. **225**:253–263. Academic Press, New York.
33. Fraser, L.R. and L. M. Drury. 1975. The relationship between sperm concentration and fertilization in vitro of mouse eggs. *Biol. Reprod.* **13**:513–518.
34. Thadani, V.M. 1982. Mice produced from eggs fertilized at very low sperm:egg ratio. *J. Exp. Zool.* **219**:277–283.
35. Bleil, J.D. and P.M. Wassarman. 1980. Mammalian sperm-egg interaction: identification of a glycoprotein in mouse egg zonae pellucidae possessing receptor activity for sperm. *Cell* **20**:873–882.
36. Behr, B.R., C.J. Stratton, W.D. Foote, V. Knutzen and G. Sher. 1990. In vitro fertilization (IVF) of mouse ova in HEPES-buffered culture media. *J. In Vitro Fertil. Embryo Transf.* **7**:9–15.
37. Bagis, H., L. Keskentepe, H. Sagirkaya, H. Odaman and S. Cetin. 2001. Influences of cumulus cells during in vitro fertilization of mouse oocytes in different mouse strains. *Turk. J. Vet. Anim. Sci.* **25**:777–782.
38. Pavlok, A. and A. McLaren. 1972. The role of cumulus cells and the zona pellucida in fertilization of mouse eggs in vitro. *J. Reprod. Fertil.* **29**:91–97.
39. Stanger, J.D. and P. Quinn. 1982. Fertilization of cumulus-free, zona-intact mouse ova in vitro at high and low sperm concentrations. *Gamete Res.* **5**:61–70.
40. Wolf, D.P. and M. Inoue. 1976. Sperm concentration dependency in the fertilization and zonae sperm binding properties of mouse eggs inseminated in vitro. *J. Exp. Zool.* **196**:27–38.
41. Gordon, J.W. and B.E. Talansky. 1986. Assisted fertilization by zona drilling: a mouse model for correction of oligospermia. *J. Exp. Zool.* **239**:347–354.
42. Talansky, B.E. and J.W. Gordon. 1988. Cleavage characteristics of mouse embryos inseminated and cultured after zona pellucida drilling. *Gamete Res.* **21**:277–287.
43. Cohen, J., H. Malter, G. Wright, H. Kort, J. Massey and D. Mitchell. 1989. Partial zona dissection of human oocytes when failure of zona pellucida penetration is anticipated. *Hum. Reprod.* **4**:435–442.
44. Depypere, H.T., K.J. McLaughlin, R.F. Seamark, G.M. Warnes and C.D. Matthews. 1988. Comparison of zona cutting and drilling as techniques for assisted fertilization in the mouse. *J. Reprod. Fertil.* **84**:205–211.
45. Tarkowski, A.K. 1961. Mouse chimeras developed from fused eggs. *Nature* **190**:857–860.
46. Mintz, B. 1965. Genetic mosaicism in adult mice of quadraparental lineage. *Science* **148**:1232–1235.
47. Mintz, B., Gearhart, J. D. and A.O. Guymont. 1973. Phytohemagglutin-mediated blastomere aggregations and development of allophenic mice. *Dev. Biol.* **31**:195–199.
48. Markert, C.L. and R.M. Petters. 1978. Manufactured hexaparental mice show that adults are derived from three embryonic cells. *Science* **202**:56–58.
49. Petters, R.M. and C.L. Markert. 1980. Production and reproductive performance of hexaparental mice. *J. Hered.* **71**:70–74.

50. McLaren, A. 1975. Sex chimerism and germ cell distribution in a series of chimeric mice. *J. Embryol. Exp. Morph.* **33**:205–216.
51. Markert, C.L. 1984. Genetic control of cell interactions in chimeras. *Dev. Genet.* **4**:267–279.
52. Mullen, R.J. and W.K. Whitten. 1971. Relationship of genotype and degree of chimerism in coat color to sex ratio and gametogenesis in chimeric mice. *J. Exp. Zool.* **178**:165–176.
53. Milet, R.G., B.B. Mukherjee and W.K. Whitten. 1972. Cellular distribution and possible mechanism of sex differentiation in XX/XY chimaeric mice. *Can. J. Genet. Cytol.* **14**:933–941.
54. Bradbury, M.W. 1987. Testes of XX in equilibrium with XY chimeric mice develop from fetal ovotestes. *Dev. Genet.* **8**:207–218.
55. Stevens, L.C. 1978. Totipotent cells of parthenogenetic origin in a chimaeric mouse. *Nature* **276**:266–267.
56. Stewart, T.A. and B. Mintz. 1982. Recurrent germ-line transmission of the teratocarcinoma genome from the METT-1 culture line to progeny in vivo. *J. Exp. Zool.* **224**:465–469.
57. Kono, T., Y. Obata, Q. Wu, K. Niwa, Y. Ono, Y. Yamamoto, E.S. Park, J.-S. Seo and H. Ogawa. 2004. Birth of parthenogenetic mice that can develop to adulthood. *Nature* **428**:860–864.
58. Stewart, C.L. 1993. Production of chimeras between embryonic stem cells and embryos. In: *Guide to Techniques in Mouse Development. Methods in Enzymology*. ed. P.M. Wassarman and M.L. DePamphilis. **25**:823–842. Academic Press, New York.
59. Brinster, R.L. 1974. The effect of cells transferred into the mouse blastocyst on subsequent development. *J. Exp. Med.* **104**:1049–1056.
60. Mintz, B. and K. Illmensee. 1975. Normal genetically mosaic mice produced from malignant teratocarcinoma cells. *Proc. Natl. Acad. Sci. USA* **72**:3585–3589.
61. Illmensee, K. and B. Mintz. 1976. Totipotency and normal differentiation of single teratocarcinoma cells cloned by injection into blastocysts. *Proc. Natl. Acad. Sci. USA* **73**:549–553.
62. Bradley, A., M. Evans, M.H. Kaufman and E. Robertson. 1984. Formation of germ-line chimeras from embryo derived teratocarcinoma cell lines. *Nature* **309**:255–256.
63. Ramirez-Sous, R., A.C. Davis and A. Bradley. 1993. Gene targeting in embryonic stem cells. In: *Guide to Techniques in Mouse Development. Methods in Enzymology*. ed. P.M. Wassarman and M.L. DePamphilis. **25**:855–878. Academic Press, New York.
64. Brinster, R.L. 1993. Stem cells and transgenic mice in the study of development. *Intl. J. Dev. Biol.* **37**:89–99.
65. Watanabe, J., M.J. Dewey and B. Mintz. 1978. Teratocarcinoma cells as vehicles for introducing specific mutant mitochondrial genes into mice. *Proc. Natl. Acad. Sci. USA* **75**:5113–5117.
66. Pinkert, C.A. and I.A. Trounce. 2002. Production of transmitochondrial mice. *Methods* **26**:348–357.
67. Thompson, S., A.R. Clarke, A.M. Pow, M.L. Hooper and D.W. Melton. 1989. Germ line transmission and expression of a corrected HPRT gene produced by gene targeting in embryonic stem cells. *Cell* **56**:313–321.
68. Zijlstra, M., F. Sajjadi, S. Subramani and R. Jaensich. 1989. Germ-line transmission of a disrupted β2-microglobulin gene produced by homologous recombination in embryonic stem cells. *Nature* **342**:435–438.

69. Polzin, V.J., D.L. Anderson, G.B. Anderson, R.H. BonDurant, J.E. Butler, R.L. Pashen, M.C.T. Penedo and J.D. Rowe. 1987. Production of sheep-goat chimeras by inner cell mass transplantation. *J. Anim. Sci.* **65**:325–330.
70. Anderson, G.B., N.A. Ruffing, R.H. BonDurant, R.H. and R.L. Pashen. 1991. Preliminary observations on reproduction in a female sheep-goat chimera. *Vet. Rec.* **129**:467–469.
71. Nagashima, H., C. Giannakis, R.J. Ashman.and M.B. Nottle. 2004. Sex differentiation and germ cell production in chimeric pigs produced by inner cell mass injection into blastocysts. *Biol. Reprod.* **70**:702–707.
72. Kimura, Y. and R. Yanagimachi. 1995. Mouse oocytes injected with testicular spermatozoa or round spermatids can develop into normal offspring. *Development* **121**:2397–2405.
73. Kimura, Y., H. Tateno, M.A. Handel and R. Yanagimachi. 1998. Factors affecting meiotic developmental competence of primary spermatocyte nuclei injected into mouse oocytes. *Biol. Reprod.* **59**:871–877.
74. Markert, C.L. 1983. Fertilization of mammalian eggs by sperm injection. *J. Exp. Zool.* **228**:195–201.
75. Kuretake, S., Y. Kimura, K. Hoshi and R. Yanagimachi. 1996. Fertilization and development of mouse oocytes injected with isolated sperm heads. *Biol. Reprod.* **55**:798–795.
76. Kishikawa, H., H. Tateno and R. Yanagimachi. 1999. Chromosome analysis of BALB/c mouse spermatozoa with normal and abnormal head morphology. *Biol. Reprod.* **61**:809–812.
77. Burrel, V.R., R. Yanagimachi and W.K. Whitten. 1996. Normal mice develop from oocytes injected with spermatozoa with grossly misshapen heads. *Biol. Reprod.* **55**:709–714.
78. Szczygiel, M.A., H. Kusakabe, R. Yanagimachi, R. and D.G. Whittingham. 2002. Intracytoplasmic sperm injection is more efficient than in vitro fertilization for generating mouse embryos from cryopreserved semen. *Biol. Reprod.* **67**:1278–1284.
79. Rijnders, P.M., E.Y. Jaarsma, H.C. Van Os, M. Tombrock-Scheffer and C.A.M. Jansen. 1990. Pregnancies from embryos without evidence of fertilization. *Hum. Reprod.* **5S**:39.
80. Perry, A.C.F., T. Wakayama, H. Kishikawa, T. Kasai, M. Okabe, Y. Toyoda and R. Yanagimachi. 1999. Mammalian transgenesis by intracytoplasmic sperm injection. *Science* **284**:1180–1183.
81. Chan, A.W., C.M. Luetjens, T. Dominko, J. Ramacho-Santos, C.R. Simerly, L. Hewitson and G. Schatten. 2000. Foreign DNA transmission by ICSI: injection of spermatozoa bound with exogenous DNA results in embryonic GFP expression and live rhesus monkey births. *Mol. Hum. Reprod.* **6**:20–33.
82. Kim, N.-H. and H. Shim. 2000. Intracytoplasmic sperm injection in pigs. *ETS Newsletter* **8**:10–13.
83. Shim, SW, Y.H. Kim, S.H. Jun, J.M. Lim, H.M. Chung, J.J. Ko, H.T. Lee and H. Shim. 2000. Transgenesis of porcine embryos using intracytoplasmic sperm injection. *Theriogenology* **53**:521 (abstr.).
84. Bittner, J. and C.C. Little. 1937. The transmission of breast and lung cancer in mice. *J. Heredity* **28**:117–121.
85. McLaren, A and D. Michie. 1959. Studies on the transfer of fertilized mouse eggs to uterine foster mothers. 2. The effect of transferring large numbers of eggs. *J. Exp. Biol.* **36**:40–50.
86. McLaren, A. and Michie, D. 1959. Superpregnancy in the mouse 1. Implantation and fetal mortality after induced superovulation in females of various ages. *J. Exp. Biol.* **36**:281–300.

87. Canseco, R.S., A.E.T. Sparks, R.L. Page, C.G. Russell, J.L. Johnson, W.H. Velander, R.E. Pearson, W.N. Drohan and F.C. Gwazdauskas. 1994. Gene transfer efficiency during gestation and the influence of co-transfer of non-manipulated embryos on production of transgenic mice. *Transgenic Res.* **3**:20–25.

88. Reetz, I.C., M. Wullenweber-Schmidt, V. Kraft and H.J. Hedrich. 1988. Rederivation of inbred strains of mice by means of embryo transfer. *Lab. Anim. Sci.* **38**:696–701.

89. Tsunoda, Y. and Y. Kato. 2002. Production of transgenic nonhuman primates. In: *Transgenic Animal Technology: A Laboratory Handbook.* ed. C.A. Pinkert. pp. 195–231. Academic Press, San Diego.

90. Paterson, L., W. Ritchie and I. Wilmut. 2002. Nuclear transfer technology in cattle, sheep, and swine. In: *Transgenic Animal Technology: A Laboratory Handbook.* ed. C.A. Pinkert. pp. 395–416. Academic Press, San Diego.

91. Godke, R.A., M. Sansinena and C.R. Youngs. 2002. Assisted reproductive technologies and embryo culture methods for farm animals. In: *Transgenic Animal Technology: A Laboratory Handbook.* ed. C.A. Pinkert. pp. 513–568. Academic Press, San Diego.

92. Wakayama, T., I. Rodriguez, A.C. Perry, M. Zuccotti, K.R. Johnson and R. Yanagimachi. 1998. Full-term development of mice from enucleated oocytes injected with cumulus cell nuclei. *Nature* **394**:369–374.

93. Wakayama, T. and R. Yanagimachi. 1999. Cloning of male mice from adult tail-tip cells. *Nat. Genet.* **22**:127–128.

94. Ogura, A., K. Inoue, N. Ogonuki, A. Noguchi, K. Takano, R. Nagano, O. Suzuki, J. Lee, F. Ishino and J. Matsuda. 2000. Production of male cloned mice from fresh, cultured, and cryopreserved immature Sertoli cells. *Biol. Reprod.* **62**:1579–1594.

95. Ono, Y., N. Shimozawa, K. Muguruma, S. Kimoto, K. Hioki, M. Tachibana, Y. Shinkai, M. Ito and T. Kono. 2001. Production of cloned mice from embryonic stem cells arrested at metaphase. *Reproduction* **122**:731–736.

96. Gao, S., M. McCarry, T. Ferrier, B. Pallante, B. Gasparrini, J. Fletcher, L. Harkness, P. DeSousa, J. McWhir and I. Wilmut. 2003. Effect of cell confluence on production of cloned mice using an inbred embryonic stem cell line. *Biol. Reprod.* **68**:595–603.

97. Eggan, K., H. Akutsu, J. Loring, L. Jackson-Grusby, M. Klemm, W.M. Rideout III, R. Yanagimachi and R. Jaenisch. 2001. Hybrid vigor, fetal overgrowth, and viability of mice derived from nuclear cloning and tetraploid embryo complementation. *Proc. Natl. Acad. Sci. USA* **98**:6209–6214.

98. Chung, Y.G., M.R. Mann, M.S. Bartolomei and K.E. Latham. 2002. Nuclear-cytoplasmic "tug of war" during cloning: effects of somatic cell nuclei on culture medium preferences of preimplantation cloned mouse embryos. *Biol. Reprod.* **66**:1178–1184.

99. Heindryckx, B., A. Rybouchkin, J. Van der Elst and M. Dhont. 2001. Effect of culture media on in vitro development of cloned mouse embryos. *Cloning* **3**:41–50.

100. Young, L.E., K.D. Sinclair and I. Wilmut. 1998. Large offspring syndrome in cattle and sheep. *Rev. Reprod.* **3**:155–163.

101. Takeda, K., S. Akagi, S. Takahashi, A. Onishi, H. Hanada and C.A. Pinkert. 2001. Mitochondrial activity in response to serum starvation in bovine and murine cells. *Conf. on Current Status and Perspectives in Cloning and Related Studies*, Tsukuba, Japan, p 71 (abstr.).

102. Humpherys, D., Eggan, K., Akutsu, H., Friedman, A., Hochedlinger, K., Yanagimachi, R., Lander, E.S., Golub, T.R. and R. Jaenisch. 2002. Abnormal gene expression in cloned mice

derived from embryonic stem cell and cumulus cell nuclei. *Proc. Natl. Acad. Sci. USA* **99**:12889–12894.

103. Inoue, K., Kohda, T., Lee, J., Ogonuki, N., Mochida, K., Noguchi, Y., Tanemura, K., Kaneko-Ishino, T., Ishino, F. and A. Ogura. 2002. Faithful expression of imprinted genes in cloned mice. *Science* **295**:297.
104. Lin, T.P. 1966. Microinjection of mouse eggs. *Science* **151**:333–337.
105. Martin, M.J., D.A. Dunn and C.A. Pinkert. 2004. Transgenic animals: Secreted products. In: *Encyclopedia of Animal Science*. ed. W.G. Pond and A.W. Bell. Marcel Dekker, Inc., New York (in press).
106. Kim, T. 2002. Retrovirus-mediated gene transfer. In: *Transgenic Animal Technology: A Laboratory Handbook*. ed. C.A. Pinkert. Academic Press, San Diego, pp. 173–193.
107. Chan, A.W.S., K.Y. Chong and G. Schatten. 2002. Production of transgenic nonhuman primates. In: *Transgenic Animal Technology: A Laboratory Handbook*. ed. C.A. Pinkert. pp. 359–394. Academic Press, San Diego.
108. Capecchi, M.R. 1989. Altering the genome by homologous recombination. *Science* **244**:1288–1292.
109. Joyner, A. L. 2000. *Gene Targeting: A Practical Approach*. IRL Press, New York.
110. Doetschman, T. 2002. Gene targeting in embryonic stem cells: I. History and methodology. In: *Transgenic Animal Technology: A Laboratory Handbook*. ed. C.A. Pinkert. pp. 113–141. Academic Press, San Diego.
111. Rucker, E.B. III, J.G. Thompson and J.A. Piedrahita. 2002. Gene targeting in embryonic stem cells: II. Conditional technologies. In: *Transgenic Animal Technology: A Laboratory Handbook*. ed. C.A. Pinkert. pp. 143–171. Academic Press, San Diego.
112. Matsui, Y. K. Zsebo and B.L. Hogan. 1992. Derivation of pluripotential embryonic stem cells from murine primordial germ cells in culture. *Cell* **70**:841–847.
113. Stewart, C.L., I. Gadi and H. Bhatt. 1994. Stem cells from primordial germ cells can reenter the germ line. *Dev. Biol.* **161**:626–628.
114. Keefer, C.L., S.L. Stice and D.L. Matthews 1994. Bovine inner cell mass cells as donor nuclei in the production of nuclear transfer embryos and calves. *Biol Reprod.* **50**:935–939.
115. Collas, P. and F. Barnes. 1994. Nuclear transplantation by microinjection of inner cell mass and granulosa cell nuclei. Mol. *Reprod. Dev.* **38**:264–267.
116. Cheong, H.T., Y. Takahashi and H. Kanagawa. 1993. Birth of mice after transplantation of early cell-cycle-stage embryonic nuclei into enucleated oocytes. *Biol. Reprod.* **48**:958–963.
117. Tsunoda, Y. and Y. Kato. 1993. Nuclear transplantation of embryonic stem cells in mice. *J. Reprod. Fertil.* **98**:537–540.
118. Brinster, R.L., E.P. Sandgren, R.R. Behringer and R.D. Palmiter. 1989. No simple solution for making transgenic mice. *Cell* **59**:239–241.
119. Lavitrano, M., A. Camaioni, V.M. Fazio, A. Dolci, M.G. Farace and C. Spadafora. 1989. Sperm cells as vectors for introducing foreign DNA into eggs: Genetic transformation of mice. *Cell* **57**:717–723.
120. Lavitrano, M., M.L. Bacci, M. Forni, D. Lazzereschi, C. Di Stefano, D. Fioretti, P. Giancotti, G. Marfe, L. Pucci, L. Renzi, H. Wang, A. Stoppacciaro, G. Stassi, M. Sargiacomo, P. Sinibaldi, V. Turchi, R. Giovannoni, G. Della Casa, E. Seren and G. Rossi. 2002. Efficient production by sperm-mediated gene transfer of human decay accelerating factor (hDAF) transgenic pigs for xenotransplantation. *Proc. Natl. Acad. Sci. USA* **98**:13090–13095.

121. Brinster, R.L. and M.R. Avarbock. 1994. Germline transmission of donor haplotype following spermatogonial transplantation. *Proc. Natl. Acad. Sci. USA* **91**:11303–11307.
122. Nagano, M., C.J. Brinster, K.E. Orwig, B.-Y. Rye, M.R. Avarbock and R.L. Brinster. 2001. Transgenic mice produced by retroviral transduction of male germ-line stem cells. *Proc. Natl. Acad. Sci. USA* **98**:13090–13095.
123. Wallace, D.C. 2001. Mouse models for mitochondrial disease. *Am. J. Med. Genet.* **106**:71–93.
124. McKenzie, M., I.A. Trounce, C.A. Cassar and C.A. Pinkert. 2004. Production of homoplasmic xenomitochondrial mice. *Proc. Natl. Acad. Sci. USA* **101**:1685–1690.
125. Trounce, I.A., M. McKenzie, C.A. Cassar, C.A. Ingraham, C.A. Lerner, D.A. Dunn, C.L. Donegan, K. Takeda, W.K. Pogozelski, R.L. Howell and C.A. Pinkert. 2004. Development and initial characterization of xenomitochondrial mice. *J. Bioenerg. Biomembr.* (in press).
126. Evans, M.J., C. Gurer, J.D. Loike, I. Wilmut, A.E. Schnieke and E.A. Schon. 1999. Mitochondrial DNA genotypes in nuclear transfer-derived cloned sheep. *Nat. Genet.* **23**:90–93.
127. Hiendleder, S., S.M. Schmutz, G. Erhardt, R.D. Green and Y. Plante. 1999. Transmitochondrial differences and varying levels of heteroplasmy in nuclear transfer cloned cattle. *Mol. Reprod. Devel.* **54**:24–31.
128. Hiendleder, S., V. Zakhartchenko, H. Wenigerkind, H.D. Reichenbach, K. Bruggerhoff, K. Prelle, G. Brem, M. Stojkovic, and E. Wolf. 2003. Heteroplasmy in bovine fetuses produced by intra- and inter-subspecific somatic cell nuclear transfer: neutral segregation of nuclear donor mitochondrial DNA in various tissues and evidence for recipient cow mitochondria in fetal blood. *Biol. Reprod.* **68**:159–166.
129. Takeda, K., S. Takahashi, A. Onishi, Y. Goto, A. Miyazawa and H. Imai. 1999. Dominant distribution of mitochondrial DNA from recipient oocytes in bovine embryos and offspring after nuclear transfer. *J. Reprod. Fertil.* **116**:253–259.
130. Steinborn, R., P. Schinogl, V. Zakhartchenko, R. Achmann, W. Schernthaner, M. Stojkovic, E. Wolf, M. Muller and G. Brem. 2000. Mitochondrial DNA heteroplasmy in cloned cattle produced by fetal and adult cell cloning. *Nat. Genet.* **25**:255–257.
131. Cohen, J., R. Scott, T. Schimmel, J. Levron, and S. Willadsen. 1997. Birth of an infant after transfer of anucleate donor oocyte cytoplasm into recipient eggs. *Lancet* **350**:186–187.
132. Brenner, C.A., J.A. Barritt, S. Willadsen and J. Cohen. 2000. Mitochondria DNA heteroplasmy after human ooplasmic transplantation. *Fertil. Steril.* **74**:573–578.
133. Perez, G.I., A.M. Trbovich, R.G. Gosden and J.L. Tilly. 2000. Mitochondria and the death of oocytes. *Nature* **403**:500–501.
134. Barritt, J.A., C.A. Brenner, H. Malter and J. Cohen. 2001. Mitochondria in human offspring derived from ooplasmic transplantation. *Hum. Reprod.* **16**:513–516.
135. Hawes, S.M., C. Sapienza and K.E. Latham. 2002. Ooplasmic donation in humans: the potential for epigenic modifications. *Hum. Reprod.* **17**:850–852.
136. St. John, J. 2003. The potential risks of abnormal transmission of mtDNA through assisted reproductive technologies. *Reprod. BioMed. Online* **8**:34–44.
137. Whittingham, D.G., S.P. Leibo and P. Mazur. 1972. Survival of mouse embryos frozen to -195°C and -269°C. *Science* **178**:411–414.
138. Wilmut, I. 1972. The effect of cooling rate, warming rate, cryoprotective agent and stage of development on survival of mouse embryos during freezing and thawing. *Life Sci.* **11**:1071–1079.

139. Leibo, S.P., and K. Oda. 1993. High survival of mouse zygotes and embryos cooled rapidly or slowly in ethylene glycol plus polyvinylpyrrolidone. *Cryoletters* **14**:133–144.
140. Songsasen, N. and S.P. Leibo. 1998. Live mice from cryopreserved embryos derived in vitro with cryopreserved ejaculated spermatozoa. *Lab. Anim. Sci.* **48**:275–281.
141. Liebermann, J., F. Nawroth, V. Isachenko, E. Isachenko, G. Rahimi and M.J. Tucker. 2002. Potential importance of vitrification in reproductive medicine. *Biol. Reprod.* **67**:1671–1680.
142. Rall, W.F. and G.M. Fahy. 1985. Ice-free cryopreservation of mouse embryos at –196 C by vitrification. *Nature* **313**:573–575.
143. Ali, J. and N. Shelton. 1993. Vitrification of preimplantation stages of mouse embryos. *J. Reprod. Fertil.* **98**:459–465.
144. Bagis, H., H. Odaman, H. Sagirkaya and A. Dinnyes. 2002. Production of transgenic mice from vitrified pronuclear-stage embryos. *Mol. Reprod. Dev.* **61**:173–179.
145. Otsuka, J., A. Takahashi, M. Nagaoka and H. Funahashi. 2002. Optimal equilibration conditions for practical vitrification of two-cell mouse embryos. *Comp. Med.* **52**:342–346.
146. Tharnprisarn, W., S. Suwajanakorn, W. Sereepapong, K. Pruksananonda, W. Boonyakasemsanti, P. Virutamasen, V. Ahnonkitpanit, D. Chompurat and P. Numchaisrika. 2003. Mouse blastocyst vitrification compared with the conventional slow-freezing method. *J. Med. Assoc. Thai.* **86**:666–671.
147. Zhu, L., C. Lou, M.Z. Huang, H. Li and F.Q. Xing. 2003. Vitrification of mouse blastocysts using two kinds of straw systems. *Di Yi Jun Yi Da Xue Xue Bao* **23**:992–995.
148. Huxley, A. 1932. *Brave New World*. Chatto and Windus, London.
149. Hübner, K., G. Fuhrmann, L.K. Christenson, J. Kehler, R. Reinbold, R. De La Fuente, J. Wood, J. F. Strauss III, M. Boiani and H.R. Schöler. 2003. Derivation of oocytes from mouse embryonic stem cells. *Science* **300**:1251–1256.
150. Toyooka, Y., N. Tsunekawa, R. Akasu and T. Noce. 2003. Embryonic stem cells can form germ cells in vitro. *Proc. Natl. Acad. Sci. USA* **100**:11457–11462.
151. Geijsen, N., M. Horoschak, K. Kim, J. Gribnau, K. Eggan and G.Q. Daley. 2004. Derivation of embryonic germ cells and male gametes from embryonic stem cells. *Nature* **427**:148–154.
152. Pinkert, C.A. 2003. Transgenic animal technology: Alternatives in genotyping and phenotyping. *Comp. Med.* **53**:126–139.

CHAPTER 14

TRANSGENICS

J. D. Murray[1,2,3] and E. A. Maga[1,4]

Departments of Animal Science[1] and Population Health and Reproduction[2]
University of California, Davis, CA, USA
[3]jdmurray@ucdavis.edu
[4]eamaga@ucdavis.edu

1. Introduction

Since the initial demonstration in 1982 of profound phenotypic effects stemming from the expression of a single transgene, genetic engineering techniques have revolutionized fundamental biological and biomedical research. While a variety of methods for producing transgenic mice have been developed, little progress has been realized for increasing the efficiency of producing transgenic animals or in developing an understanding of the mechanism of DNA integration. Even so, work in the mouse has led the way both in terms of technique development and applications. Since the generation of the first transgenic mice over-expressing a gene product, techniques have been developed to allow the disruption (knockouts), mutation, or replacement of a gene. In addition, RNAi technology will allow the targeted down-regulation of a specific gene. In the era following the sequencing of the human, mouse, and other genomes, transgenic mice will remain at the forefront of elucidating the function and developmental regulation of gene expression.

The scientific breakthroughs that enabled the genetic engineering of animals slowly developed over the past century, beginning with the first attempts to culture and transfer embryos in the late 1800s.[1] While recent progress has been extremely rapid, it is still difficult to believe that following the first published report of a microinjection method using mouse zygotes,[2] 15 years passed before the first transgenic mice were produced.[3] This delay was caused by the need to be able to manipulate DNA, and with the advent of recombinant DNA technology in the late 1970s, transgenic animal production became a reality. Today, transgenic mice, including both over-expression models and knockouts, are routinely used to help dissect the genetic control of development throughout the mammalian life cycle.

The first technological shift toward transgenic mouse production occurred in 1977, when Gurdon[4] transferred mRNA and DNA into *Xenopus* embryos and observed that the transferred nucleic acids could function in an appropriate manner. Then, in 1980, Brinster *et al.*[5] reported similar studies in the mouse, where they demonstrated that an appropriate translation product was produced following transfer of a globin mRNA into mouse embryos. Sequentially, these studies laid the groundwork for the development of the first gain-of-function transgenic mouse models.

From late 1980 through 1981, five research groups reported successful genetic engineering in mice.[6–10] To describe animals carrying new genes (integrating foreign or novel DNA into their genome), Gordon and Ruddle[6] coined the term ''transgenic.'' This definition has since been extended to include animals that result from the molecular manipulation of endogenous genomic DNA, including all techniques ranging from DNA microinjection to embryonic stem (ES) cell transfer and ''knockout'' mouse production.

Palmiter, Brinster, and their colleagues published a series of papers in the early 1980s on the dramatic phenotypic effects in mice that resulted from the over-expression of growth hormone.[11,12] This research was the first demonstration of the phenotypic effects resulting from the expression of a transgene and demonstrated the developmental plasticity of the mouse in response to the over-expression of a single gene. These studies ushered in a new era in research, i.e., the targeted over-expression of single genes in order to study the resulting biochemical and physiological processes. In addition, these studies demonstrated the viability of using the mouse to study genetic control and structural elements such as promoters, enhancers, and introns. In the 24 years that have passed since the publication of the generation of the first transgenic animals, transgenic mice have been used extensively for a variety of purposes. These include the identification of important gene sequences, as models for biomedical research, as models for populations following the integration of a "new" gene, and as predictors of transgene function for applied purposes such as the production of transgenic livestock. This wide variety of applications has been made possible, as a number of alternative methods to the original method of pronuclear microinjection have since been developed. While, in general, the overall efficiency of these methods is low (1 to 3% of manipulated eggs), we have now progressed to being able to carry out a wide range of different genetic manipulations (Table 1).

Table 1. Genetic manipulations in the mouse
Adding a new gene
Adding additional copies of an existing gene
Deleting a gene (knockouts)
Mutating an existing gene
Replacing the coding region of an existing gene
Down-regulating expression of a gene (siRNA)

The development of transgenic technology, concomitantly with the sequencing of the human and mouse genomes, places the mouse as the premier research animal, as we seek to understand how the genetic program regulates vertebrate development. An in-depth review of transgenic work in the mouse is not possible as over 32,000 papers citing transgenic mouse work, based on thousands of transgenic mouse lines, were in the literature at the time of writing. Nor will we cover in-depth methods for producing transgenic mice, as there are excellent laboratory manuals currently available.[13,14] Therefore, this review will briefly cover some of the methods for producing transgenic mice, including our perceptions of their strengths and limitations, with examples of the types of manipulations that are possible.

2. Genetic Engineering Methods

Through the early 1980s, the production of transgenic mice by microinjection of DNA into the pronucleus of zygotes was the most productive and widely used technique. Using transgenic technology in the mouse, it is possible to add a new gene to the genome, increase the level of expression or change the tissue specificity of expression of a gene, or decrease the level of synthesis of a specific protein through the use of short interfering RNA (siRNA) encoding transgenes.[15] Removal or alteration of an existing gene also can be effected via homologous recombination,[16] using either ES cells[17,18] or somatic cell nuclear transfer-based cloning procedures.[19] A wide variety of methods have been developed that allow genetic engineering in the mouse (Table 2). Some methods will be briefly reviewed, while descriptions of these and other methods may also be found in this volume (see Ch. 13).

Table 2. Principle gene transfer methods
Pronuclear microinjection
Viral vectors
Sperm-mediated
Embryonic stem (ES) cell transfer
Nuclear transfer-based cloning
Spermatogonial cell-mediated transfer

2.1. *Pronuclear Microinjection*

Pronuclear microinjection was the first method developed for the genetic engineering of mammals and remains a useful technique, even though no significant advances in efficiency have been achieved yet in the mouse. Pronuclear microinjection is principally used to produce over-expression models, since targeting a specific insertion site has not been achieved with this method. Transgenes insert at random locations in the genome and usually integrate as tandem arrays,[20] which can pose problems with expression if integration occurs in a transcriptionally inactive region of the genome. For research models these negative aspects are mitigated by the ability to produce a number of different transgenic lines, each with different copy numbers and insertion sites. Thus, only those lines yielding appropriate expression patterns or levels are then used for study. Furthermore, different lines with different levels of transgene expression may provide useful information. In addition to generating multiple lines, the probability of obtaining lines with appropriate expression levels and patterns can be further increased by the use of insulator elements.[21] For instance, matrix attachment regions (MARs), which act as gene domain boundaries and bind to the nuclear matrix, can be positioned at the 5′ and 3′ ends of a transgene to result in position-independent expression of the transgene.[22]

Pronuclear microinjection can be used to insert gene constructs or isolated genomic DNA fragments, up to several hundred kb in size, into the mouse genome. While most transgene constructs fall in the range of a few kb to 25 or 30 kb in size, bacterial artificial chromosome (BAC) inserts, ranging from 200 to 300 kb, have been successfully microinjected.[23] The microinjection of BAC and yeast artificial chromosome inserts allows the expression of unknown genes, thus facilitating their characterization.

Over-expressing transgenic mouse lines produced by pronuclear microinjection are useful for studying the effects of altering the level or pattern of expression of a coding region to determine the genetic, biochemical, or physiological consequences. The over-expression model is also appropriately used to study the role of enhancer or promoter elements. Constructs consisting of a complete promoter of interest, or an enhancer element coupled to a minimal promoter which is joined to a reporter gene such as Green Fluorescent Protein (GFP), can be used to determine tissue specificity or developmental time of expression controlled by the promoter or enhancer under study. In both of these cases, the over-expression model has the benefit of not intentionally disrupting expression of an endogenous gene, particularly if care is taken in the choice of lines being studied.

Transgenic mice are also often used to model transgene effects in other animals. Much of the preliminary work relevant to commercially important livestock species such as pigs and cattle was conducted first in mice. For instance, pharmaceutical production in the milk using the mammary gland as a bioreactor[24] and work in the field of xenotransplantation[25] were both first carried out in the mouse before moving to the larger animals. The rapid reproduction rate and lower cost of mice make them an attractive model for the large animal. In most instances, results obtained in the mouse models have been predictive of what occurs in the transgenic large animal.

2.2. *Viral Vectors*

Shortly after the initial reports of the production of transgenic mice by pronuclear microinjection, Jahner *et al.*[26] reported the production of transgenic mice using retroviral vectors. The use of viral vectors had the advantage of using the natural infectivity of the virus to ensure reasonably efficient production of transgenic animals after infection of early embryos. Furthermore, using retroviral vectors resulted in integration of single copies of the construct, although the site of integration was still random. However, retroviral vectors were limited in the size of insert DNA that could be packaged into the viral capsid and suffered from high recombination rates, and obtaining reliable, high level expression of the transgene was difficult. Problems with expression were caused both by the need to retain certain elements of the retrovirus within the construct, which interfered with the transgene promoter, and by the tendency for retroviral inserts to be inactivated (methylated) in the embryo. Current retroviral vectors have been designed to be self-inactivating and, through selective deletions of *cis* retroviral sequences, problems with promoter interference have been largely overcome.[27] However, packaging limitations on the size of the transgene insert still exist. With the relative certainty of obtaining trangenic mice by pronuclear microinjection, and the subsequent development of ES cells as routes for the production of transgenic mice, retroviral vectors are not widely developed for general use in the production of transgenic mice for research models.

Recently, lentivirus vectors have been proposed as high efficiency vectors for use in the production of over-expression transgenic mouse models.[28] Lentiviral vectors can be used to infect ES cells, primordial germ cells, or germ cells *in vivo*. At this time, the general applicability of such vectors for use in the routine production of transgenic mice is unknown, but the high degree of efficiency[29] indicates they may find widespread use.

2.3. *Spermatogonial Cell and Sperm-Mediated Gene Transfer*

In contrast to progress in embryo manipulation, a completely different approach was taken with the advent of sperm-mediated gene transfer procedures. In 1989, sperm-mediated gene transfer was reported but hotly disputed when many laboratories around the world were unable to duplicate the procedures. However, it is now clear that this system does work in a variety of species, including the mouse.[30] An alternative potential approach to sperm-mediated gene transfer is the development of spermatogonial cell transplantation procedures for gene transfer experimentation.[31] While both of these routes for the production of transgenic mice work, they are unlikely to replace pronuclear microinjection and embryonic stem cell methods.

2.4. *Embryonic Stem Cell-Mediated Gene Transfer*

The development of ES cell technologies emanated from efforts of the early cell biologists. Teratocarcinoma cell transfer and cell aggregation studies in the 1970s evolved from the earlier characterization and studies of these cells in mice.[32–34] This led to work with the 129 mouse strain and pluripotent teratocarcinoma cells and, ultimately, to the basis for work with ES cells in 1981.[17,18] Embryonic stem cells, when returned to the environment of the blastocyst, resume a correct developmental pathway, leading to the production of germline-competent chimeric mice.[35] The efficiency of producing chimeras was enhanced by the development of co-culture techniques, where blastocyst injection was not the only route for ES cell transfer. With co-culture, host embryos could be cultured on a lawn of ES cells, with the ES cells preferentially incorporating into the embryo proper.

To the extent that the ES cells contribute to the germline, these cells provide a direct route for the gene insertion. For gene transfer, the importance of ES cells is the ability to genetically engineer cells in culture, followed by selection and characterization, prior to the production of chimeric mice. When gametes derived from the genetically engineered ES cells participate in fertilization, a transgenic animal is obtained. The isolation of ES cells allowed gene targeting via homologous recombination, and the resulting production of chimeric knockout mice ushered in a new era of genetically engineered loss-of-function mutants.[33,36,37]

3. Homologous Recombination or "Knockouts"

Gene transfer has been used to produce both random and targeted insertion of discrete DNA fragments into the mouse genome. For targeted insertions, where the integration of foreign genes is based on a recombination-based gene insertion event with specific homology to cellular sequences (termed homologous recombination), the efficiency at which DNA microinjection is effective is extremely low.[38] In contrast, the use of ES cell transfer into mouse embryos has been quite effective in allowing an investigator to pre-select a specific genetic modification, via homologous recombination, at a precise chromosomal position.[36] This pre-selection has led to the fairly routine production of mice (a) incorporating a novel foreign gene at a predetermined site in their genome, (b) carrying a modified endogenous gene, or (c) lacking a specific endogenous gene following gene deletion or ''knockout'' procedures.[33,37] As of this writing, there were over 3,400 reviews in the literature involving data from gene knockout experiments in mice, a fact which clearly demonstrates the integral role of this technology in modern biology.

The great power of homologous recombination-generation of loss-of-functions mutations through the production of a knockout mouse is lost, however, if the heterozygous or homozygous knockout is lethal. Furthermore, there is an increasing desire to obtain a tissue specific knockout. As our knowledge of the multiplicity of functions of some genes has grown, it has become clear that often a knockout at a specific time during development is required. The Cre-loxP and FLP-FRT site-specific recombination systems have been developed for use in murine ES cells for the production of tissue specific gene targeting.[39,40] The Cre-loxP system is the most commonly used system and allows for the directed deletion or insertion of material into a targeted area marked by loxP sites, depending on whether the loxP sites are in a *cis* or *trans* relationship relative to each other. For a detailed discussion of the Cre-loxP system the reader is referred to Rucker *et al.*[41]

4. Strain Considerations

In mice, differences in reproductive productivity, behavior, related husbandry requirements, and responses to various experimental procedures that affect overall production efficiency are well documented. However, strain differences can have significant influences on modifying gene expression; e.g., gene expression and tumor formation in lines of transgenic mice harboring human oncogenes (or with tumor suppressor genes knocked out) vary when these mice are backcrossed to different inbred or outbred strains.[42–45] The same consideration applies with some over-expression transgenic models in mice. For example, Eisen *et al.*[46,47] demonstrated with lines expressing three different growth hormone transgene constructs, that the growth phenotype depended, in part, on the genetic background. Furthermore, the effect of selection for increased 8-week body weight differed between lines carrying the oMt1a-oGH transgene depending on genetic background.[48,49]

5. Integration of a Transgene into a Breeding Population

The work outlined above clearly illustrates that in addition to transgenic mice, transgenic animals potentially useful in production agriculture can be produced. However, the production and characterization of a useful transgenic line are only the first step towards introducing the transgene into a production population. Following integration of a transgene into nucleus breeding herds, the nucleus herds for each breed into which the transgene is integrated will need to undergo selection to optimize the performance for the production traits of interest since it cannot be assumed that the transgene will yield the same phenotype when placed on different genetic backgrounds. This factor will be particularly important for transgenes affecting quantitative traits such as growth, body composition, or reproduction and for those transgenes which modify intermediary metabolism. A number of theoretical papers are available,[50,51] but research in this area using transgenic mouse lines as a model has only just begun. The introgression of a transgene into a nucleus herd has a number of associated costs, e.g., the cost of breeding and the cost associated with the loss of gain if selection had not been interrupted. Gama *et al.*[50] suggest the best strategy for introgressing a transgene into a nucleus swine herd would involve three generations of backcrossing prior to initiating selection of a herd and characterization of the phenotype. Thus, the net economic merit of the transgene would need to be sufficiently high to compensate for these costs. Recent research with growth hormone transgenic mice has shown that the heritabilities of various body composition traits are altered,[52] that the phenotypic response to transgene expression may be dependent on the genetic background of the line,[47] and, thus, that the response to selection is also background-dependent.[48,49]

6. Conclusions

The mouse was the first transgenic animal produced over 20 years ago and remains today a critical tool for basic, medical and applied research. The production of transgenic mice has progressed from random gene addition to making very specific endogenous DNA modifications. Transgenic mice can be used to predict the function of unknown genes, identify important gene regulatory elements, study physiological and phenotypic effects of gene over-expression or knockout, as models for medical research of disease and immune function, and as applied models for other species such as commercially important livestock. The availability of the mouse genome sequence,[53] and other genomic resources, such as the efforts to catalogue all of the cDNAs for the mouse,[54] opens up the potential for understanding the genetic control of development, but only through gaining detailed knowledge of what function each gene plays across developmental time and the complete range of tissues that makes a mammal. Furthermore, we need to understand the allelic interactions that occur in diverse genetic backgrounds. We are well placed to carry forward this research due to our ability to genetically engineer mice to over-express, reduce expression, or eliminate a gene product completely, either systemically or tissue specifically. The mouse thus has a central role in our quest to understand our biology.

References

1. Heape, W. 1891. Preliminary note on the transplantation and growth of mammalian ova within a uterine foster mother. *Proc. R. Soc. London* **48**:457–458.
2. Lin, T.P. 1966. Microinjection of mouse eggs. *Science* **151**:333–337.
3. Gordon, J.W., G.A. Scangos, D.J. Plotkin, J.A.Barbosa and F.H. Ruddle. 1980. Genetic transformation of mouse embryos by microinjection of purified DNA. *Proc. Natl. Acad. Sci. USA* **77**:7380–7384.
4. Gurdon, J.B. 1977. Nuclear transplantation and gene injection in amphibia. *Brookhaven Symposia in Biology* **29**:106–115.
5. Brinster, R.L., H.Y. Chen, M.E. Trumbauer, and M.R. Avarbock. 1980. Translation of globin messenger RNA by the mouse ovum. *Nature* **283**:499–501.
6. Gordon, J.W. and F.H. Ruddle. 1981. Integration and stable germ line transmission of genes injected into mouse pronuclei. *Science* **214**:1244–1246.
7. Harbers, K., D. Jahner and R. Jaenisch. 1981. Microinjection of cloned retroviral genomes into mouse zygotes; integration and expression in the animal. *Nature* **293**:540–542.
8. Costantini, F. and E. Lacy. 1981. Introduction of a rabbit β-globin gene into the mouse germ line. *Nature* **294**:92–94.
9. Wagner, E.F., T.A. Stewart and B. Mintz. 1981. The human ß-globin gene and a functional viral thymidine kinase gene in developing mice. *Proc. Natl. Acad. Sci. USA* **78**:5016–5020.
10. Wagner, T.E., P.C. Hoppe, J.D. Jollick, D.R. Scholl, R.L. Hodinka and J.B. Gault. 1981. Microinjection of a rabbit ß-globin gene into zygotes and its subsequent expression in adult mice and their offspring. *Proc. Natl. Acad. Sci. USA* **78**:6376–6380.
11. Palmiter, R.D., R.L. Brinster, R.E. Hammer, M.E. Trumbauer, M.G. Rosenfeld, N.C. Birnberg, and R.M. Evans. 1982. Dramatic growth of mice that develop from eggs microinjected with metallothionein-growth hormone fusion genes. *Nature* **300**:611–615.
12. Palmiter, R.D., G. Norstedt, R.E. Gelinas, R.E. Hammer and R.L. Brinster. 1983. Metallothionein-human GH fusion genes stimulate growth of mice. *Science* **222**:809–814.
13. Hogan, B., R. Beddington, F. Costantini and E. Lacy. 1994. *Manipulating the Mouse Embryo: A Laboratory Manual.* Cold Spring Harbor Press, Cold Spring Harbor, NY.

14. Pinkert, C.A. 2002. *Transgenic Animal Technology: A Laboratory Handbook.* Second Edition. Academic Press, San Diego, CA.
15. Shingawa, T. and S. Ishii. 2003. Generation of *Ski*-knockdown mice by expressing a long double-strand RNA from an RNA polymerase II promoter. *Genes & Development* **17**:1340–1345.
16. Cappechi, M.R. 1989. Altering the genome by homologous recombination. *Science* **244**:1288–1292.
17. Evans, M.J. and M.H. Kaufman. 1981. Establishment in culture of pluripotential cells from mouse embryos. *Nature* **292**:154–156.
18. Martin, G.R. 1981. Isolation of a pluripotent cell line from early mouse embryos cultured in medium conditioned by teratocarcinoma stem cells. *Proc. Natl. Acad. Sci. USA* **78**:7634–7638.
19. Wakayama, T., A.C. Perry, M. Zuccotti, K.R. Johnson and R. Yanagimachi. 1998. Full-term development of mice from enucleated oocytes injected with cumulus cell nuclei. *Nature* **394**:369–374.
20. Brinster, R.L., H.Y. Chen, M.E. Trumbauer, M.K. Yagle and R.D. Palmiter. 1985. Factors affecting the efficiency of introducing foreign DNA into mice by microinjecting eggs. *Proc. Natl. Acad. Sci. USA* **82**:4438–4442
21. Li, Q., K.R. Peterson, X. Fang and G. Stamatoyannopoulos. 2002. Locus control regions. *Blood* **100**:3077–3086.
22. Whitelaw, C.B.A., S. Grolli, P. Accornero, G. Donofrio, E. Farini, and J. Webster. 2000. Matrix attachment region regulates basal β-lactoglobulin transgene expression. *Gene* **244**:73–80.
23. Smith, D.J., Y. Zhu, J. Zhang, J.F. Cheng and E.M. Rubin. 1995. Construction of a panel of transgenic mice containing a contiguous 2-Mb set of YAC/P1 clones from human chromosome 21q22.2. *Genomics* **10:**425–34.
24. Gorden, K., E. Lee, J.A. Vitale, A.E. Smith, H. Westphal and L. Hennighausen. 1987. production of human tissue plasminogen activator in transgenic mouse milk. *Bio/Technology* **5**:1183–1187.
25. Tearle R.G., M.J. Tange, Z.L. Zannettino, M. Katerelos, T.A. Shinkel, B.J. Van Denderen, A.J. Lonie, I. Lyons, M.B. Nottle, T. Cox, C. Becker, A.M. Peura, P.L. Wigley, R.J. Crawford, A.J. Robins, M.J. Pearse and A.J. d'Apice. 1996. The alpha-1,3-galactosyltransferase knockout mouse. Implications for xenotransplantation. *Transplantation* **61**:13–9.
26. Jahner, D., K. Haase, R. Mulligan and R. Jaenisch. 1985. Insertion of the bacterial gpt gene into the germ line of mice by retroviral infection. *Proc. Natl. Acad. Sci. USA*. **82**:6927–6931
27. Kim, T. 2002. Retrovirus-mediated gene transfer. In: *Transgenic Animal Technology: A Laboratory Handbook.* Second Edition. ed. C.A. Pinkert. pp. 173–193. Academic Press, San Diego, CA.
28. Lois, C., E.J. Hong, S. Pease, E.J. Brown and D. Baltimore. 2002. Germline transmission and tissue-specific expression of transgenes delivered by lentiviral vectors. *Science* **295**:5556, 868–872.
29. Punzon I., L.M. Criado, A. Serrano, F. Serrano and A. Bernad. 2004. Highly efficient lentiviral-mediated human cytokine transgenesis on the NOD/scid background. *Blood* **103**:580–582.
30. Maione, B., M. Lavitrano, C. Spadafora, A.A. Kiessling. 1998. Sperm-mediated gene transfer in mice *Mol. Reprod. Dev.* **50**:406–409.
31. Brinster R.L. and J.W. Zimmermann. 1994. Spermatogenesis following male germ-cell transplantation. *Proc. Natl. Acad. Sci. USA*. **91**:11298–11302.
32. Pierce, G.B. 1975. Teratocarcinoma and perspectives. In: *Teratomas and Differentiation*. ed. M.I. Sherman, and D. Solter. pp. 3–12. Academic Press, New York.

33. Brinster, R.L. 1993. Stem cells and transgenic mice in the study of development. *Int. J. Dev. Biol.* **37**:89–99.
34. Pinkert, C.A. 1997 The history and theory of transgenic animals. *Lab. Anim.* **26**:29–34.
35. Nagy, A., J. Rossant, R. Nagy, W. Abramow-Newerly, and J.C. Roder. 1993. Derivation of completely cell culture-derived mice from early-passage embryonic stem cells. *Proc. Natl. Acad. Sci. USA* **90**:8424–8428.
36. Thomas, K.R. and M.R. Capecchi. 1987. Site-directed mutagenesis by gene-targeting in mouse embryo-derived stem cells. *Cell* **51**:503–512.
37. Capecchi M.R. 1989. Altering the genome by homologous recombination. *Science* **244**:1288–1292.
38. Brinster, R.L., E.P. Sandgren, R.R. Behringer, and R.D. Palmiter. 1989. No simple solution for making transgenic mice. *Cell* **59**:239–241.
39. Kos C.H. 2004. Cre/loxP system for generating tissue-specific knockout mouse models. *Nutr. Rev.* **62**:243–246.
40. Vooijs, M., M. van der Valk, H. te Riele and A. Berns. 1998. Flp-mediated tissue-specific inactivation of the retinoblastoma tumor suppressor gene in the mouse. *Oncogene* **17**:1–12.
41. Rucker III, E.B, J.G. Thomson and J.A. Piedrahita. 2002. Gene targeting in embryonic stem cells: II. Conditional Technologies. In: *Transgenic Animal Technology: A Laboratory Handbook.* Second Edition. ed. C.A. Pinkert. pp. 143–171 Academic Press, San Diego, CA.
42. Harris, A.W., C.A. Pinkert, M. Crawford, W.Y. Langdon, R.L. Brinster and J.M. Adams. 1988. The Eμ-*myc* transgenic mouse: a model for high-incidence spontaneous lymphoma and leukemia of early B cells. *J. Exp. Med.* **167**:353–371.
43. Chisari, F.V., K. Klopchin, T. Moriyama, C. Pasquinelli, J.A. Dunsford, S. Sell, C.A. Pinkert, R.L. Brinster and R.D. Palmiter. 1989. Molecular pathogenesis of hepatocellular carcinoma in hepatitis B virus transgenic mice. *Cell* **59**:1145–1156.
44. Cho, H.J., M. Seiberg, I. Georgoff, A.K. Teresky, J.R. Marks and A.J. Levine. 1989. Impact of the genetic background of transgenic mice upon the formation and timing of choroid plexus papillomas. *J. Neurosci. Res.* **24**:115–122.
45. Donehower, L.A., M. Harvey, H. Vogel, M.J. McArthur, C.A. Montgomery, Jr., S.H. Park, T. Thompson, R.J. Ford and A. Bradley. 1995 Effects of genetic background on tumorigenesis in p53-deficient mice. *Molec. Carcinogen.* **14**:16–22.
46. Eisen E.J., M. Fortman, W.Y. Chen and J.J. Kopchick. 1993. Effect of genetic background on growth of mice hemizygous for wild-type or dwarf mutated bovine growth hormone transgenes. *Theor. Appl. Genet.* **87**:161–169.
47. Eisen, E.J., J.D. Murray and T.J. Schmitt. 1995. An ovine-growth-hormone transgene model suitable for selection experiments for growth in mice. *J. Anim. Breed. Genet.* **112**:401–413.
48. Siewerdt, F., E.J. Eisen, J.S. Conrad-Brink and J.D. Murray. 1998. Gene action of the oMt1a-oGH transgene in two lines of mice with distinct selection backgrounds. *J. Anim. Breed. Genet.* **115**:211–226.
49. Siewerdt, F., E.J. Eisen and J.D. Murray. 2000. Correlated changes in fertility and fitness traits in lines of oMt1a-oGH transgenic mice selected for increased 8-week body weight. *J. Anim. Breed. Genet.* **117**:83–95.
50. Gama, L.T., C. Smith and J.P. Gibson. 1992. Transgene effects, introgression strategies and testing schemes in pigs. *Anim. Prod.* **54**:427–440.

51. Hoeschele I. 1990. Potential gain from insertion of major genes into dairy cattle. *J. Dairy Sci.***73**:2601–2618.
52. Clutter A.C., D. Pomp and J.D. Murray. 1996. Quantitative genetics of transgenic mice: components of phenotypic variation in body weights and weight gain. *Genetics* **143**:1753–1760.
53. Mouse Genome Sequencing Consortium. 2002. Initial sequencing and comparative analysis of the mouse genome. *Nature* **420**, 520–562.
54. Anonymous. 2004. The Status, Quality, and Expansion of the NIH Full-Length cDNA Project: The Mammalian Gene Collection (MGC). *Genome Res.* **14**:2121–2127.

CHAPTER 15

THE MOUSE IN BIOMEDICAL RESEARCH

R. B. Roberts[1] and D. W. Threadgill[2]

Department of Genetics and Carolina Center for Genome Sciences
University of North Carolina, Chapel Hill, NC, USA
[1]*reade_roberts@med.unc.edu*
[2]*dwt@med.unc.edu*

1. Introduction

Since the beginning of genetics research using the mouse as an experimental model, science has been driven by a desire to understand basic mammalian biology. A major goal has been to investigate the causes of, and ultimately develop therapies for, human disease. The first several decades of experimental mouse genetics revolved around understanding tumor biology, yet the results of early pioneering studies impacted fields far beyond cancer biology, influencing virtually all areas of biomedical research. Since its humble beginnings, mice have become the model of choice to investigate every aspect of normal and abnormal human pathology, from hair loss to psychiatric disorders. Indeed, over a dozen Nobel Prizes in Physiology or Medicine have been awarded for discoveries directly utilizing the mouse (Table 1), a fact that serves as a testament to the essential role that mice serve in biomedical research. Mouse-based research is destined to support many more such prizes, especially since research incorporating mouse models has expanded exponentially in the last two decades with the advent of genetic engineering technologies, and more recently, with the sequencing and annotation of the mouse and human genomes.

The increasing appreciation that most human diseases are the result of highly complex interactions between genetics and environmental influences requires a similarly complex experimental analysis. Furthermore, most pathologies, including heart disease, cancer, and degenerative neurological disorders, result from interactions between different cell types or organ systems and cannot be accurately investigated using *in vitro* studies. Consequently, the mouse, being biologically homologous to humans while allowing a vast array of experimental manipulations at the whole organism level, has become widely recognized as the preeminent experimental model supporting biomedical research into complex disease processes. This chapter provides representative examples illustrating the impact of mouse models, past, present and future, on biomedical research and the breakthroughs that these models have provided in understanding and treating human disease.

Table 1. Nobel Prizes in Physiology or Medicine awarded for discoveries utilizing the mouse

Year	Scientists	Contribution
1928	Nicolle	Pathogenesis of typhus
1939	Domagk	Antibacterial effects of prontosil
1943	Dam, Doisy	Discovery of function of vitamin K
1945	Fleming, Chain, Florey	Curative effect of penicillin in bacterial infections
1951	Theiler	Development of yellow fever vaccine
1954	Enders, Weller, Robbins	Culture of poliovirus that led to development of vaccine
1975	Baltimore, Dulbecco, Temin	Interaction between tumor viruses and genetic material
1980	Benacerraf, Dausset, Snell	Identification of histocompatibility antigens and mechanism of action
1984	Milstein, Kohler, Jerne	Techniques of monoclonal antibody formation
1986	Levi-Montalcini, Cohen	Nerve growth factor and epidermal growth factor
1987	Tonegawa	Basic principles of antibody synthesis
1996	Doherty, Zinkernagel	Immune-system detection of virus-infected cells
1997	Prusiner	Discovery of prions
2000	Carlsson, Greengard, Kandel	Signal transduction in the nervous system

Table adapted from[1].

2. Tissue Transplantation

At the start of the twentieth century, E. E. Tyzzer likely made the first genetic examination of a disease process with his transplantation of tumors from a line of genetically homogeneous Japanese fancy mice.[2] Tumors survived transplantation only between members of the Japanese line, but not to a genetically heterogeneous house mouse stock. While F_1 progeny of the Japanese and house mouse stocks supported tumor survival, F_2 progeny were resistant to tumor grafts. Although the resulting conclusion that survival of tumor grafts was not inherited as a simple Mendelian trait may not have had a major impact on cancer biology at the time, it did serve as the impetus for seminal developments in the mouse genetics field. Based upon Tyzzer's initial experimental observations, C. C. Little recognized the need for genetically homogeneous mouse lines and developed the first inbred mouse strain, DBA. Through subsequent tumor transplantation studies with the new inbred strain, Little and Tyzzer demonstrated that the ability to support tumor survival was inherited as a highly polygenic trait.[3]

Within the next two decades, the inheritance of susceptibility to tumor transplantation would be confirmed, and the underlying immunological mechanism elucidated by P. Gorer. He was the first to derive antiserum from the mouse, and using such antisera was able to show cellular antigen differences between inbred strains. Gorer found that the single gene inheritance for one of these antigens co-segregated with one of the tumor graft resistance genes.[4] He had discovered the major histocompatibility complex (*H2* or *Mhc*), though it was later work by G. Snell that conclusively demonstrated that *H2* had the most significant (major) effect on the tumor graft phenotype and that the locus contained multiple genes (complex). Snell produced the first congenic strains by selectively backcrossing *H2* loci onto various inbred backgrounds, producing the tools for the first fine genetic structure analysis in vertebrates.[5] Utilizing these congenic strains in conjunction with a visible marker mutation and an elegant breeding strategy, Snell genetically dissected the *H2* locus, demonstrating that most mouse strains carry distinct *H2* alleles.[6] From these interbreeding experiments Snell also suggested that the *H2* locus was actually two closely linked loci, an interpretation that was formally demonstrated two decades later.[7] Importantly, Snell's findings with *H2* and tumor transplants held true for studies using the transplantation of normal tissue grafts[8] and thus provided the foundation to prevent the rejection of tissue and organ transplants in humans. Snell's mouse studies led to the discovery of the human major histocompatibility locus (*HLA*) and were seminal to the basic understanding of adaptive immunity. For his many contributions, Snell received the Nobel Prize in 1980.

3. Immunology

Similar to the pivotal role played during investigation into the genetics of tissue transplantation, mouse models were essential for elucidating the biological basis of the immune response. A central dogma, held during most of the twentieth century, hypothesized that each protein is encoded by one gene, producing a long-standing dilemma in immunology. When applied to the production of antibodies, the one gene-one protein model could not explain the tremendous gene diversity required for response to each different antigen; it was known that the immune system produces many more antibodies than there are potential genes.[9] A second dilemma resulted from observations suggesting that reactivity of antibodies is responsive to the structure of their cognate antigens. This puzzle created a situation by which biological information appeared to flow in the opposite direction. Both of these dilemmas were resolved by S. Tonegawa who won the Nobel Prize in 1987 for discovering the genetic principle for the generation of antibody diversity.

Tonegawa noted that immunoglobulin genes are far apart in mouse embryonic cells but are adjacent in B-lymphocytes,[10] suggesting that genes are rearranged in the genome during differentiation of the antibody-producing B-lymphocytes. In subsequent experiments, he showed how the genome is rearranged and even that part of the intervening DNA is lost.[11] This research gave rise to the model of VDJ recombination that leads to antibody diversity,[12] with a similar mechanism having since been shown for the T-cell receptor genes.[13,14] Tonegawa's seminal discovery using mice explained how a limited number of genes could dramatically increase the complexity of the immune response, which had major implications in fields ranging from the response to infectious organisms and vaccinations to providing insights into autoimmunity. Furthermore, Tonegawa showed that each B-lymphocyte only produces its own unique antibody through allelic exclusion.[15] Only recently has the heritable nature of the monoclonality of B-lymphocyte-produced antibodies been conclusively demonstrated in whole animals using mice cloned from mature B-lymphocytes that produce only a single type of antibody.[16]

Concurrent with the discovery of how antibodies are generated, G. J. F. Kohler and C. Milstein developed a method of fusing B-lymphocytes with a myeloma cell line, an immortal lymphocyte that does not produce antibodies, to produce a hybridoma cell line that is both immortal and a producer of a specific antibody.[17] They further showed that mice could be immunized with an antigen to recover spleen-derived B-lymphocytes, expressing antibodies specific to the immunizing antigen, for fusion with myeloma cells.[18] For their development of hybridomas, Kohler and Milstein were awarded the Nobel Prize in 1984. Their technological development using mice permitted the unlimited production of monoclonal antibodies with pre-determined specificity, providing powerful biological reagents. This technology revolutionized diagnostics and permitted the development of immunotherapy for the treatment of cancer, as exemplified by Herceptin, a monoclonal antibody targeting the ERBB2 trans-membrane receptor that is frequently over-expressed or otherwise mis-regulated in breast cancer.[19]

Mice were also critical to the discovery of how the cellular immunity, particularly T-lymphocytes, distinguishes healthy cells from infected cells. Previous work by Snell and others described above showed that T-lymphocytes could recognize and kill cells from unrelated mice during tumor transplantation studies. Subsequent work by H. O. McDevitt linked immune responses to the *Mhc* by showing that the *Mhc* produced the proteins recognized by T-lymphocytes and that T-lymphocytes would only kill virus-infected cells from MHC-matched mice.[20] However, these results could not explain how T-lymphocytes distinguish infected cells from uninfected cells of the same mouse strain. This question was answered by P. Doherty and R. Zinkernagel who discovered killer T-cells and the mechanism by which they function only against infected cells of the same mouse strain from which the T-lymphocytes are derived but which cannot be transmitted to unrelated mice of different strains. For their discovery of the specificity of cell mediated immune defense, a discovery which has had major impacts on many types of diseases from autoimmune diseases like arthritis, multiple sclerosis and diabetes to infectious agents, Doherty and Zinkernagel were awarded the Nobel Prize in 1996.

Doherty and Zinkernagel used mice infected with lymphocytic choriomeningitis virus (LCMV) that causes meningitis to discover that the T-lymphocytes recognize not only the virus but also the MHC histocompatibility antigens.[21–23] To demonstrate this, they mixed brain fluid from infected mice, containing T-lymphocytes, with mouse cells of like or different strains that were separately infected with LCMV. This experiment showed that for T-lymphocytes to function, they must recognize both the viral antigen and proteins encoded by the *Mhc*, explaining why the response cannot be transferred to unrelated strains. They further determined that the pathological damage resulting from LCMV infection was due to T-lymphocytes responding to the virus and not to the virus directly.[24] This discovery has supported more recent research to develop vaccines against cancer and HIV-infected cells and methods to halt the aberrant responses causing autoimmune diseases.

4. Tumor Biology

Like the discovery of the immunological basis of tumor transplantation, research examining spontaneous neoplasia in mice had its beginnings early in the twentieth century. Indeed, many common inbred strains were selectively inbred for susceptibility or resistance to spontaneous formation of specific tumor types; a continually updated summary of these strain susceptibilities can be found at

The Jackson Laboratory's Mouse Tumor Biology Database (http://tumor.informatics.jax.org/straingrid.html).

In addition to his work on tumor transplantation, Little was involved in early spontaneous tumor biology studies, experiments that would again have repercussions in multiple areas of human biology. Under his leadership The Jackson Laboratory found that susceptibility to spontaneous mammary tumor formation in the C3H strain was maternally inherited.[25] Transfer of susceptibility was later traced to the dam's milk through fostering experiments,[26] leading to the discovery of the mouse mammary tumor virus (MMTV) and provirus inheritance in the genome.[27,28] Subsequent mouse experiments with similar viruses created the basis for our understanding of mammalian retroviral biology, and culminated in the 1975 Nobel Prize being awarded to D. Baltimore, R. Dulbecco and H. Temin. Importantly, MMTV insertion sites were used to discover the *Int* (MMTV integration site) genes (now called *Wnt* genes), normal host genes that act as oncogenes when transcriptionally activated by nearby proviral insertions.[29,30] Not only was it shown that misregulation of specific genes was responsible for tumor growth, the number of *Wnt* genes demonstrated that multiple genetic alterations could lead to the same tumor phenotype. Though a few *Wnt* genes were found to be in other pathways, most were eventually identified as ligands for the WNT pathway, a key pathway during embryonic development and for the growth of many tumor types.

Subsequently, a mutation in the *Apc* gene, the central regulator of the WNT pathway, arose in a chemical mutagenesis screen.[31] The resulting multiple intestinal neoplasia phenotype, caused by the Apc^{Min} allele, was similar to the human cancer syndrome familial adenomatous polyposis (FAP). Additionally, as *APC* is mutated in nearly all human colon cancers, the Apc^{Min} mouse serves as a valuable colorectal tumorigenesis model, and as genetic proof that loss of APC function is sufficient for neoplasia in the gut. Studies using the Apc^{Min} model also led to the identification of the Modifier of Min (*Mom1*) locus, containing the first identified cancer modifier gene, *Pla2g2a*, a phospholipase A_2 family gene.[32] While previous characterization of inbred strain tumor susceptibility showed the presence of such modifiers, the discovery of a specific polymorphism in *Pla2g2a* that was at least partially responsible for a major difference in tumor burden gave new credence to the idea that similar loci may be utilized as prognostic factors in humans, and whose gene products may serve as targets for therapeutic intervention. Although no association has been found between human cancer and *PLA2G2A* variants, the action of PLA2G2A in the inflammatory prostaglandin pathway corroborates the use of anti-inflammatory drugs to treat colon cancer.[33]

Compared to genetically-initiated tumor models, carcinogen-based models may more accurately model human cancers, the majority of which are sporadic, resulting from differential susceptibility to accumulated mutations over a lifetime, rather than inheritance as single gene traits. Use of carcinogens also has the advantage of immediate application to many inbred strains, including recombinant inbred or recombinant consomic strains. Such studies utilizing treatment of recombinant congenic strains with 1,2-dimethylhydrazine (DMH) or azoxymethane (AOM) have identified multiple Susceptibility to Colon Cancer loci (*Scc*).[34,35] The modifier gene responsible for the *Scc1* locus was identified as the tyrosine phosphatase *Ptprj*, and subsequent analysis showed frequent loss of *PTPRJ* in multiple human tumor types, which would have otherwise gone unnoted in the absence of the mouse studies.[36]

More recently, microarray technology has allowed large-scale expression analysis of both human and mouse tissue as a means to identify cancer modifiers. Of particular interest has been the identification of molecular expression profiles associated with specific tumor pathologies and therapeutic outcomes, and thus carrying diagnostic and prognostic value. One molecular profile study of human primary tumors revealed a seventeen-gene expression "signature" associated with the potential of the primary tumor to metastasize to distant tissue sites.[37] This finding was not readily explainable based upon current models of cancer progression, in part because acquisition of metastatic potential was considered a rare event, and thus a metastasis gene expression signature should not be found in the bulk of the primary tumor but only in those few cells within the larger tumor acquiring metastatic potential. Concurrent mouse studies, utilizing a transgenic metastatic tumor model on a number of inbred genetic backgrounds,[38,39] helped resolve the controversy by elucidating the nature of the metastasis signature. Expression analysis using tissues obtained from the mouse model revealed that the metastasis signature is probably not tumor-specific or the result of specific mutations, but rather is a characteristic of the individual's genetic background.[39] The mouse and human metastasis signatures matched for twelve out of the thirteen genes studied, confirming the ability of the transgenic model to recapitulate complex host-tumor interactions found in humans.[40]

5. Cardiovascular Biology

The discovery of cancer modifier genes and expression profiles through mouse genetic studies illustrates the importance of the mouse to understanding human genetic factors contributing to the progression and treatment of disease.

However, the relevance of mouse models is in no way confined to cancer studies, as a similarly extensive body of work has been accomplished in the study of clinically relevant cardiovascular pathologies, including cardiac hypertrophy, atherosclerosis, and hypertension. Since these physiological phenotypes are a result of complex interactions between multiple tissues and organs, a proper study of cardiac pathology requires an entire organism. Additionally, a hallmark of cardiac disease is compensatory processes that produce systemic changes, and whole organism studies permit the analysis of these secondary and tertiary effects. Aside from the obvious size difference, the anatomy, development, and maturation of the mouse heart are remarkably similar to that of the human heart; thus, the mouse can be used to model both developmental defects and adult heart disease.[41]

The rat has classically been the standard model organism for cardiac studies, due in part to the ease of surgical and other manipulation afforded by its relatively large size. However, the mouse is gaining ground in cardiac studies given the extensive genetic tools available, the ease of genetic manipulation, and recent technological advances permitting measurements and manipulation of the small mouse cardiovascular system. Presently, comprehensive examinations of questions in cardiovascular biology are relatively easy in the mouse through the utilization of interstrain variation, transgenics, surgical models, and pharmacological treatments to modulate the cardiovascular system. Analysis of these various forms of perturbation can in turn be accomplished both *ex vivo* (including culture of intact heart and isolated cell types) and *in vivo* (including magnetic resonance imaging, echocardiography (Figure 1), doppler, catheterization, telemetry, and measurements of heart rate, oxygen consumption, and blood pressure). Non-invasive *in vivo* techniques such as echocardiography provide the additional benefit of permitting multiple measurements in the same individual mouse over time or through different treatments, allowing more accurate analysis of progressive conditions.

Currently there are no fewer than fifty genetically engineered mouse models of cardiac hypertrophy and failure.[42] In some cases, transgenes mimicking potential therapeutic treatment have been crossed to these hypertrophy models, decreasing the extent of pathology and demonstrating efficacy.[43] Fortunately for modifier mapping studies, cardiac disease model phenotypes vary with genetic background,[44] as do most baseline cardiovascular phenotypes,[45] permitting the identification of loci that may contain therapeutically relevant target genes. In a recent survey of human and mouse atherosclerosis-related modifiers, roughly eighty percent of the human modifier loci overlapped with mapped loci in conserved syntenic regions of the mouse genome.[46] The extensive overlap

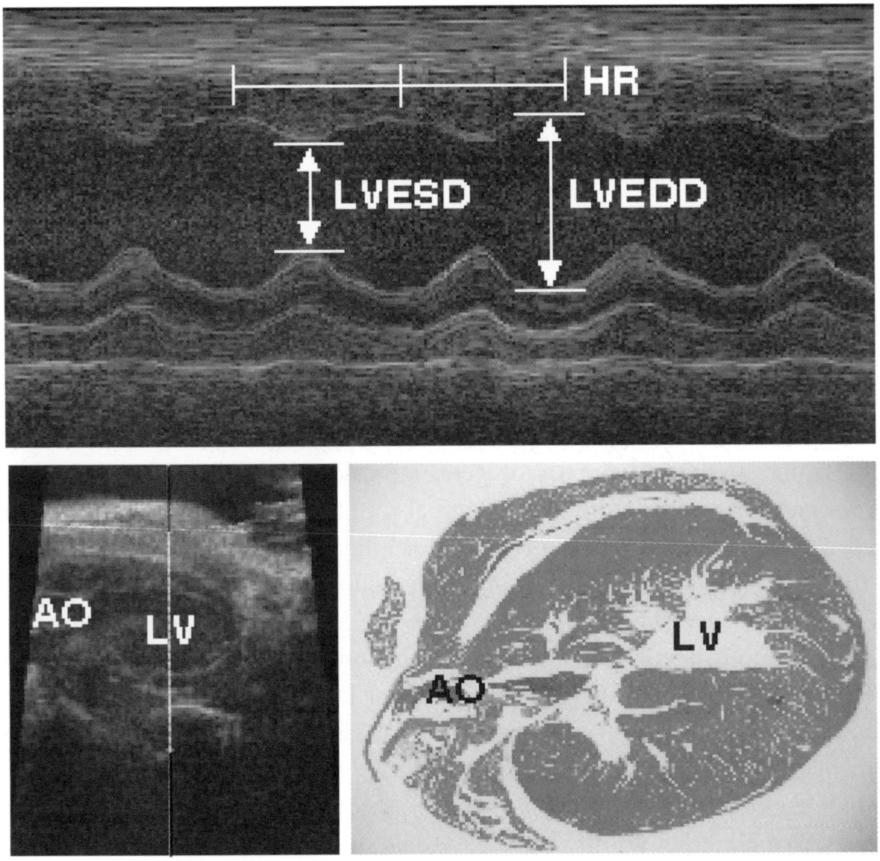

Figure 1. Comparison of mouse heart histology to echocardiography (ECG). Top, ECG showing representative measurements that can be obtained. HR, heart rate; LVESD, left ventricle end systolic diameter; LVEDD, left ventricle end diastolic diameter. Lower left and right panels, ultrasound and histology, respectively, of heart shown in top image. LV, left ventricle; AO, aorta. Images courtesy of D. Barrick, UNC.

suggests a set of modifiers that may represent conserved regulatory genes in cholesterol management and thus may serve as points of therapeutic intervention within the complex network of genes modulating cholesterol levels.

Atherosclerosis and other forms of heart disease have also been examined extensively with genetically engineered mice. The first engineered atherosclerosis model was the apolipoprotein E (*Apoe*) knockout mouse.[47,48] *Apoe* was chosen as a prime candidate due to its central role in clearance of low density lipoproteins from circulation. Indeed, *Apoe* null mice develop profound atherosclerotic lesions early in life, even when on a low-fat, low-cholesterol diet; and these lesions progress in a way that recapitulates human disease.[49,50] Without genetic manipulation, mice rarely develop aspects of atherosclerosis on a normal rodent diet. However, high-fat, high-cholesterol "atherogenic" diets have been shown to produce atherosclerosis in the mouse,[51] permitting cross-strain analysis. Diet-based atherosclerosis models may more accurately model human disease since genetically engineered models often carry profound physiological perturbations.

Engineered mice have also given insight to the biology of hypertension, with perhaps the most studied pathway being the renin-angiotensin system. Angiotensinogen (AGT) is a secreted circulating molecule and is processed by renin and angiotensin converting enzyme (ACE) to angiotensin II, a potent vasoconstrictor which can ultimately lead to cardiac hypertrophy if chronically dysregulated. The transgenic strategy of "gene titration," pioneered by O. Smithies, one of the developers of gene knockout technology in mice, reveals roles for genes by correlating physiological measurements to gene copy number, altered through engineered deletions and duplications to produce a range from one to four gene copies. Interestingly, increasing gene copy number of *Agt* correlated to an increase in blood pressure, while increasing *Ace* dosage had no effect on blood pressure.[52,53] The conclusion was that the mouse renin-angiotensin system is unable to compensate for increased AGT levels, but can do so in response to chronically increased ACE activity, even though acute treatment with pharmacological ACE inhibitors modulates blood pressure.[54] This seeming paradox demonstrates an important caveat to studies in genetically engineered mice, the ability of biological systems to equilibrate to normal phenotypes despite profound genetic manipulations that often mask the role of the manipulated gene.

Strategies to overcome masking by homeostatic adjustment seek to maintain a constant level of a normally variable metabolite, a technique known as "clamping". One traditional clamping method utilizes controlled infusion of a metabolite to prevent physiological compensation to normal levels, a technique commonly used in the study of insulin-related physiology.[55] *In vivo* studies of the role of rennin, encoded by the *Ren1* gene, have been hampered, in part, by the strong homeostatic compensation in the renin-angiotensin system. Recently, "genetic clamping" was achieved by ectopically expressing *Ren1* in the liver, an organ where this factor cannot be regulated.[56] Mice with high, genetically clamped, levels of circulating renin develop hypertension and hypertrophy,[56,57] phenotypes not found in a conventional transgenic *Ren1* over-expression model.[58] Thus, genetic engineering strategies in the mouse can be developed to circumvent physiological control of molecular systems.

6. Birth Defects

Complex, systemic disease analysis requires the complexity and potency of mouse models, as do complex syndromes resulting from multigenic or chromosome-level changes. Human trisomy 21, the cause of Down Syndrome, is the most common viable human chromosomal abnormality and is characterized by developmental delays, mental retardation, language deficits, hypotonic musculature, and eventual Alzheimer's disease pathology.[59] Fortuitously, centromeric fusions, or Robertsonian translocations, provided the necessary hybrid chromosomes to permit the production of the first Down Syndrome model with trisomy 16 (Ts16) mice. Mouse Chromosome 16 (Chr 16) includes a region of conserved synteny with the distal arm of human Chr 21, and Ts16 mouse fetuses, though dying *in utero*, share a number of characteristics with Down Syndrome patients, including heart defects and developmental delays.[60,61] However, the relatively more severe phenotypes of the Ts16 model were likely due to trisomy of the rest of the chromosome, the bulk of which does not show syntenic conservation with human Chr 21. A separate, reciprocal translocation between mouse Chr 16 and 17 allowed for the production of segmental trisomy 16 (Ts16Dn) mice, which are trisomic only for the segment of human Chr 21 showing syntenic conservation with mouse Chr 16, and thus serves as a more accurate Down Syndrome model.[62] Ts16Dn mice are viable and have a range of Down Syndrome-like phenotypes, including reduced birth weight, male sterility, muscular trembling, and developmental and age-related neurological defects, such as deficient memory and Alzheimer's disease symptoms.[63,64] The Ts16Dn mice have extra copies of an estimated 108 genes,[65] demonstrating the ability of mouse models to recapitulate human syndromes that are highly complex at both the genetic and the phenotypic level. Subsequently, mice trisomic for even smaller portions of mouse Chr 16 have been developed and characterized, aiding in the identification of chromosomal regions responsible for specific phenotypes.[66,67]

While availability and controllability of the Ts16Dn model has permitted detailed molecular analysis of Down Syndrome-related processes, the model has more importantly allowed the identification of therapeutic interventions to lessen certain aspects of the syndrome. Although the neuroprotective compound piracetam was unable to modulate the Ts16Dn phenotype, both estrogen treatment and environmental enrichment were shown to individually increase cognition and memory in the Ts16Dn mouse.[68–70] Though the developmental aspects of Down Syndrome cannot be overcome, studies using the Ts16Dn mouse model suggest that certain interventions may increase the quality of life for individuals with Down Syndrome.

7. Infectious Diseases

Nobel-prize winning research has utilized the mouse to study many types of infectious diseases and their resulting host response (Table 1), in part because such studies require a whole organism to accurately model infection and pathogenesis. Recently, genetic engineering technologies have combined with infectious disease studies to resolve a long-standing controversy surrounding the prion-based diseases. Prions (PrP) were found by S. Prusiner to be the transmissible agents responsible for the spongiform encephalopathies, including scrapie in sheep, bovine spongiform encephalopathy ("mad cow disease"), and Creutzfeldt-Jacob disease in humans; for this and subsequent discoveries on the biological properties of prions, Prusiner was awarded the Nobel Prize in 1997. Prusiner's biochemical work on prions led to the controversial "protein-only hypothesis," whereby normal host PrP^{C} protein is converted by the abnormal isomer PrP^{Sc} into an identical PrP^{Sc} isomer. The PrP^{Sc} isomer is resistant to protease degradation and aggregates in the brain of infected animals with pathological consequences.[71]

The protein-only hypothesis of prion disease received its strongest confirmation in studies using engineered *Prnp* knockout mice, which lack PrP^{C} protein. Mice nullizygous for *Prnp* are completely resistant to prion propagation and related pathology out to two years post-infection, while mice wild-type for *Prnp* quickly succumb to prion infection, dying around six months of age.[72] These results proved the necessity of host PrP protein for the disease process. Interestingly, mice heterozygous for a *Prnp* null allele remain symptom-free for roughly twice as long as homozygous wild-type mice, suggesting that gene dosage effects the timing but not the ultimate pathology of the disease.[73] Since heterozygous mice harbor high levels of infectivity for most of their symptom-free lives, these results have strong implications for the monitoring of livestock or transplant tissue, since overtly healthy mammals can live for long periods with high levels of infectious prion in their tissues.

Additional experiments revealed other aspects of prion biology. Transgenic replacement of *Prnp* to the *Prnp* nullizygous background restored susceptibility to prion infection and pathology. This replacement experiment confirmed the central role of host PrP to prion disease, and by varying the dosage of the *Prnp* trangene, showed that the speed of prion disease progression is directly related to the levels of host PrP^C.[74] A similar series of experiments utilized the addition of PrP^C from different species (hamster, sheep, pig, human) to the *Prnp* nullizygous mouse, followed by inoculation of infectious prion from the various species.[75,76] These experiments demonstrated that species-specific sequence differences in host *Prnp* account for much of the species barrier to infection, or in other cases, resistance to progression of pathology. Similarly, a series of engineered alleles containing deletions were used to determine the domain responsible for prion pathogenicity; and targeting PrP^C expression to specific cell types identified those cells capable of harboring prion infectivity.[77] The study of prion biology serves as an example of the variety of questions that can be answered through the creation of an allelic series (Table 2). Interestingly, despite the extensive research described above, the normal biological function of PrP^C proteins remains unknown.

Table 2. Engineered *Prnp* alleles for the study of prion disease

Class	Phenotype	Conclusion
Conservative knockout	Normal; resistant to prion disease	Normal role of PrP^C unknown; host PrP^C is required for prion disease
Radical knockout	Severe ataxia and Purkinje cell loss	Targeting strategy induced ectopic expression of downstream *Doppel* gene
Transgenic *Prnp* replacement	Restoration of susceptibility to prion disease	Confirmation of host PrP^C requirement for prion disease; host PrP^C level correlates with disease
Prnp deletion series	Varying prion disease susceptibility	Identification of PrP^C domains necessary for susceptibility
Varied species *Prnp* replacement	Varying prion disease susceptibility when challenged with varied species PrP^{Sc}	Identification of specific species barriers preventing prion infection
Human inherited *PRNP* disease alleles	General failure to produce pathology or infectivity	Lack of success in developing inherited prion disease mouse model
Cell-specific *Prnp* replacement	PrP^{Sc} infectivity varies with cell types	PrP^C is required in proper cellular context for prion disease susceptibility

Reviewed in.[77]

8. Neurological Diseases

The mouse has been central to the study of prion biology in part due to the requirement for the whole organism to accurately model stages of disease from inoculation, through the accumulation of prion aggregates, to late stage behavioral symptoms. For similar reasons, genetically engineered mice have been invaluable to the study of many neurological diseases like Alzheimer's disease, also a progressive neurodegenerative disorder. Alzheimer's symptoms result in part from the buildup of amyloid plaques and neurofibrillary tangles.[78] Depending on the action of proteases and other factors, amyloid precursor protein (APP) is processed to amyloid-β (Aβ), which can oligomerize if the Aβ fragments are not degraded, resulting in plaque formation. Many of the genes in the pathways modulating the formation of these pathogenic structures have been activated and inactivated, and the resulting genetically engineered mice crossed to Alzheimer disease models (Figure 2). The resulting differences in Alzheimer-related pathology and behavior in these mice have confirmed the function of the

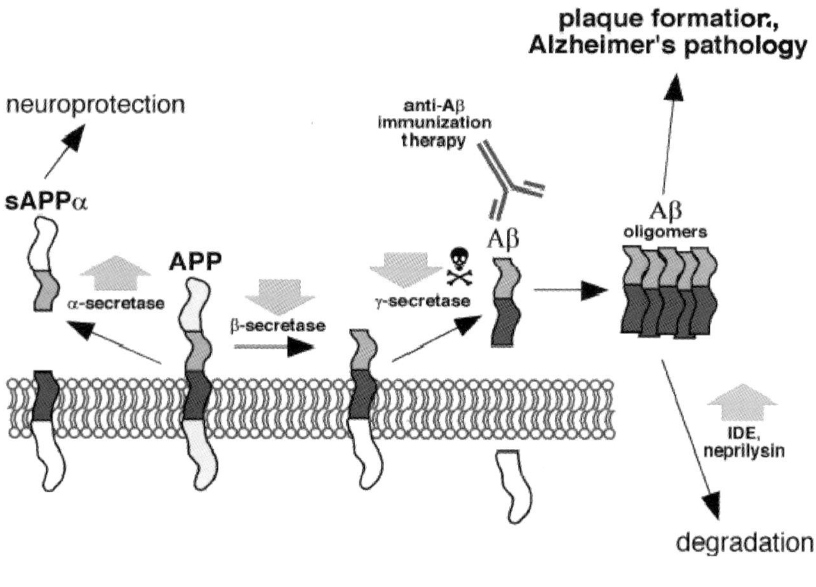

Figure 2. Mouse models reveal potential therapeutic intervention points for Alzheimer's disease treatment. Studies in genetically engineered mice have demonstrated the prevention of Alzheimer's pathology by the genetic activation (up arrows) or inhibition (down arrows) of specific proteases. Pharmacological activation or inhibition of corresponding proteins would likely have therapeutic efficacy, though the severe defects found in γ-secretase deficient mice suggest a strong likelihood of toxicity as a result of inhibiting that specific protease.

APP pathway *in vivo* and suggested important targets for potential therapeutic intervention. The direction of targeted therapy is revealed by the nature of the genetic modification; for example, if ablation of a gene reduces disease in a model, then inhibitors to the protein product of that gene are likely to have clinical efficacy in treating the disease. Specifically, removal of β-secretase or γ-secretase activity via gene knockouts prevents plaque formation and behavioral deficits in Alzheimer's disease models, suggesting pharmacological inhibition of these secretases as therapy.[79–81] Conversely, activation of α-secretase, insulin-degrading enzyme (IDE), or neprilysin by transgenic over-expression in neurons blocks plaque formation, suggesting therapeutic activation is required for these proteins to prevent disease pathology.[82,83] In some cases engineered alleles have demonstrated activity that drives pathogenesis; in this way transgenic over-expression of β-secretase, or knockout of IDE or neprilysin, results in enhanced Aβ build-up.[84–86] Finally, mouse models can be used to test novel therapeutic strategies that require metabolic or immunological functions. For example, plaque formation has been prevented by immunization against Aβ in an Alzheimer's disease mouse model.[87]

Genetically engineered mice also serve as valuable predictors of targeted therapy toxicity.[88] Defects in genetically engineered mice often mirror side-effects appearing in humans as a result of pharmacological modulation of the corresponding gene product. As such, the severe developmental and degenerative defects of γ-secretase deficient mice suggest potential toxicity from pharmacological inhibition of γ-secretase.[89–91] The β-secretase knockout mice are essentially normal; thus, β-secretase inhibitors will likely carry less toxicity.[80] Consequently, β-secretase inhibitors are currently in development to treat Alzheimer's disease.[92]

9. Conclusions

As described herein, the mouse has a proven track record of supporting seminal discoveries in virtually all fields of biomedical research. Because of this successfully history, the mouse is poised to make even greater contributions to existing and emerging fields. For example, studies in mice have recently rebutted the basic tenet of reproductive biology that females are born with their entire complement of eggs. Rather, recent observations revealed that female mice can produce new eggs throughout their adult life.[93] Likewise, mice completely maternally derived, without a sperm or paternal contribution, have recently been produced through cloning.[94] Both of these discoveries open new possibilities for fertility treatments. Other developing areas that the mouse will undoubtedly have a major impact on are the dissection of gene-environment interactions, developing interventions to curb the growing obesity epidemic, and in the search for genes that can alter the aging process.

Revealing and testing modes of therapeutic intervention have been and will continue to be the overarching goal of biomedical research using the mouse. One step towards achieving this goal is the engineering of accurate models of human disease, to serve as tools in the development of therapies. More recently, genetic engineering in the mouse has become appreciated as a means to directly model both the intended and unintended effects of therapy.[88,95] Crossing mouse models of disease with models of therapy is a potent approach to reveal gene actions underlying pathology, and thus therapeutic potential, all with the specificity of genetics. Mouse studies are essential to the continued elucidation of the interplay of disease, pathology, and human genetic background that will make the dream of individualized therapy a reality.

References

1. Foundation for Biomedical Research. 2002. Nobel Prizes: The Payoff from Animal Research. Foundation for Biomedical Research.
2. Tyzzer, E.E. 1909. Heritable difference in the rejection of transplantable tumor. *J. Med. Res.* **21**:519.
3. Little, C.C. and E.E. Tyzzer. 1916. A Mendelian explanation of rejection and susceptibility. *J. Med. Res.* **33**:393–425.
4. Gorer, P.A. 1937. Further studies on antigenetic differences in mouse erythrocytes. *Br. J. Exp. Pathology* **18**:31–36.
5. Snell, G.D. 1948. Methods for the study of histocompatibility genes. *J. Genetics* **49**:87–108.
6. Snell, G.D. and G.F. Higgins. 1951. Alleles at the histocompatibility-2 locus in the mouse as determined by tumor transplantation. *J. Natl. Cancer Inst.* **11**:1299–1305.
7. Klein, J. and D.C. Schreffler. 1971. The H-2 model for the major histocompatibility systems. *Transplant Rev.* **3**: 3–29.
8. Counce, S., P. Smith, R. Barth, and G.D. Snell. 1956. Strong and weak histocompatibility gene differences in mice and their role in the rejection of homografts of tumors and skin. *Ann. Surg.* **144**: 198–204.
9. Tonegawa, S. 1976. Reiteration frequency of immunoglobulin light chain genes: further evidence for somatic generation of antibody diversity. *Proc. Natl. Acad. Sci. USA* **73**:203–207.
10. Hozumi, N. and S. Tonegawa. 1976. Evidence for somatic rearrangement of immunoglobulin genes coding for variable and constant regions. *Proc. Natl. Acad. Sci. USA* **73**: 3628–3632.
11. Brack, C., M. Hirama, R. Lenhard-Schuller, and S. Tonegawa. 1978. A complete immunoglobulin gene is created by somatic recombination. *Cell* **15**:1–14.

12. Sakano, H., R. Maki, Y. Kurosawa, W. Roeder, and S. Tonegawa. 1980. Two types of somatic recombination are necessary for the generation of complete immunoglobulin heavy-chain genes. *Nature* **286**:676–683.
13. Malissen, M., K. Minard, S. Mjolsness, M. Kronenberg, J. Goverman, T. Hunkapiller, M.B. Prystowsky, Y. Yoshikai, F. Fitch, T.W. Mak, and et al. 1984. Mouse T cell antigen receptor: structure and organization of constant and joining gene segments encoding the beta polypeptide. *Cell* **37**:1101–1110.
14. Chien, Y.H., N.R. Gascoigne, J. Kavaler, N.E. Lee, and M.M. Davis. 1984. Somatic recombination in a murine T-cell receptor gene. *Nature* **309**:322–326.
15. Perry, R.P., D.E. Kelley, C. Coleclough, J.G. Seidman, P. Leder, S. Tonegawa, G. Matthyssens, and M. Weigert. 1980. Transcription of mouse kappa chain genes: implications for allelic exclusion. *Proc. Natl. Acad. Sci. USA* **77**: 937–1941.
16. Hochedlinger, K. and R. Jaenisch. 2002. Monoclonal mice generated by nuclear transfer from mature B and T donor cells. *Nature* **415**:1035–1038.
17. Kohler, G. and C. Milstein. 1975. Continuous cultures of fused cells secreting antibody of predefined specificity. *Nature* **256**:495–497.
18. Kohler, G. and C. Milstein. 1976. Derivation of specific antibody-producing tissue culture and tumor lines by cell fusion. *Eur. J. Immunol.* **6**:511–519.
19. Baselga, J., L. Norton, J. Albanell, Y.M. Kim, and J. Mendelsohn. 1998. Recombinant humanized anti-HER2 antibody (Herceptin) enhances the antitumor activity of paclitaxel and doxorubicin against HER2/neu overexpressing human breast cancer xenografts. *Cancer Res.* **58**:2825–2831.
20. Meruelo, D., B. Deak, and H.O. McDevitt. 1977. Genetic control of cell-mediated responsiveness to an AKR tumor-associated antigen: mapping of the locus involved to the I region of the H-2 complex. *J. Exp. Med.* **146**:1367–1379.
21. Zinkernagel, R.M. and P.C. Doherty. 1974. Restriction of in vitro T cell-mediated cytotoxicity in lymphocytic choriomeningitis within a syngeneic or semiallogeneic system. *Nature* **248**: 701–702.
22. Zinkernagel, R.M. and P.C. Doherty. 1974. Immunological surveillance against altered self components by sensitised T lymphocytes in lymphocytic choriomeningitis. *Nature* **251**:547–548.
23. Doherty, P.C. and R.M. Zinkernagel. 1975. H-2 compatibility is required for T-cell-mediated lysis of target cells infected with lymphocytic choriomeningitis virus. *J. Exp. Med.* **141**:502–507.
24. Doherty, P.C., M.B. Dunlop, C.R. Parish, and R.M. Zinkernagel. 1976. Inflammatory process in murine lymphocytic choriomeningitis is maximal in H-2K or H-2D compatible interactions. *J. Immunol.* **117**:187–190.
25. Roscoe B. Jackson Memorial Laboratory. 1933. The existence of non-chromosomal influence in the incidence of mammary tumors in mice. *Science* **78**:465–466.
26. Bittner, J.J. 1936. Some possible effects of nursing on the mammary gland tumor incidence in mice. *Science* **84**:162.
27. Muhlbock, O. and P. Bentvelzen. 1968. The transmission of the mammary tumor viruses. *Perspect. Virol.* **6**:75–87.
28. Varmus, H.E., J.M. Bishop, R.C. Nowinski, and N.H. Sarker. 1972. Mammary tumor virus specific nucleotide sequences in mouse DNA. *Nat. New Biol.* **238**:189–191.

29. Nusse, R., A. van Ooyen, D. Cox, Y.K. Fung, and H. Varmus. 1984. Mode of proviral activation of a putative mammary oncogene (int-1) on mouse chromosome 15. *Nature* **307**: 131–136.
30. Peters, G., A.E. Lee, and C. Dickson. 1984. Activation of cellular gene by mouse mammary tumour virus may occur early in mammary tumour development. *Nature* **309**:273–275.
31. Su, L.-K., K.W. Kinzler, B. Vogelstein, A.C. Preisinger, A.R. Moser, C. Luongo, K.A. Gould, and W.F. Dove. 1992. Multiple intestinal neoplasia caused by a mutation in the murine homolog of the APC gene. *Science* **256**:668–670.
32. Cormier, R.T., K.H. Hong, R.B. Halberg, T.L. Hawkins, P. Richardson, R. Mulherkar, W.F. Dove, and E.S. Lander. 1997. Secretory phospholipase Pla2g2a confers resistance to intestinal tumorigenesis. *Nat. Genet.* **17**:88–91.
33. Sheng, H., J. Shao, S.C. Kirkland, P. Isakson, R.J. Coffey, J. Morrow, R.D. Beauchamp, and R.N. DuBois. 1997. Inhibition of human colon cancer cell growth by selective inhibition of cyclooxygenase-2. *J. Clin. Invest.* **99**:2254–2259.
34. Moen, C.J.A., M. Snoek, A.A.M. Hart, and P. Demant. 1992. *Scc-1*, a novel colon cancer susceptibility gene in the mouse: linkage to *CD44* (*Ly-24*, *Pgp-1*) on chromosome 2. *Oncogene* **7**:563–566.
35. Ruivenkamp, C.A., T. van Wezel, C. Zanon, A.P. Stassen, C. Vlcek, T. Csikos, A.M. Klous, N. Tripodis, A. Perrakis, L. Boerrigter, P.C. Groot, J. Lindeman, W.J. Mooi, G.A. Meijjer, G. Scholten, H. Dauwerse, V. Paces, N. van Zandwijk, G.J. van Ommen, and P. Demant. 2002. Ptprj is a candidate for the mouse colon-cancer susceptibility locus Scc1 and is frequently deleted in human cancers. *Nat. Genet.* **31**:295–300.
36. Ruivenkamp, C., M. Hermsen, C. Postma, A. Klous, J. Baak, G. Meijer, and P. Demant. 2003. LOH of PTPRJ occurs early in colorectal cancer and is associated with chromosomal loss of 18q12-21. *Oncogene* **22**:3472–3474.
37. Ramaswamy, S., K.N. Ross, E.S. Lander, and T.R. Golub. 2003. A molecular signature of metastasis in primary solid tumors. *Nat. Genet.* **33**:49–54.
38. Hunter, K.W. 2003. Allelic diversity in the host genetic background may be an important determinant in tumor metastatic dissemination. *Cancer Lett.* **200**:97–105.
39. Hunter, K., D.R. Welch, and E.T. Liu. 2003. Genetic background is an important determinant of metastatic potential. *Nat. Genet.* **34**:23–24.
40. Qiu, T.H., G.V. Chandramouli, K.W. Hunter, N.W. Alkharouf, J.E. Green, and E.T. Liu. 2004. Global expression profiling identifies signatures of tumor virulence in MMTV-PyMT-transgenic mice: correlation to human disease. *Cancer Res.* **64**:5973–5981.
41. Wessels, A. and D. Sedmera. 2003. Developmental anatomy of the heart: a tale of mice and man. *Physiol. Genomics* **15**:165–176.
42. Takeishi, Y. and R.A. Walsh. 2001. Cardiac hypertrophy and failure: lessons learned from genetically engineered mice. *Acta Physiol. Scand.* **173**:103–111.
43. Rockman, H.A., K.R. Chien, D.J. Choi, G. Iaccarino, J.J. Hunter, J. Ross, Jr., R.J. Lefkowitz, and W.J. Koch. 1998. Expression of a beta-adrenergic receptor kinase 1 inhibitor prevents the development of myocardial failure in gene-targeted mice. *Proc. Natl. Acad. Sci. USA* **95**: 7000–7005.
44. Suzuki, M., K.M. Carlson, D.A. Marchuk, and H.A. Rockman. 2002. Genetic modifier loci affecting survival and cardiac function in murine dilated cardiomyopathy. *Circulation* **105**: 1824–1829.

45. Nadeau, J.H., L.C. Burrage, J. Restivo, Y.H. Pao, G. Churchill, and B.D. Hoit. 2003. Pleiotropy, homeostasis, and functional networks based on assays of cardiovascular traits in genetically randomized populations. *Genome Res.* **13**:2082–2091.
46. Wang, X. and B. Paigen. 2002. Quantitative trait loci and candidate genes regulating HDL cholesterol: a murine chromosome map. *Arterioscler. Thromb. Vasc. Biol.* **22**:1390–1401.
47. Zhang, S.H., R.L. Reddick, J.A. Piedrahita, and N. Maeda. 1992. Spontaneous hypercholesterolemia and arterial lesions in mice lacking apolipoprotein E. *Science* **258**:468–471.
48. Plump, A.S., J.D. Smith, T. Hayek, K. Aalto-Setala, A. Walsh, J.G. Verstuyft, E.M. Rubin, and J.L. Breslow. 1992. Severe hypercholesterolemia and atherosclerosis in apolipoprotein E-deficient mice created by homologous recombination in ES cells. *Cell* **71**:343–353.
49. Reddick, R.L., S.H. Zhang, and N. Maeda. 1994. Atherosclerosis in mice lacking apo E. Evaluation of lesional development and progression. *Arterioscler. Thromb.* **14**:141–147.
50. Nakashima, Y., A.S. Plump, E.W. Raines, J.L. Breslow, and R. Ross. 1994. ApoE-deficient mice develop lesions of all phases of atherosclerosis throughout the arterial tree. *Arterioscler. Thromb.* **14**:133–140.
51. Ishida, B.Y., P.J. Blanche, A.V. Nichols, M. Yashar, and B. Paigen. 1991. Effects of atherogenic diet consumption on lipoproteins in mouse strains C57BL/6 and C3H. *J. Lipid Res.* **32**:559–568.
52. Krege, J.H., J.B. Hodgin, J.R. Hagaman, and O. Smithies. 1995. A noninvasive computerized tail-cuff system for measuring blood pressure in mice. *Hypertension* **25**:1111–1115.
53. Smithies, O. and H.S. Kim. 1994. Targeted gene duplication and disruption for analyzing quantitative genetic traits in mice. *Proc. Natl. Acad. Sci. USA* **91**:3612–3615.
54. Krege, J.H., H.S. Kim, J.S. Moyer, J.C. Jennette, L. Peng, S.K. Hiller, and O. Smithies. 1997. Angiotensin-converting enzyme gene mutations, blood pressures, and cardiovascular homeostasis. *Hypertension* **29**:150–157.
55. Radziuk, J. 2000. Insulin sensitivity and its measurement: structural commonalities among the methods. *J. Clin. Endocrinol. Metab.* **85**:4426–4433.
56. Caron, K.M., L.R. James, H.S. Kim, S.G. Morham, M.L. Sequeira Lopez, R.A. Gomez, T.L. Reudelhuber, and O. Smithies. 2002. A genetically clamped renin transgene for the induction of hypertension. *Proc. Natl. Acad. Sci. USA* **99**:8248–8252.
57. Caron, K.M., L.R. James, H.S. Kim, J. Knowles, R. Uhlir, L. Mao, J.R. Hagaman, W. Cascio, H. Rockman, and O. Smithies. 2004. Cardiac hypertrophy and sudden death in mice with a genetically clamped renin transgene. *Proc. Natl. Acad. Sci. USA* **101**:3106–3111.
58. Veniant, M., J. Menard, P. Bruneval, S. Morley, M.F. Gonzales, and J. Mullins. 1996. Vascular damage without hypertension in transgenic rats expressing prorenin exclusively in the liver. *J. Clin. Invest.* **98**:1966–1970.
59. Epstein, C.J., J.R. Korenberg, G. Anneren, S.E. Antonarakis, S. Ayme, E. Courchesne, L.B. Epstein, A. Fowler, Y. Groner, J.L. Huret, and et al. 1991. Protocols to establish genotype-phenotype correlations in Down syndrome. *Am. J. Hum. Genet.* **49**:207–235.
60. Miyabara, S., A. Gropp, and H. Winking. 1982. Trisomy 16 in the mouse fetus associated with generalized edema and cardiovascular and urinary tract anomalies. *Teratology* **25**:369–380.
61. Epstein, C.J., D.R. Cox, and L.B. Epstein. 1985. Mouse trisomy 16: an animal model of human trisomy 21 (Down syndrome). *Ann. N. Y. Acad. Sci.* **450**:157–168.

62. Davisson, M.T., C. Schmidt, R.H. Reeves, N.G. Irving, E.C. Akeson, B.S. Harris, and R.T. Bronson. 1993. Segmental trisomy as a mouse model for Down syndrome. *Prog. Clin. Biol. Res.* **384**:117–133.
63. Reeves, R.H., N.G. Irving, T.H. Moran, A. Wohn, C. Kitt, S.S. Sisodia, C. Schmidt, R.T. Bronson, and M.T. Davisson. 1995. A mouse model for Down syndrome exhibits learning and behaviour deficits. *Nat. Genet.* **11**:177–184.
64. Holtzman, D.M., D. Santucci, J. Kilbridge, J. Chua-Couzens, D.J. Fontana, S.E. Daniels, R.M. Johnson, K. Chen, Y. Sun, E. Carlson, E. Alleva, C.J. Epstein, and W.C. Mobley. 1996. Developmental abnormalities and age-related neurodegeneration in a mouse model of Down syndrome. *Proc. Natl. Acad. Sci. USA* **93**:13333–13338.
65. Reeves, R.H. 2001. Down's syndrome. A complicated genetic insult. *Lancet* **358 Suppl**: S23.
66. Sago, H., E.J. Carlson, D.J. Smith, J. Kilbridge, E.M. Rubin, W.C. Mobley, C.J. Epstein, and T.T. Huang. 1998. Ts1Cje, a partial trisomy 16 mouse model for Down syndrome, exhibits learning and behavioral abnormalities. *Proc. Natl. Acad. Sci. USA* **95**:6256–6261.
67. Sago, H., E.J. Carlson, D.J. Smith, E.M. Rubin, L.S. Crnic, T.T. Huang, and C.J. Epstein. 2000. Genetic dissection of region associated with behavioral abnormalities in mouse models for Down syndrome. *Pediatr. Res.* **48**:606–613.
68. Moran, T.H., G.T. Capone, S. Knipp, M.T. Davisson, R.H. Reeves, and J.D. Gearhart. 2002. The effects of piracetam on cognitive performance in a mouse model of Down's syndrome. *Physiol. Behav.* **77**:403–409.
69. Granholm, A.C., K.A. Ford, L.A. Hyde, H.A. Bimonte, C.L. Hunter, M. Nelson, D. Albeck, L.A. Sanders, E.J. Mufson, and L.S. Crnic. 2002. Estrogen restores cognition and cholinergic phenotype in an animal model of Down syndrome. *Physiol. Behav.* **77**:371–385.
70. Martinez-Cue, C., C. Baamonde, M. Lumbreras, J. Paz, M.T. Davisson, C. Schmidt, M. Dierssen, and J. Florez. 2002. Differential effects of environmental enrichment on behavior and learning of male and female Ts65Dn mice, a model for Down syndrome. *Behav. Brain Res.* **134**:185–200.
71. Aguzzi, A. and M. Polymenidou. 2004. Mammalian prion biology: one century of evolving concepts. *Cell* **116**:313–327.
72. Bueler, H., A. Aguzzi, A. Sailer, R.A. Greiner, P. Autenried, M. Aguet, and C. Weissmann. 1993. Mice devoid of PrP are resistant to scrapie. *Cell* **73**:1339–1347.
73. Bueler, H., A. Raeber, A. Sailer, M. Fischer, A. Aguzzi, and C. Weissmann. 1994. High prion and PrPSc levels but delayed onset of disease in scrapie-inoculated mice heterozygous for a disrupted PrP gene. *Mol. Med.* **1**:19–30.
74. Fischer, M., T. Rulicke, A. Raeber, A. Sailer, M. Moser, B. Oesch, S. Brandner, A. Aguzzi, and C. Weissmann. 1996. Prion protein (PrP) with amino-proximal deletions restoring susceptibility of PrP knockout mice to scrapie. *EMBO J.* **15**:1255–1264.
75. Telling, G.C., M. Scott, K.K. Hsiao, D. Foster, S.L. Yang, M. Torchia, K.C. Sidle, J. Collinge, S.J. DeArmond, and S.B. Prusiner. 1994. Transmission of Creutzfeldt-Jakob disease from humans to transgenic mice expressing chimeric human-mouse prion protein. *Proc. Natl. Acad. Sci. USA* **91**:9936–9940.
76. Scott, M.R., R. Will, J. Ironside, H.O. Nguyen, P. Tremblay, S.J. DeArmond, and S.B. Prusiner. 1999. Compelling transgenetic evidence for transmission of bovine spongiform encephalopathy prions to humans. *Proc. Natl. Acad. Sci. USA* **96**:15137–15142.

77. Weissmann, C. and E. Flechsig. 2003. PrP knock-out and PrP transgenic mice in prion research. *Br. Med. Bull.* **66**:43–60.
78. Mattson, M.P. 2004. Pathways towards and away from Alzheimer's disease. *Nature* **430**:631–639.
79. Luo, Y., B. Bolon, S. Kahn, B.D. Bennett, S. Babu-Khan, P. Denis, W. Fan, H. Kha, J. Zhang, Y. Gong, L. Martin, J.C. Louis, Q. Yan, W.G. Richards, M. Citron, and R. Vassar. 2001. Mice deficient in BACE1, the Alzheimer's beta-secretase, have normal phenotype and abolished beta-amyloid generation. *Nat. Neurosci.* **4**:231–232.
80. Roberds, S.L., J. Anderson, G. Basi, M.J. Bienkowski, D.G. Branstetter, K.S. Chen, S.B. Freedman, N.L. Frigon, D. Games, K. Hu, K. Johnson-Wood, K.E. Kappenman, T.T. Kawabe, I. Kola, R. Kuehn, M. Lee, W. Liu, R. Motter, N.F. Nichols, M. Power, D.W. Robertson, D. Schenk, M. Schoor, G.M. Shopp, M.E. Shuck, S. Sinha, K.A. Svensson, G. Tatsuno, H. Tintrup, J. Wijsman, S. Wright, and L. McConlogue. 2001. BACE knockout mice are healthy despite lacking the primary beta-secretase activity in brain: implications for Alzheimer's disease therapeutics. *Hum. Mol. Genet.* **10**:1317–1324.
81. De Strooper, B., P. Saftig, K. Craessaerts, H. Vanderstichele, G. Guhde, W. Annaert, K. Von Figura, and F. Van Leuven. 1998. Deficiency of presenilin-1 inhibits the normal cleavage of amyloid precursor protein. *Nature* **391**:387–390.
82. Postina, R., A. Schroeder, I. Dewachter, J. Bohl, U. Schmitt, E. Kojro, C. Prinzen, K. Endres, C. Hiemke, M. Blessing, P. Flamez, A. Dequenne, E. Godaux, F. van Leuven, and F. Fahrenholz. 2004. A disintegrin-metalloproteinase prevents amyloid plaque formation and hippocampal defects in an Alzheimer disease mouse model. *J. Clin. Invest.* **113**:1456–1464.
83. Leissring, M.A., W. Farris, A.Y. Chang, D.M. Walsh, X. Wu, X. Sun, M.P. Frosch, and D.J. Selkoe. 2003. Enhanced proteolysis of beta-amyloid in APP transgenic mice prevents plaque formation, secondary pathology, and premature death. *Neuron* **40**:1087–1093.
84. Bodendorf, U., S. Danner, F. Fischer, M. Stefani, C. Sturchler-Pierrat, K.H. Wiederhold, M. Staufenbiel, and P. Paganetti. 2002. Expression of human beta-secretase in the mouse brain increases the steady-state level of beta-amyloid. *J. Neurochem.* **80**:799–806.
85. Farris, W., S. Mansourian, Y. Chang, L. Lindsley, E.A. Eckman, M.P. Frosch, C.B. Eckman, R.E. Tanzi, D.J. Selkoe, and S. Guenette. 2003. Insulin-degrading enzyme regulates the levels of insulin, amyloid beta-protein, and the beta-amyloid precursor protein intracellular domain in vivo. *Proc. Natl. Acad. Sci. USA* **100**:4162–4167.
86. Iwata, N., S. Tsubuki, Y. Takaki, K. Shirotani, B. Lu, N.P. Gerard, C. Gerard, E. Hama, H.J. Lee, and T.C. Saido. 2001. Metabolic regulation of brain Abeta by neprilysin. *Science* **292**: 1550–1552.
87. Schenk, D., R. Barbour, W. Dunn, G. Gordon, H. Grajeda, T. Guido, K. Hu, J. Huang, K. Johnson-Wood, K. Khan, D. Kholodenko, M. Lee, Z. Liao, I. Lieberburg, R. Motter, L. Mutter, F. Soriano, G. Shopp, N. Vasquez, C. Vandevert, S. Walker, M. Wogulis, T. Yednock, D. Games, and P. Seubert. 1999. Immunization with amyloid-beta attenuates Alzheimer-disease-like pathology in the PDAPP mouse. *Nature* **400**:173–177.
88. Roberts, R.B., C.L. Arteaga, and D.W. Threadgill. 2004. Modeling the cancer patient with genetically engineered mice: prediction of toxicity from molecule-targeted therapies. *Cancer Cell* **5**:115–120.
89. Shen, J., R.T. Bronson, D.F. Chen, W. Xia, D.J. Selkoe, and S. Tonegawa. 1997. Skeletal and –NS defects in Presenilin-1-deficient mice. *Cell* **89**:629–639.

90. Feng, R., C. Rampon, Y.P. Tang, D. Shrom, J. Jin, M. Kyin, B. Sopher, M.W. Miller, C.B. Ware, G.M. Martin, S.H. Kim, R.B. Langdon, S.S. Sisodia, and J.Z. Tsien. 2001. Deficient neurogenesis in forebrain-specific presenilin-1 knockout mice is associated with reduced clearance of hippocampal memory traces. *Neuron* **32**:911–926.
91. Dewachter, I. and F. Van Leuven. 2002. Secretases as targets for the treatment of Alzheimer's disease: the prospects. *Lancet Neurol.* **1**:409–416.
92. John, V., J.P. Beck, M.J. Bienkowski, S. Sinha, and R.L. Heinrikson. 2003. Human beta-secretase (BACE) and BACE inhibitors. *J. Med. Chem.* **46**:4625–4630.
93. Johnson, J., J. Canning, T. Kaneko, J.K. Pru, and J.L. Tilly. 2004. Germline stem cells and follicular renewal in the postnatal mammalian ovary. *Nature* **428**:145–150.
94. Kono, T., Y. Obata, Q. Wu, K. Niwa, Y. Ono, Y. Yamamoto, E.S. Park, J.S. Seo, and H. Ogawa. 2004. Birth of parthenogenetic mice that can develop to adulthood. *Nature* **428**:260–864.
95. Zambrowicz, B.P. and A.T. Sands. 2003. Knockouts model the 100 best-selling drugs--will they model the next 100? *Nat. Rev. Drug Discov.* **2**: 38–51.

CHAPTER 16

THE MOUSE GENOME SEQUENCING PROJECT: AN OVERVIEW

Michael C. Wendl[1], Robert S. Fulton[2], Tina Graves[3],
Elaine R. Mardis[4], and Richard K. Wilson[5]

Genome Sequencing Center
Washington University School of Medicine, St. Louis, MO, USA
[1]*mwendl@wustl.edu,* [2]*bfulton@wustl.edu,* [3]*tgraves@wustl.edu,*
[4]*emardis@wustl.edu,* [5]*rwilson@wustl.edu*

1. Introduction

The role of the mouse as a formalized biological model system dates to the 19th century and became firmly established with the re-discovery of Mendel's Inheritance Laws around 1900.[1–3] Variation in domestic mice was soon being examined in the context of Mendelian genetics, and breeding programs for inbred strains were systematically expanded. The first genetic mapping work was undertaken by John Haldane and colleagues shortly thereafter[4] and resulted in a two-marker linkage group that would later be localized to chromosome 7. By the 1990s, a map having roughly 6,600 microsatellite markers had been developed.[5] Although somewhat simplistic, this short historical perspective illustrates the enormous progress in characterizing the mouse genome that had been made by the closing years of the 20th century.

At about this same time, the Human Genome Project (HGP) was getting underway. From its earliest planning phases, investigators recognized the necessity of obtaining companion sequence from model organisms, specifically the yeasts, *E. coli*, the roundworm *C. elegans*, the common fruit fly *D. melanogaster*, and the mouse.[6] Insurmountable difficulties were anticipated in analyzing the human genome sequence unless data from model organisms were to be available for comparison. The sequence of the mouse genome was considered especially important in the contexts of mammalian experimental biology and human disease modeling. HGP leaders thus resolved to quickly follow-up human sequencing with a mouse genome sequencing effort. The initial phase of that project is now complete.[7]

The mouse genome project is a landmark. Its design and execution benefited from the collective experience of all large-scale sequencing projects before it, especially the HGP.[8,9] In turn, this project has set the stage for succeeding mammalian projects, notably for the rat and chimpanzee genomes.[10,11] Here, we provide an overview of the project in terms of its architecture, results, and legacy.

2. Genome Sequencing and Assembly: An Evolving Process

Before delving specifically into the mouse project, it will be helpful to briefly review the process of DNA sequencing and its evolution over roughly the last three decades. Fred Sanger and colleagues invented the chain-termination method of sequencing in the late 1970s.[12] Sanger sequencing allows one to deduce the sequence of relatively short fragments of DNA. Specifically, a single chain-terminating reaction can yield a "read" of up to about 800 base pairs. Consequently, numerous such reads are required to cover an entire genome.

Improved genome fragmentation and cloning methods led to a "shotgun" approach, whereby genome coverage could be efficiently generated by randomly subcloning and selecting fragments for sequencing without regard to their origin within the genome.[13–15] The basic idea is that over-sampling a genome yields mutually overlapping sequence reads that can be assembled, enabling one to infer the original genome sequence. Although the shotgun technique had led to notable megabase-sized sequencing accomplishments by the mid-1990s,[16,17] controversy arose regarding how to best extend the technique into the gigabase realm necessary for human sequencing. Debate focused to some degree on the tremendous ramp-up in sequence generation rate that would be required. However, this hurdle could, in principle, be addressed by factory-style economies of scale that would be realized with proper software and hardware automation and integration.

The more fundamental issue would be how best to handle sequence repeats in the assembly process. Mammalian genomes have long been known to contain a significant percentage of repeated sequence.[18] Demonstrating the "repeat problem" is a matter of simple statistics. Suppose a target genome sequence is completely random. If we consider a specific sub-sequence of length L, the genome would have to be at least 4^L bases long before we would expect the presence of another copy of this sub-sequence. The actual assembly procedure would be relatively straightforward for this case. For example, if a 35 base-pair sub-sequence appears in two reads, we would conclude that these reads actually overlap. Our degree of confidence in this decision would be quite high because we would only expect this sub-sequence about once every $4^{35} \approx 10^{21}$ base pairs. It is, therefore, almost a certainty that this sub-sequence would be unique within the genome. Conversely, if this genome were to contain many repeated copies of the 35 base sub-sequence, the likelihood of these two reads representing different parts of the genome would be appreciably higher. In this case, the assembly problem would be dramatically more difficult because of the increased likelihood of falsely joining different copies. The primary debate focused on just how much of a difficulty this high repeat content would pose for sequence assembly.

The initial approach to this problem, taken by the Human Genome Project consortium, was termed Hierarchical Shotgun Sequencing (HSS). The HSS strategy is a two-phase process based mainly on large-insert clones, especially BACs (bacterial artificial chromosomes).[19] One phase involves creating a physical map of a clone library to determine spatial relationships of the clones to one another based on shared restriction digest fragment lengths.[20] In the second phase, a subset of large-insert clones covering the genome is selected and each one is sequenced and assembled as a self-contained shotgun project. Ideally, these phases are conducted in series, so that the clone subset is minimally overlapping. The phases can also be conducted in parallel to a limited degree, although this complicates the clone selection process. Clone overlaps, and thus duplicated work will be slightly higher. The main advantages of this approach are that large-insert clones localize repeat problems by isolating and encapsulating repeat structures, and that the physical map provides an independent method of confirming the correct genome assembly. Moreover, the physical map provided a reference framework for organizing the HGP across multiple participating laboratories. Proponents of the HSS method felt that its success in pilot projects, especially for the multi-cellular *C. elegans* genome,[21] made it the best approach for human sequencing. Studies also indicated that an HSS draft sequence would be well-positioned to act as a substrate for the downstream task of completing the human sequence.[22]

A second strategy called Whole Genome Shotgun (WGS) sequencing was proposed a few years later for human sequencing.[23] This approach relies on a simple scale-up of the basic shotgun technique to capture the entire genome in the form of subclones of various insert sizes, which are then sequenced from both ends. These paired end reads create extra "linking" information because the distance between each mating read pair is known. As reads merge into contiguous segments (contigs), the mate-pair information begins linking contigs at a larger scale. The theory is that repeat-related mistakes in an assembly could then be detected both by standard coverage-based statistical tests[24] *and* by any inconsistencies in the topology of links. The WGS method had been successfully applied to a number of low-repeat microbial genomes, e.g., *H. influenzae,*[17] and proponents argued via studies[25] and a pilot shotgun project for the fruit fly *D. melanogaster* [26] that its feasibility would extend to the human genome.

3. Designing the Mouse Project

Human sequencing proceeded in the form of two separate projects based on the above two methods. A private venture attempted to apply the WGS strategy, while the publicly funded HGP consortium utilized the HSS method. Draft sequences were announced simultaneously in 2001.[8,9] These results resolved a number of strategic issues pertaining to mammalian projects:

(a) The WGS technique is straightforward to implement and scale-up, leading to rapid coverage in the early phase of a project.
(b) Even with ancillary linking information, the WGS strategy can realize formidable assembly problems when long near-perfect repeats are present.
(c) The HSS approach effectively handles the repeat problem by isolating repetitive sequence within large-insert clones.
(d) Clone-picking for the HSS strategy is most effective when based on an established physical map.
(e) Physical and genetic maps facilitate project completion by providing anchors to specific genomic locations for clones and sequences.

In short, the two strategies are somewhat complementary; WGS provides better coverage performance early in a project, while HSS promotes repeat resolution and downstream map-based finishing advantages.

The mouse project was formalized by the establishment of the Mouse Genome Sequencing Consortium (MGSC). This organization consisted originally of the Whitehead Institute for Biomedical Research/Massachusetts Institute of Technology, the Washington University Genome Sequencing Center, and the Wellcome Trust Sanger Institute. MGSC investigators resolved to ultimately produce a completed, so-called "base-perfect" mouse sequence as a canonical reference for mammalian biology. However, the scientific community at-large also considered intermediate "draft" sequences valuable for research work. Consequently, the MGSC decided to pursue a novel hybrid strategy having elements of both the WGS and HSS approaches. This strategy consisted of four components:

(a) Whole Genome Shotgun Sequence: The plan called for an initial phase of whole genome shotgun sequence generation to a redundancy of 7X. This means that the average base position in the genome would be covered by 7 shotgun reads. Experience from past projects indicated that 7X sequencing would be sufficient to produce a reasonable assembly and draft sequence relatively early in the mouse project.
(b) Physical Map: A BAC-based physical map would simultaneously be constructed from the Roswell Park Cancer Institute (RPCI) large-insert mouse libraries.[27] This map would serve as a framework for better assembling the amalgam of shotgun reads, for organizing the finishing of the genome, and for eventually anchoring finished sequence to specific chromosomes. It would be based on both the fingerprint technique of detecting clone overlaps[20] and on BAC end sequences.[28]
(c) Hierarchical Shotgun Sequence: In addition to the mapping component, the plan included sequencing each of the BAC clones covering the mouse genome at low (4X) coverage to foster the creation of an enhanced BAC-WGS assembly.
(d) Finishing Component: Following established procedures for HSS, the BAC clones would then serve as final templates for completing the sequence. Specifically, remaining gaps are targeted for closure using directed reads rather than shotgun reads.

Based on discussions with the science community at-large, MGSC investigators decided to sequence a single mouse strain, specifically C57BL/6, rather than multiple strains to avoid polymorphism issues. C57BL/6 was chosen primarily because of its widespread use, favorable breeding attributes, and well-characterized phenotype.[29] The investigators opted further to sequence a female, so that there would be statistically equal coverage for the autosomes and the X chromosome. This plan intentionally omitted the Y chromosome, which was anticipated to be too repeat-laden and complicated for anything but a pure HSS-based approach. Like its counterpart in the human genome, the sequencing of mouse Y would largely be separate from the main genome project.[30]

4. Sequencing and Assembling the Mouse Genome

A draft mouse sequence was reported in 2002 by Waterston and colleagues.[7] The assembly was based on a collection of 41.4 million sequencing reads, derived predominantly (87.7%) from small-insert plasmids of 2, 4, and 6 kb lengths. About 6% of these reads originated from 10 kb plasmids, while the remaining ones were generated from large-insert fosmid and mapped BAC clone templates. Almost 30 million of the reads were mate-paired, i.e. derived from both ends of the same insert. Varying the insert sizes in this fashion, rather than using inserts of uniform size, resulted in the creation of links over different length scales. This feature is exploited by algorithms used for assembling the data.[31]

As discussed above, the initial assembly utilized whole genome shotgun reads without the benefit of mapping information. Actual redundancy of the data was about 7.7X for the euchromatic portion of the genome. The resulting assembly consisted of about 225,000 contiguous elements (contigs). Linking information connected these into roughly 7,400 supercontigs of at least 2 kb in length. Most assembled reads were found in a relatively small number of large supercontigs. Specifically, the 200 largest supercontigs accounted for more than 98% of the assembly.

The whole genome shotgun assembly, as we have just described it, does not actually yield much information regarding the location of contigs and supercontigs within the genome itself. A follow-on step of *anchoring* sequence contigs and supercontigs is required. For the mouse project, MGSC investigators used the extensive mouse map resources[5,32,33] for anchoring sequences in their proper order and orientation and for obtaining additional contiguity. Comparison with publicly available finished sequence and mouse cDNA sequences suggested that the draft sequence covered about 96% of the mouse genome.

The quality of the assembly was assessed at several levels. On a global level, no large-scale mis-assemblies were found in the course of anchoring the sequence supercontigs to the mouse map. On a local level, the positions of several thousand well-studied markers were compared between the genetic map[5] and the sequence. There were conflicts in fewer than 2% of the cases. Eleven specific discrepancies were further tested by remapping them in a mouse cross and 10 of these supported the position assignment given by the draft sequence. From this and a number of similar comparisons, it was estimated that less than 0.3% of the assembly was affected by local errors.

One shortcoming of applying the WGS method to eukaryotic genomes is that a non-trivial percentage of the raw data usually cannot be assembled or anchored. This phenomenon was first realized in a significant way for the fruit fly pilot project.[34] In the case of the mouse genome, about 5 million of the WGS reads (16%) could not be integrated into the assembly, remaining instead as singletons. This value lies in the 10 to 20% range typical of eukaryotic genome projects.[35] These so-called "chaff" reads are often associated with genomic regions having various repeat structures.

The final step in obtaining the draft sequence was integrating all available finished BAC sequences. In all, about 48 Mb of sequence from 210 finished BAC clones was modified in this way. The resulting draft, designated MGSCv3, was deposited in the major public databases, specifically GenBank[36] and Ensembl[37] The GenBank accession number is CAAA01000000.

The draft sequence provided an unprecedented opportunity to examine mouse biology. MGSC project leaders convened a group of scientists representing several dozen institutions from around the world to undertake initial analysis of the mouse sequence. This group studied the general composition of the genome, repeat structures, genes, the mouse proteome, and genome evolution. Comparative studies with respect to human sequence were especially notable. Among the discoveries were:

(a) At about 2.5 Gb, the mouse genome is roughly 14% smaller than the 2.9 Gb human genome, which suggests faster deletion in the mouse lineage.
(b) Gene order is conserved across over 90% of the mouse and human genomes. About 40% of the human genome can be aligned to mouse sequence.
(c) The mouse genome contains on the order of 30,000 protein-coding genes. About 80% of mouse genes have an ortholog in the human genome.
(d) There has been notable expansion in gene families related to reproduction, immunity, and olfaction over evolutionary time. Specific classes of secreted proteins related to reproduction and immuno-defense appear to be under positive selection.
(e) Despite fundamental differences in transposon activity, there is appreciable correlation between mouse and human repeat structures and their locations in their respective genomes.

Moreover, additional sequence from other mouse strains enabled about 80,000 single nucleotide polymorphisms (SNPs) to be identified. MGSC investigators expect to broaden the SNP collection with more data and the inclusion of sequence from additional mouse strains.

5. The Mouse Project Legacy

The mouse genome project is a significant milestone in the genomic era and has already established an important research legacy. In particular, the mouse sequence has quickly become one of the cornerstones for post-genomic research, as casual browsing of the literature will reveal. In the few years that it has been available to researchers, the genome sequence has played a fundamental role in quite a few notable discoveries, for example:

(a) Humans have roughly twice the recombination rate per generation as mice.[38] The mouse X chromosome has a significantly lower recombination rate than the autosomes.
(b) Sequence analysis is expanding and clarifying gene families. Recent work has shown that gene families related to odor detection are larger than previously thought.[39]
(c) About 2% of the mouse genome constitutes recent segmental duplications.[40] This is about half the value for the human genome.

The mouse project itself was the first real endeavor to utilize a "hybrid" architecture, combining elements of both the WGS and HSS approaches. As such, it served as a proof-of-concept for hybrid projects and now stands as a prototype for further refining subsequent large-scale eukaryotic genome efforts. Specifically, the rat project employed a hybrid WGS-HSS strategy that was even more highly integrated than that for mouse.[10] Likewise, the current effort to sequence the chimpanzee genome relies on a mixture of WGS plasmid reads and large-insert clone paired-end reads.[11] It is likely that such hybrid strategies will be the basis of most eukaryotic projects for the foreseeable future.

The mouse genome sequence is also having profound implications related to laboratory mice. For example, there are thousands of mutants having heritable Mendelian phenotypes,[41] yet the molecular basis for only a fraction of these has been characterized via positional cloning. Mouse sequence enables the process to be accelerated via PCR-amplifying and sequencing of candidate regions. This approach should enhance the speed with which the molecular basis of such mutants can be determined.[42] Identifying quantitative trait loci (QTL) is another area in which mouse sequence is expected to accelerate progress. There is an ongoing effort to characterize phenotypes for a standard set of inbred mouse lines.[43] Crosses between lines, followed by genotyping, enable the mapping of QTL and subsequent positional cloning. However, QTL are somewhat more difficult than Mendelian defects to clone because the underlying genomic regions are larger and the boundaries are more troublesome to identify. The DNA sequence will clearly improve the prospects for such endeavors. Lastly, the mouse sequence has obvious implications for creating knockout and transgenic mice via homologous recombination in embryonic stem cells.

The methods by which many other biological factors are studied, for example gene expression, will also be profoundly impacted by the mouse sequence. Only a small portion of the mammalian genome is under selection pressure and investigators believe few of these regions code for proteins. Rather, most are thought to be related to regulating gene expression. Cross-species sequence comparison will provide powerful tools for finding and describing such regulatory elements.[44] Of course, the mouse sequence is considered to be one of the most important instruments in this context because of its experimental capacity for testing hypotheses regarding function.

6. Conclusions

Waterston and colleagues[7] called the mouse sequence "a unique lens through which we can view ourselves" and this is quite an appropriate analogy. The availability of this sequence is a superlative informational complement to the mouse's primary and long-established role as an experimental system. At the same time, it functions merely as one of the stepping stones in a broader research program to fundamentally understand biological function and the differences that make species unique.

References

1. Correns, C. 1899. Untersuchungen über die Xenien bei *Zea mays*. *Ber. Deutschen Bot. Ges.* **17**:410–418.
2. Tschermak, E. 1900. Über künstliche Kreuzung bei *Pisum sativum*. *Ber. Deutschen Bot. Ges.* **18**:232–239.
3. DeVries, H. 1900. Sur la loi de disjonction des hybrides. *Compt. Rend. Acad. Sci. (Paris)* **130**:845–847.
4. Haldane, J. B. S., A. D. Sprunt and N. M. Haldane. 1915. Reduplication in mice. *J. Genet.* **5**:133–135.
5. Dietrich, W. F., J. Miller, R. Steen, M. A. Merchant, D. Damron-Boles, Z. Husain, R. Dredge, M. J. Daly, K. A. Ingalls, T. J. O'Connor et al. 1996. A comprehensive genetic map of the mouse genome. *Nature* **380**:149–152.
6. Watson, J. D. 1990. The Human Genome: Past, present and future. *Science* **248**:44–49.
7. Waterston, R. H., K. Lindblad-Toh, E. Birney, J. Rogers, J. F. Abril, P. Agarwal, R. Agarwala, R. Ainscough, M. Alexandersson, P. An et al. 2002. Initial sequencing and comparative analysis of the mouse genome. *Nature* **420**:520–562.

8. Lander, E. S., L. M. Linton, B. Birren, C. Nusbaum, M. C. Zody, J. Baldwin, K. Devon, K. Dewar, M. Doyle, W. FitzHugh et al. 2001. Initial sequencing and analysis of the human genome. *Nature* **409**:860–921.
9. Venter, J. C., M. D. Adams, E. W. Myers, P. W. Li, R. J. Mural, G. G. Sutton, H. O. Smith, M. Yandell, C. A. Evans, R. A. Holt et al. 2001. The sequence of the human genome. *Science* **291**:1304–1351.
10. Gibbs, R. A., G. M. Weinstock, M. L. Metzker, D. M. Muzny, E. J. Sodergren, S. Scherer, G. Scott, D. Steffen, K. C. Worley, P. E. Burch et al. 2004. Genome sequence of the Brown Norway rat yields insights into mammalian evolution. *Nature* **428**:493–521.
11. Chimpanzee Sequencing Consortium. 2004. Initial sequencing and analysis of the chimpanzee genome. *Nature* :In Press.
12. Sanger, F., S. Nicklen and A. R. Coulson. 1977. DNA sequencing with chain-terminating inhibitors. *Proc. Natl. Acad. Sci. USA* **74**:5463–5467.
13. Sanger, F., A. R. Coulson, B. G. Barrell, A. J. Smith and B. A. Roe. 1980. Cloning in single-stranded bacteriophage as an aid to rapid DNA sequencing. *J. Mol. Biol.* **143**:161–178.
14. Anderson, S. 1981. Shotgun DNA sequencing using cloned DNase I-generated fragments. *Nucleic Acids Res.* **9**:3015–3027.
15. Deininger, P. L. 1983. Random subcloning of sonicated DNA: Application to shotgun DNA sequence analysis. *Anal. Biochem.* **129**:216–223.
16. Wilson, R., R. Ainscough, K. Anderson, C. Baynes, M. Berks, J. Burton, M. Connell, J. Bonfield, T. Copsey, J. Cooper et al. 1994. 2.2 Mb of contiguous nucleotide sequence from chromosome III of *C. elegans*. *Nature* **368**:32–38.
17. Fleischmann, R. D., M. D. Adams, O. White, R. A. Clayton, E. F. Kirkness, A. R. Kerlavage, C. J. Bult, J. F. Tomb, B. A. Dougherty, J. M. Merrick et al. 1995. Whole-genome random sequencing and assembly of *H. influenzae* rd. *Science* **269**:496–512.
18. Britten, R. J. and D. E. Kohne. 1968. Repeated sequences in DNA. *Science* **161**:529–540.
19. Shizuya, H., B. Birren, U. J. Kim, V. Mancino, T. Slepak, Y. Tachiiri and M. Simon. 1992. Cloning and stable maintenance of 300-kilobase-pair fragments of human DNA in Escherichia coli using an F-factor-based vector. *Proc. Natl. Acad. Sci. USA* **89**:8794–8797.
20. Marra, M. A., T. A. Kucaba, N. L. Dietrich, E. D. Green, B. Brownstein, R. K. Wilson, K. M. McDonald, L. W. Hillier, J. D. McPherson and R. H. Waterston. 1997. High throughput fingerprint analysis of large-insert clones. *Genome Res.* **7**:1072–1084.
21. *C. elegans* Sequencing Consortium. 1998. Genome sequence of the nematode *C. elegans*: A platform for investigating biology. *Science* **282**:2012–2018.
22. Green, P. 1997. Against a whole-genome shotgun. *Genome Res.* **7**:410–417.
23. Venter, J. C., M. D. Adams, G. G. Sutton, A. R. Kerlavage, H. O. Smith and M. Hunkapiller. 1998. Shotgun sequencing of the human genome. *Science* **280**:1540–1542.
24. Wendl, M. C., J. W. Wallis, S. P. Yang, A. T. Chinwalla and L. W. Hillier. 2004. Genome Assembly, volume 2 of *Encyclopedia of Genetics, Genomics, Proteomics, and Bioinformatics*. John Wiley & Sons, New York NY, In Press.
25. Weber, J. L. and E. W. Myers. 1997. Human whole–genome shotgun sequencing. *Genome Res.* **7**:401–409.
26. Adams, M. D., S. E. Celniker, R. A. Holt, C. A. Evans, J. D. Gocayne, P. G. Amanatides, S. E. Scherer, P. W. Li, R. A. Hoskins, R. F. Galle et al. 2000. The genome sequence of *Drosophila melanogaster*. *Science* **287**:2185–2195.

27. Osoegawa, K., M. Tateno, P. Y. Woon, E. Frengen, A. G. Mammoser, J. J. Catanese, Y. Hayashizaki and P. J. de Jong. 2000. Bacterial artificial chromosome libraries for mouse sequencing and functional analysis. *Genome Res.* **10**:116–128.
28. Mahairas, G. G., J. C. Wallace, K. Smith, S. Swartzell, T. Holzman, A. Keller, R. Shaker, J. Furlong, J. Young, S. Zhao et al. 1999. Sequence-tagged connectors: A sequence approach to mapping and scanning the human genome. *Proc. Natl. Acad. Sci. USA* **96**:9739–9744.
29. Battey, J., E. Jordan, D. Cox and W. Dove. 1999. An action plan for mouse genomics. *Nat. Genetics* **21**:73–75.
30. Skaletsky, H., T. Kuroda-Kawaguchi, P. J. Minx, H. S. Cordum, L. Hillier, L. G. Brown, S. Repping, T. Pyntikova, J. Ali, T. Bieri et al. 2003. The male-specific region of the human Y chromosome is a mosaic of discrete sequence classes. *Nature* **423**:825–837.
31. Jaffe, D. B., J. Butler, S. Gnerre, E. Mauceli, K. Lindblad-Toh, J. P. Mesirov, M. C. Zody and E. S. Lander. 2003. Whole-genome sequence assembly for mammalian genomes: Arachne 2. *Genome Res.* **13**:91–96.
32. Hudson, T. J., D. M. Church, S. Greenaway, H. Nguyen, A. Cook, R. G. Steen, W. J. Van Etten, A. B. Castle, M. A. Strivens, P. Trickett et al. 2001. A radiation hybrid map of mouse genes. *Nat. Genetics* **29**:201–205.
33. Gregory, S. G., M. Sekhon, J. Schein, S. Y. Zhao, K. Osoegawa, C. E. Scott, R. S. Evans, P. W. Burridge, T. V. Cox, C. A. Fox et al. 2002. A physical map of the mouse genome. *Nature* **418**:743–750.
34. Myers, E. W., G. G. Sutton, A. L. Delcher, I. M. Dew, D. P. Fasulo, M. J. Flanigan, S. A. Kravitz, C. M. Mobarry, K. H. J. Reinert, K. A. Remington et al. 2000. A whole-genome assembly of *Drosophila*. *Science* **287**:2196–2204.
35. Wendl, M. C. and S. P. Yang. 2004. Gap statistics for whole genome shotgun DNA sequencing projects. *Bioinformatics* **20**:1527–1534.
36. Benson, D. A., I. Karsch-Mizrachi, D. J. Lipman, J. Ostell and D. L. Wheeler. 2003. GenBank. *Nucleic Acids Res.* **31**:23–27.
37. Birney, E., T. D. Andrews, P. Bevan, M. Caccamo, Y. Chen, L. Clarke, G. Coates, J. Cuff, V. Curwen, T. Cutts et al. 2004. An overview of Ensembl. *Genome Res.* **14**:925–928.
38. Jensen-Seaman, M. I., T. S. Furey, B. A. Payseur, Y. T. Lu, K. M. Roskin, C. F. Chen, M. A. Thomas, D. Haussler and H. J. Jacob. 2004. Comparative recombination rates in the rat, mouse, and human genomes. *Genome Res.* **14**:528–538.
39. Zhang, X. M., I. Rodriguez, P. Mombaerts and S. Firestein. 2004. Odorant and vomeronasal receptor genes in two mouse genome assemblies. *Genomics* **83**:802–811.
40. Bailey, J. A., D. M. Church, M. Ventura, M. Rocchi and E. E. Eichler. 2004. Analysis of segmental duplications and genome assembly in the mouse. *Genome Res.* **14**:789–801.
41. Bult, C. J., J. A. Blake, J. E. Richardson, J. A. Kadin and J. T. Eppig. 2004. The Mouse Genome Database (MGD): Integrating biology with the genome. *Nucleic Acids Res.* **32**:D476 D481.
42. Loftus, S. K., D. M. Larson, L. L. Baxter, A. Antonellis, Y. D. Chen, X. F. Wu, Y. Jiang, M. Bittner, J. A. Hammer and W. J. Pavan. 2002. Mutation of melanosome protein RAB38 in chocolate mice. *Proc. Natl. Acad. Sci. USA* **99**:4471–4476.
43. Paigen, K. and J. T. Eppig. 2000. A mouse phenome project. *Mamm. Genome* **11**:715–717.

44. Loots, G. G., R. M. Locksley, C. M. Blakespoor, Z. E. Wang, W. Miller, E. M. Rubin and K. A. Frazer. 2000. Identification of a coordinate regulator of interleukins 4, 13, and 5 by cross-species sequence comparisons. *Science* **288**:136–140.

Index